Relativistic Electrodynamics
and Differential Geometry

Stephen Parrott

Relativistic Electrodynamics and Differential Geometry

With 37 Illustrations

Springer-Verlag
New York Berlin Heidelberg
London Paris Tokyo

Stephen Parrott
Department of Mathematics and Computer Science
University of Massachusetts
Harbor Campus
Boston, MA 02125, USA

AMS Subject Classifications: 78–A25, 83–C50

Library of Congress Cataloging in Publication Data
Parrott, Stephen.
  Relativistic electrodynamics and differential
geometry.
  Bibliography: p.
  Includes index.
  1. Electrodynamics. 2. Special relativity
(Physics) 3. Geometry, Differential. I. Title.
QC631.P34 1986   537.6   86-24793

© 1987 by Springer-Verlag New York Inc.

All rights reserved. No part of this book may be translated or reproduced in any form without written permission from Springer-Verlag, 175 Fifth Avenue, New York, New York 10010, USA.

Printed and bound by Quinn-Woodbine, Woodbine, New Jersey.
Printed in the United States of America.

9 8 7 6 5 4 3 2 1

ISBN 0-387-96435-5 Springer-Verlag New York Berlin Heidelberg
ISBN 3-540-96435-5 Springer-Verlag Berlin Heidelberg New York

To those who have gone before

and

those who will come after

Preface

The aim of this book is to provide a short but complete exposition of the logical structure of classical relativistic electrodynamics written in the language and spirit of coordinate-free differential geometry. The intended audience is primarily mathematicians who want a bare-bones account of the foundations of electrodynamics written in language with which they are familiar and secondarily physicists who may be curious how their old friend looks in the new clothes of the differential-geometric viewpoint which in recent years has become an important language and tool for theoretical physics. This work is not intended to be a textbook in electrodynamics in the usual sense; in particular no applications are treated, and the focus is exclusively the equations of motion of charged particles. Rather, it is hoped that it may serve as a bridge between mathematics and physics.

Many non-physicists are surprised to learn that the correct equation to describe the motion of a classical charged particle is still a matter of some controversy. The most mentioned candidate is the Lorentz-Dirac equation [†]. However, it is experimentally unverified, is known to have no physically reasonable solutions in certain circumstances, and its usual derivations raise serious foundational issues. Such difficulties are not extensively discussed in most electrodynamics texts, which quite naturally are oriented toward applying the well-verified part of the subject to concrete problems. Some authors claim that the supposed difficulties are irrelevant or easily resolved, others mention them briefly in passing, and others simply ignore them. This book focuses on them but takes no position. Rather, it attempts to present the basic issues as clearly and precisely as possible so that the reader can draw his own conclusions.

As to background, it is assumed that the reader is familiar with the language of modern mathematics and has an elementary acquaintance with electromagnetic theory, at least at the level of a good freshman physics course. In addition, a working knowledge of special relativity and elementary abstract differential geometry are highly desirable prerequisites which will in any event have to be acquired along the way. The necessary concepts from each are presented in the first two chapters, the first oriented toward mathematicians and the second toward physicists and mathematicians who are not experts in differential geometry. However, these are intended as refresher courses to establish a framework within which to develop the theory rather than as texts for beginners, and the reader who has never studied special relativity or is totally unfamiliar with differential geometry may find it hard going. If so, the obvious remedy is to spend a few weeks or months with one of the many good texts on these subjects and begin again.

Chapter 3 formulates the part of electrodynamics which deals with continuous distributions of charge, while Chapter 4 treats radiation and presents the usual motivation leading to the Lorentz-Dirac equation. These two chapters are an expository synthesis of standard material. Only Section 4.4, which applies differential-geometric ideas to clarify radiation calculations usually done in less transparent ways, has any claim to novelty of content as opposed to exposition.

---

† Not to be confused with the quantum-mechanical Dirac wave equation for the electron.

The last chapter, which is of a more specialized and speculative nature, explores further difficulties with the usual formulation of electrodynamics and discusses alternate approaches. Much of this chapter is drawn from the research literature, and some of it appears to be new. Particularly noteworthy is Section 5.5, which presents a proof (based on ideas of Hsing and Driver) of Eliezer's Theorem on nonexistence of physical solutions of the Lorentz-Dirac equation for one-dimensional symmetric motion of opposite charges. This theorem implies that if the Lorentz-Dirac equation holds, then two oppositely charged point particles released at rest can never collide, and in fact will eventually flee from each other at velocities asymptotic to that of light. Since no one seems to believe that real particles will actually behave this way, many (but by no means all) interpret this result as casting serious doubt on the Lorentz-Dirac equation. Surprisingly, few physics texts even mention this result, though it has been extensively discussed in the research literature.

Mathematicians often complain that physics texts are hard to read because of frequent looseness of language and lack of careful definitions. Physicists grumble about the insufficient attention to motivation, excessive concern with generality, and plodding definition-theorem-proof-corollary-definition style too common in mathematics texts. I have tried to avoid all of these, but style is largely a matter of taste and compromise, and it would be miraculous if my taste were to everyone's liking. No doubt some physicists will consider the style overly cautious and pedantic while some mathematicians will find it too loose.

The decision to do electrodynamics in the general context of a Lorentzian manifold without explicitly introducing general relativity (i.e. the Einstein equation) was also largely a matter of taste. It certainly is not necessary to leave Minkowski space to present all the main ideas of electrodynamics, and, unfortunately, some of the important ideas and methods extend to general spacetimes only at the expense of considerable mathematical complication or physical obscurity. On balance, however, it seemed that enough does extend easily to make the relatively small extra effort worthwhile. More importantly, I feel that some of the ideas are actually clearer if one temporarily forgets the vector space structure of Minkowski space. In the end, I did it the way I should have liked to have seen it when I was learning it for the first time.

I have tried to make the notation as coordinate-free as practical while keeping it close to traditional physics notation. The only real departure from the latter is the use of the superscript "$*$" to indicate the operation which identifies a vector with a linear form: relative to traditional physics notation, if $u = u^i$, then $u* = u_i$. The physicist who feels comfortable with abstract differential geometry and wants to dive right in to Chapter 4 or 5 should be able to do so after scanning the table of notations.

I sincerely thank all who helped, directly or indirectly, with this book. In particular, I am grateful for the hospitality of the Mathematics Department of the University of California, Berkeley, where much of the writing was done. I appreciated helpful conversations with J. D. Jackson and R. Sachs and am indebted to H. Cendra and M. Mayer for many useful suggestions. Naturally, I alone am responsible for any errors, and notification of any such will be gratefully received.

Stephen Parrott

July 11, 1986

Contents

Preface	vii
Chapter 1. Special Relativity	**1**
1.1 Coordinatizations of spacetime	1
1.2 Lorentz coordinatizations	3
1.3 Minkowski space	4
1.4 Lorentz transformations	6
1.5 Orientations	10
1.6 Spacetime diagrams and the metric tensor	12
1.7 Proper time and four-velocity	14
1.8 Mass and relativistic momentum	18
Exercises 1	24
Chapter 2. Mathematical Tools	**32**
2.1 Multilinear algebra	32
2.2 Alternating forms	38
2.3 Manifolds	43
2.4 Tangent spaces and vector fields	45
2.5 Covariant derivatives	48
2.6 Stokes' Theorem	53
2.7 The metric tensor	58
2.8 The covariant divergence	59
2.9 The equations $d\beta = \gamma$ and $\delta\beta = \alpha$	69
Exercises 2	86

Chapter 3. The Electrodynamics of Infinitesimal Charges   93

3.1 Introduction   93

3.2 The Lorentz force law   94

3.3 The electromagnetic field tensor   97

3.4 The electric and magnetic fields   98

3.5 The first Maxwell equation   99

3.6 The second Maxwell equation   100

3.7 Potentials   102

3.8 The energy-momentum tensor   104

Exercises 3   119

Chapter 4. The Electrodynamics of Point Charges   126

4.1 Introduction   126

4.2 The retarded potentials and fields of a point particle   128

4.3 Radiation reaction and the Lorentz-Dirac equation   136

4.4 Calculation of the energy-momentum radiated by a point particle   145

4.5 Summary of the logical structure of electrodynamics   161

Exercises 4   163

Chapter 5. Further Difficulties and Alternate Approaches   172

5.1 The Cauchy problem for the Maxwell-Lorentz system   172

5.2 Spherically symmetric solutions of the Maxwell-Lorentz system   176

5.3 Nonexistence of global solutions of the Maxwell-Lorentz system   187

5.4 An alternate fluid model   190

5.5 Peculiar solutions of the Lorentz-Dirac equation   195

5.6 Evidence for the usual energy-momentum tensor   204

5.7 Alternate energy-momentum tensors and equations of motion   211

Appendix on Units   238

Solutions to Exercises	245
Appendix 2	292
Bibliography	296
Table of Notations	300
Index	303

Chapter 1

Special Relativity

1.1 Coordinatizations of spacetime.

Classical Newtonian physics is formulated in terms of the notions of "distance" and "time". These are taken as primitive concepts whose meaning is supposed to be self-evident and universally agreed upon. The theory of relativity does not recognize "distance" and "time" as concepts whose meaning is self-evident. Instead, they are considered as physical quantities which must be carefully defined in terms of other, still more elementary concepts.

Of course, relativity must itself be based on some primitive concepts, and the fundamental conceptual unit recognized by the theory of relativity is that of *event*. The term *event* refers to a definite happening or occurrence, such as the explosion of a bomb or the emission of a photon by an atom. Intuitively, an "event" is "something which happens at a definite place at a definite time". The set **E** of all events is called *spacetime*.

This interpretation of the term "event" makes it reasonable to try to assign to each event e a quadruple of numbers (t_e, x_e, y_e, z_e) in which the last three components x_e, y_e, z_e specify, in some sense which we choose not to make precise yet, the point in "space" at which the event occurred, and the first component t_e is associated with the "time" at which it occurred. † If such a map $e \longmapsto (t_e, x_e, y_e, z_e)$ is a bijection (i.e. a 1-1 correspondence) from **E** onto the four-dimensional real vector space R^4, it will be called a *coordinatization* of **E**. We shall also occasionally refer to *local coordinatizations* which only map a subset of **E** onto R^4.

As a concrete illustration of how such a coordinatization might be carried out, consider four observers O, A, B, and C, each equipped with a clock and a device to measure angles between light rays. Let the observers adjust their positions so that light rays sent from A, B, and C to O are measured by O as mutually orthogonal and so that the round trip time of a light ray sent from O to any of the observers A, B, C and reflected back to O depends neither on the observer to whom the beam is sent nor on the time (as measured by O's clock) that the beam is sent. Intuitively, this means that the observers A, B, and C are at rest with respect to O, are equidistant from O, and are located on the axes of an orthogonal coordinate system with origin at O. (See Figure 1.)

Suppose we are given an event e to be coordinatized, such an explosion which emits a flash of light of infinitesimal duration. We assume that the four observers are constantly exchanging light signals, so that the angles α, β, γ, and δ in the diagram can be measured. Now imagine that O, A, B, and C are points in three-dimensional Euclidean space R^3, and choose the unit of distance in R^3 so that the distance from O to A, B, or C is equal to one. Then it is a simple matter to use standard trigonometry to compute three numbers x_e, y_e, and z_e which would be the coordinates of a point in R^3 such that the lines drawn in R^3 from this point to O, A, B, and C form the same angles α, β, γ, δ shown in the diagram. Let us call these three numbers x_e, y_e, z_e the "space"

---

† Since the theory of relativity does not recognize "space" and "time" as primitive concepts, statements like this must be taken as purely poetic descriptions. A problem in introducing relativity is that the Newtonian view of the world is so intertwined with the language we speak that it is often difficult to formulate statements in everyday language which do not use Newtonian concepts in a relativistically inadmissible way. As soon as we have replaced the Newtonian concepts of "space" and "time" with the relativistic ones of "event" and "coordinatization", this linguistic difficulty will disappear.

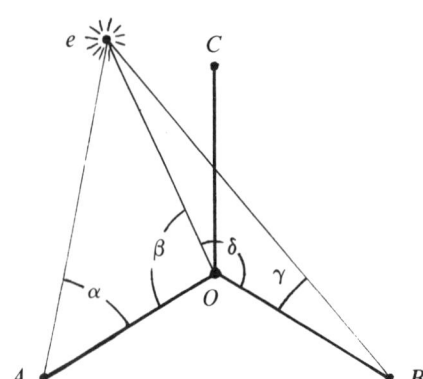

Figure 1-1. The straight lines are the paths of light rays.

coordinates of the event e. To assign a "time" coordinate to e, define a constant c as twice the reciprocal of the round-trip time of a light signal from O to A and back (so that c is the average velocity of light on this particular round-trip), let t_0 denote the time on O's clock at which the flash of light emitted by the event being coordinatized reached O, and define

$$t_e := t_0 - \frac{1}{c}(x_e{}^2 + y_e{}^2 + z_e{}^2)^{1/2},$$

That is, t_e would be the time at which the flash was emitted if the velocity of light in our imaginary Euclidean space were always c.

We have thus defined a map $e \longmapsto (t_e, x_e, y_e, z_e)$ from $\mathbf{E}$ to R^4. To postulate that this map is a a bijection is a physical assumption with far-reaching implications which we shall not examine further here, partly because we do not want to base our treatment of special relativity on this particular coordinatization, and partly because it is assumed that the reader already has at least a passing acquaintance with the physical bases of this theory. Note that this coordinatization used only *local* measurements of angles and time; the only time interval measured was measured by the clock at O, and this clock was never compared with a clock elsewhere. Thus there is no need within the above framework for the Newtonian concept of an absolute time with respect to which all clocks are synchronous; all that is needed is the operationally defined "time as measured by O's clock".

Naturally, there are a great many other reasonable ways to try to coordinatize $\mathbf{E}$, and we do not claim that particular method just described has any special merit other than its conceptual simplicity. The details of the coordinatization will not be important for our purposes; all that we shall need to begin our treatment of special relativity is the assumption that there is *some* coordinatization with special properties which we shall specify in the next section, most notably that light always travels in straight lines with constant velocity. This is a convenient and efficient way to rapidly develop special relativity, but it does slight many interesting and important physical issues, and we urge the reader who is not well acquainted with the physical ideas to read at least the first few chapters in any of the many good physics texts on the subject, some of which are recommended at the end of this chapter.

To avoid misunderstanding, we also remark that although computations using ordinary Euclidean trigonometry were used in the above coordinatization, they served only as a tool to assign coordinates, and none of the geometrical properties of Euclidean space are assumed, *a priori*, to carry over to a "space slice" of spacetime. For instance, in Euclidean space, the sum of the angles in a triangle is always 180°, but there is no reason that the angles of a triangle $OA-AB-BO$ of light rays in Figure 1 should sum to 180°.

1.2. Lorentz coordinatizations.

The theory of special relativity is based on the assumption that there exist coordinatizations of the set **E** of all events, called *Lorentz coordinatizations,* with special properties which we shall now describe. Let $e \longmapsto (t_e, x_e, y_e, z_e)$ be a coordinatization of **E**. Consider an observer who carries with him a clock, and suppose that at every instant τ of time as measured by that clock he is able to determine his coordinates $(t(\tau), x(\tau), y(\tau), z(\tau))$. (It is *not* assumed that $t(\tau) = \tau$.) We shall call the observer *stationary* at x_0, y_0, z_0 with respect to the given coordinatization if $x(\tau) = x_0$, $y(\tau) = y_0$, $z(\tau) = z_0$ for all times τ. It will be helpful to think of space as densely populated with stationary observers, each carrying a "standard clock". The "standard clocks" are conceived as identically constructed (for example, one might use excited hydrogen atoms which emit spectral lines of characteristic frequencies and can in principle serve as clocks by counting successive wavecrests). The first property which we demand of a Lorentz coordinatization is the following.

(1) We assume that the coordinatization is such that *stationary* standard clocks measure "coordinate time" $t(\tau)$. That is, for any stationary standard clock with coordinates $(t(\tau), x_0, y_0, z_0)$ as above, $t(\tau) = \tau$ for all τ.

This implies, in particular, that the rate of a stationary standard clock does not depend on its spatial location. This assumption might seem self-evident or at least innocuous, but the delicacy of the situation is shown by the fact that it is only approximately true in the real world. The theory of general relativity (which does not make this assumption) predicts, and experiment confirms, that identical clocks located at different places in a gravitational field will run at different rates. Special relativity only approximates physical reality in a sense quite closely analogous to the way that a tangent plane approximates a surface. The surface of the ocean appears flat so long as one does not have to navigate long distances, and special relativity provides a description of reality which is accurate in a laboratory small enough that differences in the gravitational field can be neglected.

Suppose we have a coordinatization satisfying (1) and for each triple of real numbers (x, y, z), a stationary observer with this triple for spatial coordinates. Imagine a pulse of light travelling through space. Think of the pulse as of infinitesimal duration and spatial extent, so that it is like a moving particle. For instance, turning on a flashlight with a very narrow beam for a very short time would approximate such a pulse. Given spatial coordinates (x, y, z), we may ask the stationary observer at (x, y, z) at what time t on his standard clock the pulse passed through his point (x, y, z), assuming that it passed through that point at all. It is reasonable to suppose that for any given number t, exactly one point, which we denote $(x(t), y(t), z(t))$ will report a passage at time t. Since all the stationary standard clocks measure coordinate time, this assumption just means that at any given coordinate time, the pulse is somewhere, and that it can't be in two different places at the same coordinate time. We now introduce the second property of a Lorentz coordinatization:

(2) Light always moves in straight lines with unit velocity. This means that if we set $\vec{r}(t) := (x(t), y(t), z(t))$, with $x(t), y(t), z(t)$ defined as above, then the function $t \longmapsto \vec{r}(t)$ is of the form $\vec{r}(t) = \vec{v}t + \vec{r}_0$, where $\vec{v}, \vec{r}_0 \in R^3$ are constant vectors, and $\vec{v}$ is a unit vector.

A coordinatization $e \longmapsto (t_e, x_e, y_e, z_e)$ of spacetime **E** satisfying (1) and (2) will be called a *Lorentz coordinatization*, or *Lorentz coordinate system*, or *Lorentz frame*. The first of two fundamental physical assumptions of special relativity is that *there exists a Lorentz coordinatization for spacetime*.

The assumption that $\vec{v}$ is a unit vector is considerably more than a normalization. The physical content of this assumption is not only that the speed $|\vec{v}|$ at which a pulse of light travels is finite, but also that it is the *same* under all conditions. For instance, $|\vec{v}|$ must be the same for a flash emitted by the headlight of a speeding motorcycle as for a signal from the flashlight of a person standing on the road. Given this, choosing the velocity of light to be unity is of course just a normalization. Physically it means that we are choosing as a unit of length the distance that light travels in a unit of time (e.g. if time is measured in years, then distance is measured in light-years); mathematically it means that we replace a coordinatization $e \longmapsto (t_e, x_e, y_e, z_e)$ in which the vectors $\vec{v}$ in (2) have length c by the new coordinatization $e \longmapsto (t_e, x_e/c, y_e/c, z_e/c)$. We shall use this normalization throughout, so the velocity of light will never enter explicitly into our formulae.

Given a Lorentz coordinatization $e \longmapsto (t_e, x_e, y_e, z_e)$ of **E**, it is often convenient to identify **E** with R^4 via the coordinatization map, and we shall do this without comment when it is not likely to cause confusion. Having fixed such a coordinatization, it is permissible to speak of the "position" (x_e, y_e, z_e) of an event or of the "time" t_e at which it occurred. The time t_e will be called *coordinate time* to distinguish it from other time measurements, such as time intervals measured by observers undergoing accelerated motion. Of course, "coordinate time" is not an absolute notion but is only defined relative to a given coordinatization.

In Newtonian physics, one describes the history of a particle by a function $t \longmapsto \vec{r}(t)$ from R^1 to R^3, where $\vec{r}(t)$ represents the position of the particle at time t. The point $(t, \vec{r}(t)) \in R^4$ corresponds to an event, so in the context of relativity theory it is natural to think of the history of the particle as the set of events

$$\{ (t, \vec{r}(t)) \mid t \in R^1 \} .$$

This set of events is called the *worldline* of the particle. It is customary to think of the worldline as a parametrized curve $s \longmapsto e(s) \in \mathbf{E}$ (so that if $e(s)$ has coordinates $(t(s), \vec{x}(s))$ relative to the coordinatization in which we are working, then $\vec{x}(s) = \vec{r}(t(s))$). Of course, we expect on physical grounds that coordinate time itself can always be used as a parameter, but there are many situations in which it is advantageous to use a different parameter, such as time as measured by a standard clock moving with the particle.

1.3. Minkowski space.

Define a symmetric bilinear form $<\cdot,\cdot>$ on R^4 as follows. If $u, v \in R^4$ have components $u = (u^0, u^1, u^2, u^3)$ and $v = (v^0, v^1, v^2, v^3)$, then

(1) $\qquad <u, v> := u^0 v^0 - u^1 v^1 - u^2 v^2 - u^3 v^3 .$

(The use of upper indices for the components of vectors is traditional in relativity theory.) This bilinear form is variously called the *Lorentz metric*, or *Minkowski metric*, or *metric tensor*, and it plays a fundamental role in relativity theory. Note that it is non-degenerate, which means that the only vector v in R^4 which satisfies $<u, v> = 0$ for all vectors u in R^4 is $v = 0$. The vector space R^4 equipped with this metric is called *Minkowski space* and will be denoted as a bold-face **M**.

We often view R^4 as $R^1 \times R^3$ in the obvious way, and we denote vectors $\vec{r} \in R^3$ by the traditional notation of vector calculus. In particular, the usual inner product of two vectors $\vec{r}$, $\vec{s}$ in R^3 is denoted as $\vec{r}\cdot\vec{s}$, so that if $\vec{r} = (r^1, r^2, r^3)$ and $\vec{s} = (s^1, s^2, s^3)$, then

$$\vec{r}\cdot\vec{s} := r^1 s^1 + r^2 s^2 + r^3 s^3 \ .$$

A vector $v \in \mathbf{M}$ is called

timelike   if $<v, v> \,>\, 0$,

null   if $<v, v> \,=\, 0$,   and

spacelike   if $<v, v> \,<\, 0$.

The physical significance of these terms will be explained in Section 6.

If we identify spacetime $\mathbf{E}$ with $\mathbf{M}$ via a given Lorentz coordinatization, assumption (2) of the previous section states that the worldline of an idealized infinitesimal pulse of light is of the form

$$t \longmapsto (t, \vec{v}t + \vec{r}_0) = (1, \vec{v})t + (0, \vec{r}_0), \qquad \text{with} \quad \vec{v}\cdot\vec{v} = 1 \ .$$

Since $<(1, \vec{v}), (1, \vec{v})> \,=\, 1 - \vec{v}\cdot\vec{v} = 0$, the worldlines of such pulses may be neatly characterized in terms of the Lorentz metric as the lines

$$t \longmapsto at + b \in \mathbf{M} \qquad \text{with} \quad <a, a> \,=\, 0 \ .$$

Such a line is called a *null line*. †

The union of all null lines passing through a given point $b \in \mathbf{M}$ is called the *light cone* at b and may be characterized as

$$\{ p \in \mathbf{M} \mid \,<p-b, p-b> \,=\, 0 \, \} \ .$$

The light cone at the origin of $\mathbf{M}$ is sometimes simply called "the light cone". The *forward* light cone at a point $b \in \mathbf{M}$, (respectively, *backward* light cone at b) is the set of $p = (p^0, p^1, p^2, p^3)$ in the light cone at b such that $p^0 - b^0 > 0$ (resp. $p^0 - b^0 < 0$).

---

† Sometimes, this remark is encapsulated by statements like "the worldlines of photons are null lines". A quantum mechanician (or should it be quantum mechanic?) might well object to the use of the term "photon" in this context. "Photons" are quantum-mechanical constructs which have no proper classical analog, and their quantum-mechanical description is very different from the classically conceivable idealization of an infinitesimal pulse of light. In particular, one can't "see", or otherwise detect, an individual photon at different times in its history in the same way that one can follow the motion of a material particle or a pulse of light. For this reason, we speak of light pulses rather than photons.

1.4. Lorentz transformations.

Having identified **E** with Minkowki space **M** via a given Lorentz coordinatization, we may seek new Lorentz coordinatizations by transforming the given coordinatization by a bijection $F: \mathbf{M} \longrightarrow \mathbf{M}$. That is, if we write $F(t, x, y, z) = (t', x', y', z')$, where t', x', y', and z' are functions of t, x, y, z, and if we assign the new coordinates (t', x', y', z') to the point whose old coordinates were (t, x, y, z), then we have produced a new coordinatization of **E**, and we may ask for which F the new coordinatization will be a Lorentz coordinatization. This question is partly mathematical and partly physical.

As noted in the last section, the worldline of a light pulse with respect to the original coordinatization is some null line $t \longmapsto at+b$, $<a,a> = 0$, in **M**, and all null lines occur in this way. With respect to the new coordinatization, the worldline is the image curve

$$t \longmapsto F(at+b) = (t'(at+b), x'(at+b), y'(at+b), z'(at+b)).$$

If the new coordinatization is to be a Lorentz coordinatization, then the new worldline must also be a null line. That is, if F defines a new Lorentz coordinatization, then it must map null lines to null lines. Moreover, it is clear that any two Lorentz coordinatizations are related in this way by some such F. Thus the mathematical part of the study of Lorentz coordinatizations may be regarded as the study of those bijections $F: \mathbf{M} \longrightarrow \mathbf{M}$ which preserve the set of all null lines. However, the definition of Lorentz coordinatization was not a purely mathematical one because it included the physical assumption that stationary "standard clocks" measure coordinate time. Even if we have an F which maps null lines to null lines, there is no guarantee that real physical clocks will measure coordinate time in the new coordinatization. We shall first examine the mathematical question and then the physical one.

Some obvious mappings which send null lines to null lines are:

(i) translations $\quad v \longmapsto v+c \quad (v \in \mathbf{M}) \quad$ by a fixed $c \in \mathbf{M}$,

(ii) multiplications $\quad v \longmapsto sv \quad$ by a nonzero scalar s, and

(iii) linear transformations L which preserve the metric in the sense that

$$<Lu, Lv> = <u, v> \quad \text{for all } u, v \in \mathbf{M}.$$

The translations and multiplications by a scalar are rather trivial, both mathematically and physically, and do not play an important role in the theory. The metric-preserving linear transformations are called *Lorentz transformations,* and we now turn to their study.

It is easy to write down examples of Lorentz transformations. For instance, if we view **M** as $R^4 = R^1 \oplus R^3$ in the obvious way, then the direct sum $R = I \oplus U$ of the identity I on R^1 with any orthogonal transformation U on R^3 will be a Lorentz transformation. We shall call such transformations *spatial rotations* with respect to the original Lorentz coordinatization which identified **E** with **M**. (Note that orthogonal transformations in R^3 include not only rotations about an axis, all of which have unit determinant 1, but also orientation-reversing orthogonal transformations such as reflections through a plane, which have determinant -1. Thus "spatial rotations" on **M** include spatial reflections.) More generally, if L is any Lorentz transformation, then $L^{-1}RL$ is of the above form with respect to the new coordinatization defined by L and is also called a spatial rotation with respect to that coordinatization.

For a more interesting example, let v be a real number with $-1 < v < 1$, define

(1) $$\gamma(v) := (1-v^2)^{-1/2},$$

and consider the linear transformation L defined by $L(t, x, y, z) = (t', x', y', z')$, where

(2) $\quad\quad\quad t' := \gamma(v)(t-vx)$

$\quad\quad\quad\quad\quad x' := \gamma(v)(x-vt)$

$\quad\quad\quad\quad\quad y' := y$

$\quad\quad\quad\quad\quad z' := z \ .$

It is routine to check that L is a Lorentz transformation, which is called a *boost* in the x-direction with velocity v.

The factor $\gamma(v)$ occurs so often that we shall permanently define $\gamma(v)$ as above, and sometimes we write simply γ when it is obvious what v is. Also, if $\vec{v}$ is a vector in R^3, we write $\gamma(\vec{v})$ for $(1-\vec{v}\cdot\vec{v})^{-1/2}$.

Inverting the above to solve for t, x, y, z yields

(3) $\quad\quad\quad t = \gamma(v)(t'+vx')$

$\quad\quad\quad\quad\quad x = \gamma(v)(x'+vt')$

$\quad\quad\quad\quad\quad y = y'$

$\quad\quad\quad\quad\quad z = z' \ ,$

so L^{-1} is a boost in the same direction with opposite velocity.

To see the physical meaning of the boost L, consider a particle at rest with respect to the new coordinatization defined by L, say at rest at the point whose new spatial coordinates are $x' = k$, $y' = 0 = z'$. Then at time t in the old coordinatization, the old x-component of position is given by the second equation of (2) as

(4) $\quad\quad\quad x = \gamma^{-1}k + vt \ .$

This shows that a particle at rest with respect to the new coordinatization is travelling at velocity v in the original coordinatization. Thus the new coordinatization defined by the boost L describes the coordinates of events as observed from a spatial coordinate system which is moving at velocity v in the x-direction with respect to the old one.

The well-known Lorentz spatial contraction follows easily from (4). To derive this result, consider two particles at rest in the new coordinate system situated at positions $x' = k_1$ and $x' = k_2$, respectively. The distance between these particles as measured in the new system is, of course, k_2-k_1, but at any time t in the old system, the difference in the old positions is given by (4) as

$$(\gamma^{-1}k_2 + vt) - (\gamma^{-1}k_1 + vt) = (k_2-k_1)\gamma^{-1}$$

If we think of the two particles as marking the ends of a measuring rod of length k_2-k_1 units in the new system, then the length of the rod in the old system is

(5) $\quad\quad\quad (k_2-k_1)\gamma^{-1} = (k_2-k_1)(1-v^2)^{1/2} \leq (k_2-k_1) \ .$

To derive the famous "time dilation", consider a clock at rest at the origin of the new system. Suppppose that at time t' in the new system the clock emits a signal. The event which is the emission of the signal has new time coordinate t' and new space coordinate

$x'=0$. The same event has time coordinate t in the old system given by

$$t = (t' + vx')\gamma = t'\gamma \ .$$

Hence if the same clock emits one signal at new time t' and another at new time $t'+1$, the time diference between these two events acording to the old coordinatization will be

(6) $\qquad (t'+1)\gamma - t'\gamma = \gamma \geq 1 \ .$

This implies that the difference in old time coordinates of two events with the same new space coordinates will differ by a factor of $\gamma(v) = (1-v^2)^{-\frac{1}{2}}$ from the difference of the new time coordinates of the same event. In other words, from the point of view of the network of stationary observers in the old system, the moving clock (which is stationary in the new system) runs slower by a factor of $\gamma(v)^{-1} = (1-v^2)^{1/2}$ relative to the stationary clocks which it is passing.

At this point this is no more than a mathematical tautology which says nothing of substance about the real world. We could obtain the same result by applying a Lorentz transformation to the coordinates of a Newtonian world. The physical content of (6) will be supplied by the postulate, or physical fact, that standard clocks at rest in the new frame do measure coordinate time in that frame. This is the second fundamental assumption of special relativity:

(7) We assume that *applying a Lorentz transformation to a Lorentz coordinatization yields a new Lorentz coordinatization.* This implies, in particular, that a standard clock moving with uniform velocity in a particular Lorentz coordinatization will also serve as a standard clock (i.e. will keep coordinate time) in a new coordinate system in which it is at rest obtained by applying a Lorentz transformation to the old system.

The first sentence of (7) also implies that the velocity of light in the new coordinate system, as measured by clocks stationary in that system, is the same as in the old system. In our treatment, this appears as a consequence of the fact that moving standard clocks keep coordinate time in the new system. Historically, the behavior of the clocks was deduced from the assumed constancy of the velocity of light rather than vice versa, since it is only fairly recently that the clock behavior could be directly verified, while the fact that the velocity of light is independent of uniform motion has been known for nearly a century.

The reader who is not familiar with special relativity may be tempted to consider as paradoxical the fact that by symmetry, a clock which is stationary with respect to the old coordinatization will be moving with respect to the new and will therefore run slow compared to the stationary clocks in the new system which it is passing. Thus it might seem that a stationary clock runs slower than a moving clock which in turn runs slower than a stationary clock, so that the stationary clock runs slower than itself! This is one of the more simple-minded versions of the so-called "clock paradox". The "solution" is the observation that statements like "clock A runs slower than clock B " have no absolute meaning in special relativity; such a statement acquires meaning only when it is specified how the clocks are to be compared, and no matter how one does this, the paradox disappears. Further discussions of such "paradoxes" can be found in nearly any text on special relativity, and the reader who is not familiar with the subject will probably have to puzzle through a few of them before he feels comfortable with it.

We add a few more words of caution for the beginner. It is tempting to try to interpret the time dilation (6) as implying that if two events occur with a time difference of t' time units in the new coordinatization, then the observed time difference in the old would be $t'\gamma$ units. This is not usually true, and our derivation of this conclusion relied on the very special hypothesis that the two events occurred *at the same space coordinate in the new system*. One must also be very careful about the use of everyday phrases such as "at the same time". It makes sense to say that two events occur at the same time in the new coordinate system (i.e. that the new time coordinates of the two events are equal) and it makes sense to say that the events occur at the same time in the old system, but the truth of one of these statements does not imply the truth of the other. In special relativity, the concept of simultaneity of events is only defined *relative to a given Lorentz coordinatization.*

The boost (2) relates two coordinate systems moving with uniform relative velocity v in the x-direction, and of course there is an analog of (2) for any spatial direction. Let $\vec{v}$ be a given vector in R^3 with $|\vec{v}| < 1$, and let

$$\vec{u} := \frac{\vec{v}}{(\vec{v} \cdot \vec{v})^{1/2}}$$

denote the unit vector in the direction of $\vec{v}$. The linear transformation B on $\mathbf{M} = R^1 \oplus R^3$ defined by $B(t, \vec{w}) = (t', \vec{w}')$, where

(8) $\qquad t' := (t - \vec{w} \cdot \vec{v})\gamma(\vec{v})$,

$\qquad \vec{w}' := ((\vec{w} \cdot \vec{u})\vec{u} - \vec{v}t)\gamma(\vec{v}) + \vec{w} - (\vec{w} \cdot \vec{u})\vec{u}$

is easily seen to be a Lorentz transformation such that a point which is stationary in the new coordinatization which it defines has velocity $\vec{v}$ in the original coordinatization. This transformation is called the *boost with velocity* $\vec{v}$ with respect to the original Lorentz coordinatization which identifies $\mathbf{E}$ with $\mathbf{M}$. Similarly, if L is any Lorentz transformation, then $L^{-1}BL$ is called a boost with respect to the new coordinatization defined by L.

The set of all Lorentz transformations is obviously a group under composition, so further Lorentz transformations can be obtained by composing boosts and spatial rotations. It is not hard to see that every Lorentz transformation L can be written as a composition $L = BRT$ of a boost B, a spatial rotation R, (both of which can be taken relative to the same original coordinatization) and a transformation T which is either the identity or multiplication by the scalar -1. (Exercise 21).

In linear algebra, linear transformations which preserve a non-degenerate real inner product are called *orthogonal,* though many texts restrict the inner product to be positive definite before making this definition. The determinant of any orthogonal transformation is always ± 1, and the same holds for indefinite inner products, though not all of the usual proofs for the positive definite case readily extend. (Exercise 2.13 outlines a proof of this fact.) In particular, Lorentz transformations have determinant ± 1. (A quick and dirty proof for this case can be obtained from the above decompositon $L = BRT$ and explicit computation that the determinant of a boost is always 1 .)

1.5. Orientations.

Any Lorentz coordinatization sets up a bijection between the set **E** of all events and **M**, and the vector space structure of **M** together with the Minkowski inner product 1.3(1) can be transferred to **E** via this bijection. Of course, the transferred structures depend on the particular Lorentz coordinatization which implements the bijection. Any two such coordinatizations are related as described in the last section by a bijection $F : \mathbf{M} \longrightarrow \mathbf{M}$ which preserves the set of null lines. If this bijection is a Lorentz transformation, scalar multiplication, translation, or some composition of these, the transferred structures are related in simple ways which we shall not bother to describe. It is an interesting mathematical question whether there are other bijections of **M** which preserve the set of null lines and which are not such a composition.

However, the question is of little importance to the physicist, who uses a particular (local) Lorentz coordinatization, established by tradition and convenience, which is quite similar in principle to the one described in Section 1. Correspondingly, we assume that there is some agreed Lorentz coordinatization which is used to transfer the vector space and inner product structure from **M** to **E**, and we then regard **E** as an abstract vector space with inner product of signature (+,-,-,-) induced by this coordinatization.

There is other structure on **M** which we wish to transfer to **E**. It is common experience that time seems to have an asymmetric character; a movie of ordinary scenes run backwards is usually easily distinguishable from the same film run forwards, and there seems to be a qualitative difference between past and future. The question of whether time asymmetry must be, or should be, built into basic physical laws is a deep and fascinating one which we do not address here. However, it *is* built into the commonly accepted formulation of the electrodynamics of massive point particles (though not into the more elementary part of the subject), and that makes it necessary to add the structure of a "time orientation" to **E**. We do this by defining a timelike vector $u = (u^0, \vec{u}) \in \mathbf{M}$ to be *forward-pointing* (or, more simply, *forward*) if $u^0 > 0$, (of course, *backward* means $u^0 < 0$), and then we use the original Lorentz coordinatization which established the vector space structure on **E** to transfer this definition to timelike vectors in **E**.

Given this new structure, it is natural to restrict our attention to those Lorentz transformations which preserve it. It is easily seen that compositions of boosts and spatial rotations preserve the time orientation in the sense that if L is such a Lorentz transformation and u a forward-pointing timelike vector, then Lu is also forward timelike, and the previous decomposition $L = BRT$ with B a boost, R a rotation and T multiplication by ± 1, shows that any L which has this property is a product of a boost and a spatial rotation. When choosing new coordinatizations, we always require them to be compatible with the given time orientation. In practice, we only use new coordinatizations induced by affine maps $z \longmapsto a + Lz$, $z \in \mathbf{M}$, with L a Lorentz transformation, and here compatibility means that L is a product $L = BR$. There is a fairly obvious sense, made precise in Exercise 13, in which this implies that all allowed coordinate times "run in the same direction".

Another element of structure on **M** which is helpful, though not strictly necessary, to transfer to **E** is its "standard" orientation. Recall that the "standard" basis for **M** is defined as the ordered basis

$$s_0 := (1, 0, 0, 0), \quad s_1 := (0, 1, 0, 0), \quad s_2 := (0, 0, 1, 0), \quad s_3 := (0, 0, 0, 1),$$

and the *standard orientation* for **M** is the orientation in which this basis is taken as positively oriented. This means that an arbitrary ordered basis e_0, e_1, e_2, e_3 is called *positively oriented* if the linear transformation T with $Te_i = s_i$ for $i = 0, 1, 2, 3$ has positive determinant and otherwise is *negatively oriented*. This orientation, unlike the time orientation, is not incorporated into the laws of electrodynamics, but merely facilitates stating them in terms of the traditional language of classical physics.

The *norm* of a vector $u \in \mathbf{M}$ is defined to be the scalar

$$\text{sign}(<u,u>)|<u,u>|^{1/2} ,$$

so that "positive norm" is a synonym for "timelike" and "negative norm" for "spacelike". We extend slightly the usual terminology of linear algebra to define a basis $\{e_i\}_{i=0}^3$ for $\mathbf{M}$ or $\mathbf{E}$ to be *orthonormal* if

(1) $\qquad <e_0, e_0> = 1 ,$

$\qquad\qquad <e_i, e_i> = -1 \qquad$ for $i = 1, 2, 3$, and

$\qquad\qquad <e_i, e_j> = 0 \qquad$ when $i \neq j$,

Similarly, we extend the meaning of the term "unit vector" to mean any vector u satisfying $<u, u> = \pm 1$.

Any orthonormal basis $\{e_i\}$ defines in a natural way an orthogonal splitting of spacetime into "space" and "time" via the decomposition

$$\mathbf{E} = [e_0] \oplus [e_1, e_2, e_3] ,$$

where [...] means "the subspace spanned by "...". Actually, we do not need the full basis to obtain the splitting but only the single timelike unit vector e_0 , since it is an easy exercise (Exercise 2) that any timelike unit vector is a member of some orthonormal basis. Notice that in the presence of the time orientation on $\mathbf{E}$, an orientation on $\mathbf{E}$ induces an orientation on the "space" part of any spacetime splitting via the definition that a basis e_1, e_2, e_3 for "space" is to be considered positively oriented if the basis e_0, e_1, e_2, e_3 is a positively oriented basis for $\mathbf{E}$, where e_0 is taken to be orthogonal to "space" and forward-pointing.

The distinction between $\mathbf{E}$ and $\mathbf{M} := (R^4, <\cdot,\cdot>)$ is , of course, mainly a matter of notational taste. We introduced $\mathbf{E}$ to clarify the physical ideas and to get away from the "standard" coordinates with which $\mathbf{M}$ comes endowed. We work in $\mathbf{M}$ when we need a coordinate system in which to do calculations and when it doesn't matter which one we use. We set our discussions in $\mathbf{E}$ when we want to emphasize the abstract structure or when a pre-existing coordinate system would only be a distraction because it's not the right one for the problem at hand. As soon as the fundamental issues are presented and understood, the passage back and forth between $\mathbf{E}$ and $\mathbf{M}$ will be as effortless as the passage between linear transformations and their matrices, and we shall then gradually phase out references to $\mathbf{E}$ and just work in $\mathbf{M}$ so as to come into harmony with standard notation in the field. (In fact, $\mathbf{E}$ is never mentioned beyond this chapter.)

Once $\mathbf{E}$ and $\mathbf{M}$ are identified, Lorentz coordinatizations induced by Lorentz transformations have a simple alternate interpretation. If L is a Lorentz transformation and we define $e_i := L^{-1}s_i$, $i = 0, 1, 2, 3$, where $\{s_i\}$ is the standard orthonormal basis for $\mathbf{M} := (R^4, <\cdot,\cdot>)$ defined above, then the line of multiples of e_0 is the t'-axis in the coordinatization defined by $(t', x', y', z') := L(t, x, y, z)$, and similarly for the other axes. Thus Lorentz coordinatizations induced by Lorentz transformations may be viewed as simply a selection of an orthonormal basis $\{e_i\}$ in $\mathbf{M}$. Of course, such a selection of an orthonormal basis also induces a Lorentz coordinatization in an obvious way without reference to Lorentz transformations: given $p \in \mathbf{M}$, just write p as a linear sum

$$p = \sum_{i=0}^{3} p^i e_i ,$$

and assign coordinates (p^0, p^1, p^2, p^3) to p .

1.6. Spacetime diagrams and the metric tensor.

In this section we shall explain the physical significance of the Lorentz metric on $\mathbf{E}$. Two events $p, q \in \mathbf{E}$ are called

> *relatively timelike*   if $<p-q, p-q> \,>0$,
>
> *relatively spacelike*   if $<p-q, p-q> \,<0$,
>
> *relatively null*   if $<p-q, p-q> \,= 0$

We shall see that the physical significance of these definitions is the following:

(a) Events p and q are relatively timelike if and only if there exists a Lorentz coordinatization of $\mathbf{E}$ such that with respect to this coordinatization,

    (i) p and q have the same space coordinates, and

    (ii) $<p-q, p-q>^{1/2}$ is the elapsed coordinate time between the events p and q,

(b) p and q are relatively spacelike if and only if there exists a Lorentz coordinatization of $\mathbf{E}$ relative to which

    (i) p and q occur simultaneously (i.e. at the same coordinate time), and

    (ii) the spatial distance between p and q is $(-<p-q, p-q>)^{1/2}$,

(c) p and q are relatively null if and only if they are "connected by a light ray". This is just a picturesque way of saying that p and q lie on some null line.

Statement (c) is, of course, trivial. The geometry behind (a) and (b) is most easily seen in a simplified spacetime which is only two-dimensional (one space dimension and one time dimension). We identify this simplified spacetime with "two-dimensional Minkowski space" R^2 via a Lorentz coordinatization, so that the Lorentz metric on $\mathbf{E}$ corresponds to the Minkowski metric

$$<(t_1, x_1), (t_2, x_2)> \,= t_1 t_2 - x_1 x_2$$

on R^2. The usual picture of R^2 as a plane is called in this context a *spacetime diagram*. (See Figure 2.) Note that the time coordinate is on the vertical axis, contrary to the conventions of elementary calculus.† Observe also that the light cone through the origin consists of the two 45° lines $x = \pm t$.

Consider the boost L given by $L(t, x) := (t', x')$ where

$$t' := (t - vx)\gamma(v)$$

$$x' := (x - vt)\gamma(v) \quad , \quad \text{with} \quad 0 < v < 1$$

---

† The latter would harmonize better with the rest of our notation, but the use of the vertical axis as time coordinate is almost universal in the relativity literature, and it was reluctantly decided that the reader would be better served by adherence to the customary notation.

1.6 Spacetime diagrams and the metric tensor

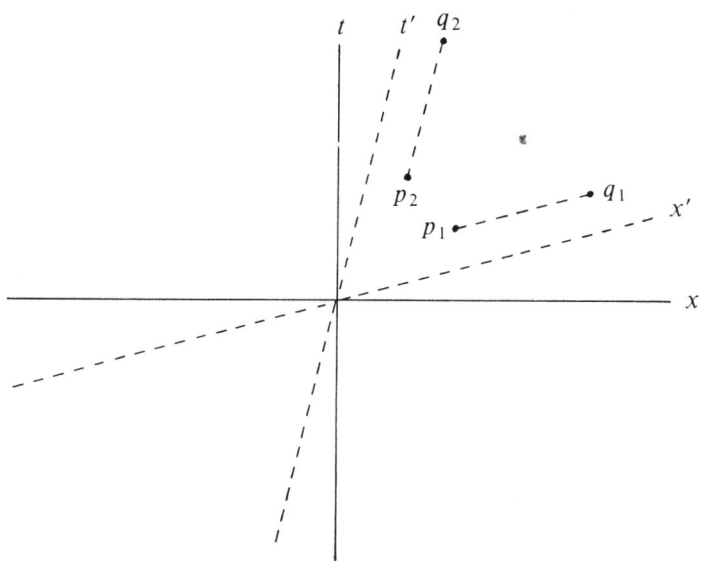

Figure 1-2. Events p_1 and q_1 are simultaneous in the primed system; p_2 and q_2 occur at the same space coordinate in that system.

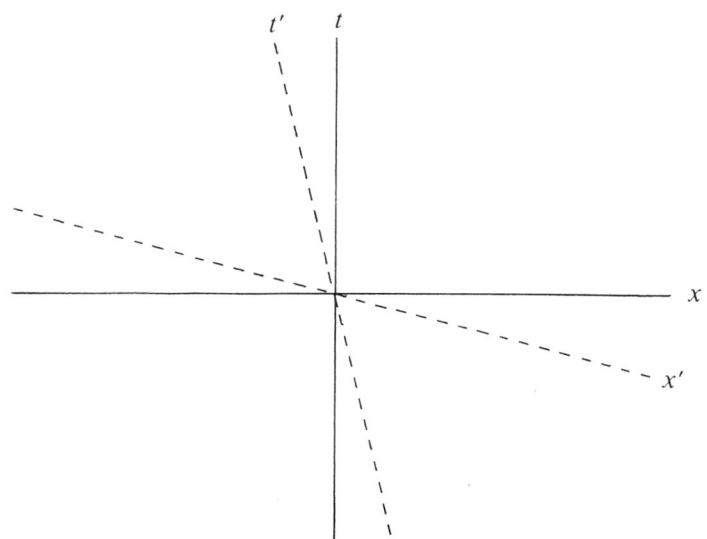

Figure 1-3. A boost 1.4(2) with $v < 0$.

With respect to the new coordinatization defined by L (that is, the coordinatization which assigns new coordinates (t', x') to the event with old coordinates (t, x)), the set of points with new spatial coordinate x' equal to 0 is the line described in the old coordinatization by the equation $x = vt$. This line is labelled t' in Figure 2, since it is natural to consider it as the "t'-axis". Similarly, the set of events with new time coodinate 0 (the "x'-axis") is the line $t = xv$ in the old coordinatization. The corresponding picture for a boost with $-1 < v < 0$ is given in Figure 3.

Two events p and q have the same new time coordinate t' (respectively, space coordinate x') if and only if the line connecting them is parallel to the x'-axis (resp., t'-axis). For instance, in Figure 2, the points p_1 and q_1 have the same new space coordinate x', while p_2 and q_2 have the same new time coordinate t'.

Now if p and q are two events which are not relatively null, the line connecting them will either be parallel to some possible t'-axis (when the line has absolute slope > 1) or to some possible x'-axis (when the absolute slope is < 1), which establishes (a) and (b) for two-dimensional spacetime. The generalization to four dimensions is left as an exercise.

For many purposes, the above two-dimensional picture of spacetime is adequate, but sometimes one needs to visualize the geometry of the other dimensions. In these cases, a three-dimensional picture (one time, two space) such as Figure 4 is usually rich enough to direct one's intuition to the correct conclusions. In this picture, forward or backward light cones look like real two-dimensional cones. Note that the intersection of a constant-time hyperplane $t = t_0$ with a light cone looks like a circle in this picture, but is actually a two-dimensional sphere in four-dimensional Minkowski space $\mathbf{M}$. A forward or backward light cone in $\mathbf{M}$ is a three-dimensional manifold (recall that the vertex is not included), which may be thought of as obtained by shrinking an infinitely large sphere to a point through the time dimension. If we go the other way and blow up a point to an infinite sphere, we get a picture of how a burst of light of infinitesimal duration can expand outward in all directions, forming a sphere at any given coordinate time and in the process of its expansion tracing out a three-dimensional light "cone" in $\mathbf{M}$.

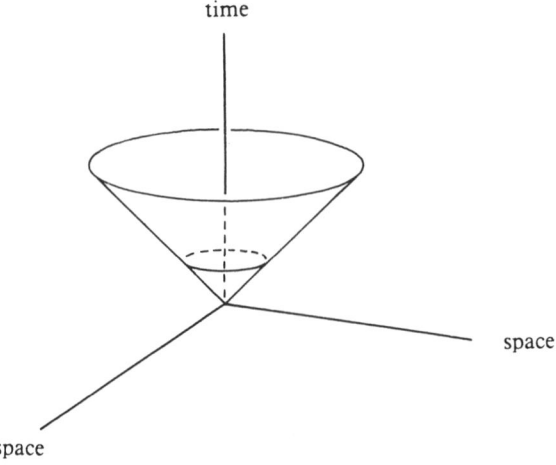

Figure 1-4. A light cone. The circles represent two-dimensional spheres.

1.7. Proper time and four-velocity.

So far we have only considered clocks moving at constant velocity with respect to a given Lorentz coordinatization. Now we shall describe how a standard clock carried by a particle moving on an arbitrary worldline behaves with respect to the standard clocks in a Lorentz coordinate system. Fix a Lorentz coordinatization, and let $\vec{x}(t)$ denote the position of the particle at time t relative to this coordinatization. Then the velocity of the particle as observed in this coordinatization is $d\vec{x}/dt$, and we shall always assume that $(d\vec{x}/dt)\cdot(d\vec{x}/dt) < 1$. That is, we assume that the velocity of any "material particle" is always strictly less than the speed of light. There are reasons to believe that particles with strictly positive mass always satisfy this condition (cf. Exercise 19); at any rate, its violation has never been observed. †

Let Δt denote a small time interval, and consider the motion of the particle between coordinate times t and $t + \Delta t$. The average velocity the particle between coordinate times t and $t + \Delta t$. The average velocity over this interval is

$$\vec{v}(t) := \frac{\vec{x}(t+\Delta t) - \vec{x}(t)}{\Delta t} ,$$

and had the particle been traveling constantly at this velocity, its standard clock would have indicated an elapsed time of

$$(1-\vec{v}(t)\cdot\vec{v}(t))^{1/2}\Delta t ,$$

according to the time dilation 1.4(6). It is therefore natural to suppose that if we set

$$\vec{v}(t) := \frac{d\vec{x}}{dt} ,$$

then

$$\int_{t_1}^{t_2} (1-\vec{v}(t)\cdot\vec{v}(t))^{1/2} \, dt$$

will be the elapsed time shown by the particle's clock corresponding to the coordinate time interval between t_1 and t_2. That is, if a stationary observer at position $\vec{x}(t_1)$ at coordinate time t_1 observes the passing clock to read, say τ_1, then a stationary observer at position $\vec{x}(t_2)$ at time t_2 will observe the same clock to read

$$\tau_1 + \int_{t_1}^{t_2} (1-\vec{v}(t)\cdot\vec{v}(t))^{1/2} \, dt .$$

---

† Hypothetical particles which travel faster than the speed of light have been named *tachyons*. Various authors have speculated on the properties they might have, but there is no evidence that they exist. Even if they do exist, it is generally believed that a particle of positive mass with velocity less than the speed of light cannot be accelerated to the velocity of light or beyond because that would necessitate giving it infinite "energy" (cf. Section 1.8).

Choose an arbitrary point $(a, \vec{x}(a))$ on the worldline $t \longmapsto z(t) := (t, \vec{x}(t))$, and define a function τ on the worldline, called the *proper time*, by

$$\tau((t, \vec{x}(t))) = \int_a^t (1 - \vec{v}(t) \cdot \vec{v}(t))^{1/2} \, dt \ . \tag{1}$$

Though the mathematical form of this definition depends on the Lorentz coordinatization with respect to which it is written, the physical quantitiy it describes (time as measured by the moving clock) is independent of the coordinatization, so we expect on physical grounds that (1) will turn out to be coordinate-independent. To prove this, note that

$$(1 - \vec{v} \cdot \vec{v})^{1/2} = \langle (1, \vec{v}), (1, \vec{v}) \rangle^{1/2} = \langle \frac{dz}{dt}, \frac{dz}{dt} \rangle^{1/2} \ . \tag{2}$$

Now (2) depends on the coordinatization only through the use of coordinate time t to parametrize the worldline. If we use a different parameter for the worldline, say $s \longmapsto z(t(s)) = (t(s), \vec{x}(t(s)))$, where we assume the relation between s and t to be strictly monotonically increasing, then we have

$$\langle \frac{dz}{ds}, \frac{dz}{ds} \rangle^{1/2} = \langle \frac{dz}{dt}, \frac{dz}{dt} \rangle^{1/2} \frac{dt}{ds} \ ,$$

and (1) written in terms of the new parameter is just a change of variables in the integral: for any b,

$$\int_b^s \langle \frac{dz}{ds}, \frac{dz}{ds} \rangle^{1/2} ds = \int_b^s \langle \frac{dz}{dt}, \frac{dz}{dt} \rangle^{1/2} \frac{dt}{ds} ds$$

$$= \int_{t(b)}^{t(s)} \langle \frac{dz}{dt}, \frac{dz}{dt} \rangle^{1/2} dt \ .$$

This shows that the more general definition

$$\tau := \int_b^s \langle \frac{dz}{ds}, \frac{dz}{ds} \rangle^{1/2} ds \tag{1'}$$

agrees with (1) and that τ is a well-defined function on the worldline given the choice of the initial point $(b, \vec{x}(b)) = (t(b), \vec{x}(t(b)))$. Since the initial point is arbitrary, τ is really only defined up to an arbitrary constant. Notice also that the definition of τ implicitly assumes the time orientation on **E**; without this additional structure the sign of τ would also be arbitrary. The definition is set up so that proper time increases with increasing coordinate time.

The reader has probably already recognized this as essentially the usual argument that shows that the arc length of a curve in a Riemannian manifold is independent of its parametrization, and the proper time is nothing more than the arc length of the worldline in the semi-Riemannian space **M**. In a Riemannian space it is often convenient to use arc length as a parameter for curves, and in terms of this parameter the tangent vector always has unit length. The same holds in the present context. Let us parametrize the worldline by proper time τ and write it (with a slight abuse of notation) as $\tau \longmapsto z(\tau) \in \mathbf{M}$. Define the *four-velocity* $u(\tau)$ as

$$u(\tau) := \frac{dz}{d\tau} \ . \tag{3}$$

Notice that $u(\tau)$ is invariantly defined; no coordinatization-dependent quantity appears in (3). Exercise 11 verifies that, just as in the Riemannian case, $u(\tau)$ is always a unit vector (of positive norm):

(4) $$\langle u(\tau), u(\tau) \rangle = 1 \, .$$

The relation between the four-velocity $u(\tau)$ and the ordinary velocity in 3-space is easily seen by passing back into a Lorentz coordinatization and writing

$$z(\tau) = (t(\tau), \vec{x}(\tau)) \, ,$$

so that

(5) $$u(\tau) = \frac{dz}{d\tau} = \left(\frac{dt}{d\tau}, \frac{d\vec{x}}{d\tau}\right) = \frac{dt}{d\tau}\left(1, \frac{d\vec{x}}{dt}\right) = \frac{dt}{d\tau}(1, \vec{v}) \, .$$

This shows that the space component of $u(\tau)$ is the ordinary velocity $\vec{v} := d\vec{x}/dt$ as measured in the given Lorentz frame multiplied by the factor $dt/d\tau$, which by (1) is

(6) $$\frac{dt}{d\tau} = \left(\frac{d\tau}{dt}\right)^{-1} = (1-\vec{v}\cdot\vec{v})^{-1/2} = \gamma(\vec{v}) \, .$$

The relation (6) occurs so often that it is worth memorizing. When the velocity $\vec{v}$ is small compared with the velocity of light, which in our units is 1, the four-velocity u is approximately equal to $(1, \vec{v})$.

It is often convenient to work in a Lorentz frame in which the particle is at rest at a particular proper time τ_0. In such a frame,

$$u(\tau_0) = (1, \vec{0}) \, ,$$

and any such frame is called a *rest frame,* or *proper frame* for the particle at proper time τ_0. Although a rest frame is only unique up to a spatial rotation, it is customary to speak of *the* rest frame at τ_0. In such a frame, differentiations with respect to proper time and coordinate time of any quantity defined on the worldline agree at τ_0.

Another important quantity is the *proper acceleration* a, also called *four-acceleration,* defined by

(7) $$a := \frac{du}{d\tau} \, .$$

Like u, a is invariantly defined. The proper acceleration $a(\tau)$ is always orthogonal to $u(\tau)$ because from (4),

$$0 = \frac{d\langle u, u\rangle}{d\tau} = 2\left\langle\frac{du}{d\tau}, u\right\rangle \, ,$$

and so $a(\tau)$ is a purely spatial vector in any rest frame at τ. In any Lorentz frame, we have

(8) $$\frac{d\gamma}{d\tau} = \frac{d(1-\vec{v}\cdot\vec{v})^{-1/2}}{d\tau} = -\frac{1}{2}(1-\vec{v}\cdot\vec{v})^{-3/2}\cdot\left(-2\frac{d\vec{v}}{d\tau}\cdot\vec{v}\right)$$

$$= \gamma^3 \frac{d\vec{v}}{d\tau}\cdot\vec{v} = \gamma^4 \frac{d\vec{v}}{dt}\cdot\vec{v} \, ,$$

and

(9) $$a = \frac{du}{d\tau} = \frac{d}{d\tau}(\gamma(v)(1, \vec{v})) = \frac{d\gamma}{d\tau} \cdot (1, \vec{v}) + \gamma \cdot (0, \frac{d\vec{v}}{d\tau})$$
$$= \gamma^4 \cdot (\frac{d\vec{v}}{dt} \cdot \vec{v}) \cdot (1, \vec{v}) + \gamma^2 \cdot (0, \frac{d\vec{v}}{dt}) \ .$$

Formula (8) occurs sufficiently frequently to make it worth remembering, and (9) shows the relation between the proper acceleration and the *coordinate acceleration* A defined by

$$\vec{A} := \frac{d\vec{v}}{dt} \ ,$$

which is usually what is actually measured in the laboratory. Setting $\vec{v} := 0$ in (9) shows that $a(\tau)$ may be viewed as the ordinary acceleration as seen from a rest frame. Note, however, that if we write

$$a = (a^0, \vec{a}) \ ,$$

then

$$\vec{a} = \gamma^4 (\vec{A} \cdot \vec{v}) \vec{v} + \gamma^2 \vec{A} \ ,$$

so that in general $\vec{a} \neq \vec{A}$. In other words, the "space" part $\vec{a}$ of the proper acceleration a is *not* usually the same as the coordinate acceleration A, though they do coincide in any rest frame.

1.8. Mass and relativistic momentum.

In Newtonian physics, the momentum $\vec{p}$ of a particle is defined as $\vec{p} := m\vec{v}$, where m denotes the mass and $\vec{v}$ the velocity of the particle. If two particles with masses m_1, m_2 and velocities $\vec{v}_1$, $\vec{v}_2$, respectively, collide, and if the velocities of the particles after collision are $\vec{v}'_1$ and $\vec{v}'_2$, respectively, then the law of conservation of momentum states that

(1) $$m_1 \vec{v}_1 + m_2 \vec{v}_2 = m_1 \vec{v}_1' + m_2 \vec{v}_2' \ .$$

This Newtonian law holds regardless of the nature of the collision (elastic or inelastic). However, a routine check reveals that if this law holds for inelastic collisions in a particular Lorentz coordinatization, then it will *not* hold in any other coordinatization related to the first by a nontrivial boost (Exercise 15). If (1) held in any particular Lorentz coordinatization, then that coordinatization, together with others related to it by spatial rotations, would constitute a "privileged" class of systems in which the laws of physics would be particulary simple. Once the striking fact that the behavior of physical clocks is consistent with Lorentz transformations is accepted, it is tempting to hypothesize that *all* the laws of physics are "invariant", in some appropriate sense, under Lorentz transformations (physicists prefer the term "covariant" in this context), and thus we are led to seek a Lorentz-invariant replacement for (1).

It is not hard to guess a likely candidate. We already know that the definition 1.7(3) of the four-velocity $u(\tau) := dz/d\tau = (dt/d\tau, d\vec{x}/d\tau)$ is coordinatization-independent, and 1.7(5) and 1.7(6) show that its space part

$$\frac{d\vec{x}}{d\tau} = \frac{d\vec{x}}{dt} \frac{dt}{d\tau} = \vec{v}(1 - \vec{v} \cdot \vec{v})^{-1/2}$$

approximates the Newtonian velocity $\vec{v} := d\vec{x}/dt$ to first order when $\vec{v}$ is small. Thus a

1.8 Mass and relativistic momentum

natural relativistic analog of (1) is:

(2) $$m_1 u_1 + m_2 u_2 = m_1 u_1' + m_2 u_2',$$

where u_1, u_2, u_1', u_2', are the four-velocities corresponding to the coordinate velocities $\vec{v}_1, \vec{v}_2, \vec{v}_1', \vec{v}_2'$, respectively. In any Lorentz coordinatization, the "space" part of (2) is

(3) $$m_1 \gamma(\vec{v}_1)\vec{v}_1 + m_2 \gamma(\vec{v}_2)\vec{v}_2 = m_1 \gamma(\vec{v}_1')\vec{v}_1' + m_2 \gamma(\vec{v}_2')\vec{v}_2',$$

and the "time" part is

(4) $$m_1 \gamma(\vec{v}_1) + m_2 \gamma(\vec{v}_2) = m_1 \gamma(\vec{v}_1') + m_2 \gamma(\vec{v}_2').$$

The space part (3) approximates (1) to first order in the velocities, and the time part (4) defines a new physical law which is roughly analogous to the Newtonian law of conservation of energy.

The latter will be discussed more fully below, but first we specify more carefully the operational definition of the "masses" m_1 and m_2, which are also called "rest masses" to avoid confusion with the concept of "relativistic mass" which will arise later. Fix a Lorentz coordinatization, arbitrarily choose a particle at rest with respect to that coordinatization, and *define* the *rest mass* of that particle to be one. That is, arbitrarily choose a "unit of mass". Now suppose we have a new particle whose rest mass we wish to define. Give the new particle a very small velocity v_1, let it collide with the stationary particle of unit mass, and measure the final velocities v_1' of the new particle and v_2' of the mass unit. (It is assumed that the motion is one-dimensional.) If (1) held, we could solve for the mass m_1 of the new particle:

(5) $$m_1 = \frac{v_2'}{v_1 - v_1'}.$$

Although we know from hindsight that (1) does not hold exactly for relativistic velocities, we expect from Newtonian physics that violations of (1) will be undetectable when v_1 is small enough, and we may therefore unambiguously define the rest mass m_1 of our test particle by the limit as $v_1 \to 0$ of the right side of (5).

In the above definition of rest mass, it is implicitly assumed that the rest mass of the standard never changes. Operationally, this means that if the above experiment is performed with two standard unit masses, then the limit on the right side of (5) is the same under all conditions. This might seem too obvious to mention, but later developments will call this assumption into question. Also, we warn the reader that despite our motivating considerations, (2) is not considered to be a universally valid relativistic law of conservation of momentum.

Nevertheless, the *relativistic momentum*, more usually called *four-momentum*, or *energy-momentum* p of a particle with rest mass m and four-velocity u is defined as

(6) $$p := mu.$$

The name "energy-momentum" arises from an analogy, to be explained later, between the "time" component of p in a given Lorentz coordinatization and the Newtonian concept of kinetic energy. Since the "space" part of p is already analogous to Newtonian momentum, the term "energy-momentum" is appropriate and more descriptive than "four-momentum".

Let the particle in question have world line $\tau \longmapsto z(\tau)$ parametrized by proper time τ, let the coordinates of $z(\tau)$ in some Lorentz frame be $z(\tau) = (t(\tau), \vec{x}(\tau))$, and write $\vec{v} := d\vec{x}/d\tau$. Then 1.7(5) and 1.7(6) show that the coordinates $(p^0(\tau), \vec{p}(\tau))$ of $p(\tau)$ with respect to this coordinatization are

$$(p^0, \vec{p}) = p(\tau) = m \frac{dz}{d\tau} = m\gamma(\vec{v})(1, \vec{v}) ,$$

so the space part $\vec{p}$ of p is

(7) $$\vec{p} := m\gamma(\vec{v})\vec{v} .$$

If we define

(8) $$M := m\gamma(\vec{v}) ,$$

then (7) looks like the usual Newtonian definition $\vec{p} := M\vec{v}$ of momentum with M playing the part of the mass, although M varies with velocity unlike a Newtonian mass. The quantity M is called the *relativistic mass* of the particle with respect to the given Lorentz coordinatization. Observe that the relativistic mass is always greater than or equal to the rest mass and that it goes to infinity as the velocity of the particle approaches the speed of light. Note also that the "time" component of p is just M, and so

$$p = (M, M\vec{v}) .$$

In particular, when (2) holds, its "time" component (4) is a law of conservation of relativistic mass.

The name "relativistic mass" tends to be misleading because the analogy between this quantity and the Newtonian concept of mass as "resistance to acceleration" (e.g. "force" divided by acceleration) is only qualitative; there is no simple and direct relation (cf. Exercise 20). We shall see shortly that relativistic mass is also related to the Newtonian concept of "kinetic energy", though again the analogy is not exact. The beginner is probably best advised to think of it as a new object not present in Newtonian physics rather than straining approximate analogies with Newtonian "mass" and "energy".

The definition (6) of relativistic momentum was motivated by the classical law (1) of conservation of momentum, so we should discuss the range of validity of the corresponding relativistic version (2). Essentially, (2) is valid in the context of collisions of *uncharged* particles whose rest masses are universal constants which can not be affected by the collision and in which neither particles nor "photons" are created or destroyed. Charged particles can "radiate" four-momentum, which is carried off by the "electromagnetic field", and it is the four-momentum of the particles *plus* that ascribed to the field which is conserved. The reason for the restriction to particles with constant rest masses (various so-called "elementary" particles, such as electrons, are thought to satisfy this condition) is discussed in the next paragraph. The role of (2) in the theory is probably best expressed by viewing it as an experimentally verifiable law for a certain more or less well-defined class of experiments (collisions of uncharged particles deemed "elementary" or whose internal structure can be neglected) and otherwise as an idealization which motivates further developments. One expects (2) to hold, and when it doesn't, one compensates by modifying in some way the definition of relativistic momentum.

As an example of the kind of trouble which can arise from an injudicious expectation that (2) will hold in all situations, consider two identical bodies of rest mass m moving toward each other with equal and opposite velocities $\vec{v}$ and $-\vec{v}$ as in Figure 5. Suppose the collision is completely inelastic (think of the bodies as lumps of soft clay, for example), so that after the collision the bodies are stuck together at the origin. This is completely consistent with the Newtonian law (1) of conservation of momentum; before and after the collision the total momentum is zero. However, it is consistent with (2) only if we assume that the rest masses of the bodies are greater after the collision than before. To see this, note that the total four-momentum p_{in} before the collision is

$$p_{in} = m\gamma(\vec{v})(1, \vec{v}) + m\gamma(-\vec{v})(1, -\vec{v}) = 2m\gamma(\vec{v})(1, \vec{0}),$$

while the total four-momentum p_{out} after the collision is

$$p_{out} = m'\gamma(\vec{0})(1, \vec{0}) + m'\gamma(\vec{0})(1, \vec{0}) = 2m'(1, \vec{0}),$$

where m' denotes the rest mass of each body *after* the collision. It is natural to initially take $m' = m$ without a thought, but *a posteriori* it becomes obvious that if we wish $p_{in} = p_{out}$, then we must have

(9) $\quad m' = m\gamma(\vec{v}) > m$.

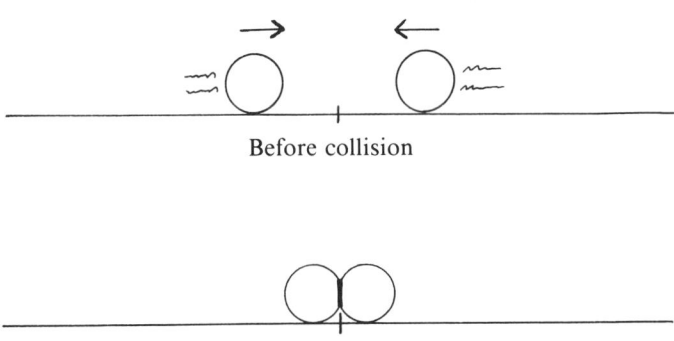

Before collision

At rest after collision

Figure 1-5. The total rest mass after collision is greater than before.

That is, assuming conservation of four-momentum, the collision has increased the rest masses of the bodies. † On the one hand this is consistent with analogies between relativistic mass and both Newtonian mass and kinetic energy; the kinetic energy of the bodies before collision has vanished and in its place has appeared additional rest mass. On the other, it shows the danger of relying on such analogies; after all, the analogy that led us from (1) to (2) did not prepare us for the possibility that the m_i on the left and right sides of (2) might not be the same.

The present example can be included within the previous framework by modifying (2) to read:

(10) $\quad m_1 u_1 + m_2 u_2 = m_1' u_1' + m_2' u_2'$,

where m_i' denotes the rest mass of particle i after the collision, and this is the generally accepted law of conservation of relativistic momentum in collisions of two uncharged particles in which nothing is created or destroyed. However, introducing the possibility of nonconstant rest masses into the theory, has the complicating side effect of calling into question the

---

† One way to "explain" this is to think of the initial kinetic energy as converted into heat energy, that is, into increased velocities of the molecules (or whatever else) of which the body is composed. The increased molecular velocities then imply increased relativistic masses.

original definition of "rest mass", which implicitly assumed the existence of a standard unit mass. If the standard mass could undergo inelastic collisions, then we would expect *its* mass to change as it underwent collisions over time. In addition, the equations analogous to (5) which would have to be solved to determine a mass m_1 in a collision experiment with a stationary standard unit mass $m_2 = 1$ would be

$$(11) \qquad m_1 \gamma(v_1)(1, v_1) + (1, 0) = m_1' \gamma(v_1')(1, v_1') + m_2' \gamma(v_2')(1, v_2') .$$

Separating into components gives only two equations for the three unknowns m_1, m_1', m_2'. Thus the theory as stated above is incomplete unless we can assume the existence of standard masses which can only undergo elastic collisions. In practice, this is not a problem because the rest masses of various elementary particles are believed to be universal constants independent of the history of the particle.

Of course, there are analogs of (10) for collisions of more than two particles and for collisions with creation and destruction, and it is also possible to include "photons" in this framework (cf. Exercise 23), but (10) conveys the main idea sufficiently clearly for our purposes, and we shall not carry the discussion further. The interested reader might consult [French] for a concrete physical account or [Sachs and Wu] for a more mathematical treatment.

Now we present the analogy between relativistic mass and classical kinetic energy summarized by the famous relation $E = Mc^2$ (which in our units simply reads $E = M$, of course). This analogy is based on the following traditional calculation, which shows that a quantity E, whose definition looks roughly like the Newtonian kinetic energy of a particle (apart from an additive constant), is equal to the relativistic mass M under the assumption that the rest mass of the particle is constant.

Classically, the rate of change of energy E of a particle moving with velocity $\vec{v}$ is given by

$$(12) \qquad \frac{dE}{dt} = \vec{F} \cdot \vec{v} ,$$

where $\vec{F}$ is the applied force. Moreover, the force is defined by

$$(13) \qquad \vec{F} := \frac{d\vec{p}}{dt} .$$

Hence

$$(14) \qquad \frac{dE}{dt} = \frac{d\vec{p}}{dt} \cdot \vec{v} ,$$

and this determines the energy E up to an additive constant.

Let us compute E for a relativistic particle with worldline $\tau \longmapsto (t(\tau), \vec{x}(\tau))$ in some Lorentz coordinatization, where τ is proper time. Recall from 1.7(5) that the four-velocity $u(\tau)$ is given in terms of the coordinate velocity $\vec{v} := d\vec{x}/dt$ by $u(\tau) = \gamma(\vec{v})(1, \vec{v})$. Let m denote the rest mass of the particle. We shall not initially assume that m is constant, though this assumption will eventually be necessary to obtain the relation $E = M$. Write the components of the four-momentum p as $p = (p^0, \vec{p}) = M(1, \vec{v})$, where $M := m\gamma(\vec{v})$ is the relativistic mass, so that

$$\vec{p} = M\vec{v} = m\gamma(\vec{v})\vec{v} .$$

This $\vec{p}$ will be taken as the proper analog of the classical 3-momentum in (14). We have

$$m^2 = m^2 <u, u> = <p, p> = M^2 - \vec{p} \cdot \vec{p}.$$

Thinking of all quantities as functions of coordinate time t and differentiating both sides with respect to t gives

$$2m\frac{dm}{dt} = 2M\frac{dM}{dt} - 2\vec{p} \cdot \frac{d\vec{p}}{dt} = 2M\frac{dM}{dt} - 2M\vec{v} \cdot \frac{d\vec{p}}{dt},$$

and substituting (14) yields

(15) $$\frac{dE}{dt} = \frac{dM}{dt} - \frac{m}{M} \cdot \frac{dm}{dt} = \frac{dM}{dt} - \frac{dm}{dt}(1-\vec{v}\cdot\vec{v})^{1/2}.$$

If the rest mass of the particle is constant, then

$$\frac{dE}{dt} = \frac{dM}{dt},$$

and hence

$$E = M + C$$

with C a constant. It is usual to take $C = 0$, thus equating relativistic mass with "energy". This equation has achieved such popularity that the analogistic character of its derivation and the possibility of nonconstant rest masses have been virtually forgotten, and it is now almost universal in the relativity literature to use the terms "energy" and "relativistic mass" interchangeably. We shall follow this convention and simply *define* the term "energy" in the relativistic context to mean "relativistic mass". When we want to emphasize the analogies between this quantity and its twin Newtonian counterparts, mass and kinetic energy, we speak of "mass-energy". The best and most descriptive way to think about this quantity is as the "time" component of four-momentum. This emphasizes that mass-energy never appears alone in relativity theory; it is always a component of a four-vector and as such is coordinatization-dependent.

The fact that the derivation of the identification $dM/dt = dE/dt$ started with the relation (12),

$$\frac{dE}{dt} = \vec{F} \cdot \vec{v},$$

might easily mislead one into thinking that the "E" which is ultimately identified with M is the classical *total* energy (kinetic plus potential). This is not correct, but a full explanation would necessitate a careful explanation of the role of "force" within the logical structure of classical Newtonian mechanics. Suffice it to say that the definition (13) of "force" in this context as $\vec{F} := d\vec{p}/dt$ makes it clear that for this type of "F", the classical quantity

$$\vec{F} \cdot \vec{v} := \frac{d\vec{p}}{dt} \cdot \vec{v} = m\frac{d\vec{v}}{dt} \cdot \vec{v} = \frac{d}{dt}(\frac{1}{2}m\vec{v}\cdot\vec{v})$$

is the time rate of change of *kinetic* energy. (For instance, the "force" which you feel your chair exerting on you right now is not of this type because your $d\vec{p}/dt$ is presumably zero.)

The relation between relativistic mass-energy and classical kinetic energy is further clarified by expanding the former in powers of the magnitude $v := |\vec{v}|$ of the velocity:

$$M = m(1-v^2)^{-1/2} = m + \frac{1}{2}mv^2 + \text{terms of order } v^4 \text{ or higher.}$$

Thus to order v^2, the relativistic mass is the rest mass plus the classical kinetic energy $mv^2/2$.

References

There are a large number of good texts on special relativity, ranging from wordy philosophical tomes to graphic masterpieces to dense, tensor-packed treatises. The representative sample mentioned below should accomodate most tastes, but the list is far from exhaustive.

[French] is elementary and physically oriented. [Taylor and Wheeler] has elaborate graphics and many problems. Further useful insights can be found in [Misner, Thorne, and Wheeler], a mammoth high-level text on general relativity produced in the same style. [Rindler] is a good, concise text on the intermediate level which may appeal to mathematicians more than some of the more wordy and diffuse physics texts. In this vein, several books on electromagnetic theory, such as [Panofsky and Phillips], contain chapters on special relativity which could serve as good mini-texts. [Synge] and [Dixon] are more advanced works which treat many difficult issues of relativistic mechanics.

Exercises 1

All vector spaces below are assumed to be real and finite dimensional, and all inner products are assumed non-degenerate (i.e. only the zero vector is orthogonal to all vectors) but not necessarily positive definite. It is a standard fact of linear algebra that any such vector space has an orthonormal basis and that any two orthonormal bases have the same number of positive norm elements (and, of course, the same number of negative norm elements) [Lang, 1965, Chapter 14]. In a vector space with indefinite inner product, a "unit vector" u is defined as one with $<u, u> = \pm 1$, and an "orthonormal basis" as a basis of orthogonal unit vectors. (cf. Section 1.5). As always, **M** stands for Minkowski space.

1.1. In Euclidean space, any nonzero vector can be embedded in an orthonormal basis. Show that this is false for Minkowski space **M**. This should only take a few seconds; the point is that many familiar properties of linear algebra on spaces with positive definite norm fail for indefinite metrics. The first few exercises are meant to accustom the reader to the peculiarities of vector spaces with indefinite inner products in general and **M** in particular.

1.2. Show that any orthonormal set $e_1, e_2, \cdots, e_m$ in an inner-product space as above can be embedded in an orthonormal basis $e_1, \cdots, e_m, e_{m+1}, \cdots, e_n$.

1.3. Show that if $n_1 \neq 0$ and n_2 are orthogonal null vectors in **M**, then n_2 is a multiple of n_1. As a corollary, conclude that any subspace of dimension two or greater contains a unit vector.

1.4. Show that every subspace of **M** of dimension two or greater contains a spacelike vector.

1.5. Give examples of one, two and three-dimensional subspaces in **M** which have no orthonormal bases.

Exercises 1

1.6. (a) Observe that any timelike unit vector $u \in \mathbf{M}$ may be written as $u = \pm\gamma(\vec{v})(1, \vec{v})$ with $|\vec{v}| < 1$.

(b) Let $u \in \mathbf{M}$ be a timelike vector written as just described. Write down a boost B such that in the new coordinatization $(t', \vec{x}') := B(t, \vec{x})$ defined by B, u has coordinates $(1, 0, 0, 0)$.

(c) Show that if a subspace of $\mathbf{M}$ contains a timelike vector, then it has an orthonormal basis.

1.7. Show that if a three-dimensional subspace S of $\mathbf{M}$ contains no null vector, then it has an orthonormal basis, and the restriction of the inner product to S is negative definite. Show that there exists a Lorentz transformation such that

$$L(S) = \{ (0, x^1, x^2, x^3) \mid (x^1, x^2, x^3) \in R^3 \}.$$

In other words, S is "space" relative to some Lorentz coordinatization.

1.8. Let us say that two subspaces S_1, S_2 of $\mathbf{M}$ are *isomorphic* if there exists a Lorentz transformation L with $LS_1 = S_2$. Show that there are exactly three isomorphism classes of three-dimensional subspaces in $\mathbf{M}$, and write down canonical examples of each.

1.9. Prove the "backward Schwartz" inequality for $\mathbf{M}$, which states that for any two timelike vectors $u, v \in \mathbf{M}$,

$$|<u, v>| \geq (<u, u><v, v>)^{1/2}.$$

Note that this implies that no two timelike vectors can be orthogonal.

1.10. Show that if two events $p = (p^0, \vec{p})$, $q = (q^0, \vec{q}) \in \mathbf{M}$ are relatively spacelike, then there is a Lorentz transformation L such that relative to the new coordinatization $(t', \vec{x}') := L(t, \vec{x})$ it defines, the new time coordinates p'^0 and q'^0 of p and q are equal.

1.11. Prove that a particle's four-velocity is always a unit vector.

1.12. Denote the boost 1.4(8) with velocity $\vec{v}$ as $B(\vec{v})$. Compute that if $R = I \oplus U$ is any spatial rotation, where U is an orthogonal transformation on R^3 and I the identity on R^1, then

$$RB(\vec{v})R^{-1} = B(U\vec{v}).$$

1.13. Let $t \longmapsto z(t) = (t, \vec{z}(t)) \in \mathbf{M}$ be a particle worldline parametrized by coordinate time t. Consider a new coordinatization defined by a Lorentz transformation L:

$$(t', \vec{x}') := L(t, \vec{x}) \ .$$

To any point $z(t) = (t, \vec{z}(t))$ on the worldline corresponds a unique new coordinate time t'; denote this relationship by writing $t' = f(t)$.

Show that $t \longmapsto f(t)$ is a strictly monotonically increasing function of t if and only if L is of the form BR with B a boost and R a rotation, and is strictly monotonically decreasing if and only if $L = -BR$. This makes precise a sense in which Lorentz transformations BR preserve the "direction of time", while those of the form $-BR$ reverse it.

1.14. Taking space as one-dimensional for simplicity, consider two Lorentz frames related as in 1.4(2) by

$$t' := \gamma(v)(t - vx)$$

$$x' := \gamma(v)(x - vt) \ .$$

Suppose a particle is moving with instantaneous coordinate velocity $w := dx'/dt'$ as seen from the primed frame. Derive a formula for its coordinate velocity in the original frame in terms of v and w. (It may be easier to think first about the case of motion with constant velocity.)

1.15. Consider two particles of rest mass unity moving in a straight line with equal and opposite velocities v_1 and $v_2 = -v_1$ in a given Lorentz frame. Suppose that they collide and move apart with velocities v_1' and v_2', respectively. Suppose also that the rest masses are not altered by the collision and that the Newtonian momentum conservation law

(1) $\qquad v_1 + v_2 = v_1' + v_2'$

holds in the given frame, so that $v_2' = -v_1'$. Transfer to a frame

$$\bar{t} := \gamma(v)(t - vx)$$

$$\bar{x} := \gamma(v)(x - vt) \ ,$$

and show that the corresponding law

(2) $\qquad \bar{v}_1 + \bar{v}_2 = \bar{v}_1' + \bar{v}_2'$,

for the barred frame, where $\bar{v}_1$ denotes the velocity corresponding to v_1 in the barred frame, etc. (cf. Exercise 14), is equivalent to

(3) $\qquad vv_1^2 = v(v_1')^2$.

In particular, if $v \neq 0$ and (2) holds, then $v_1' = \pm v_1$. Physically, this says that the Newtonian law of conservation of momentum can hold in the barred frame only if all collisions are "elastic" (no change in total kinetic energy).

1.16. (Relativistic Doppler shift) Suppose a rocket carries a standard clock which emits a flash of light every second. A stationary observer on the ground measures the time interval between flashes as Δt with his standard clock. Assume that the rocket is coming directly toward the observer with uniform velocity $v > 0$. Show that Δt is independent of time until the rocket passes the observer and is given by

$$\Delta t = \left(\frac{1-v}{1+v}\right)^{1/2} .$$

(The same formula with v replaced by $-v$ applies when the rocket is receding from the observer.) Note that Δt is *not* given by the time dilation formula 1.4(6). We are not concerned with time intervals measured by a network of stationary clocks on the ground which the rocket is passing as in 1.4(6), but rather with time intervals as measured by a *particular* stationary clock.

The periodic flashes may be considered as a signal of frequency 1, and the observed signal has frequency $1/\Delta t$. Of course, the result scales to other frequencies, so it may be viewed as a calculation of the ratio of the transmitted frequency of a periodic signal to the received frequency. When the rocket is coming toward the observer, the received frequency is higher than the transmitted frequency. This phenomenon occurs nonrelativistically as well for any sort of signal travelling at finite speed through a medium, though the formula is different and also depends on whether the observer or source is moving with respect to the medium. For example, the whistle of an approaching train sounds higher than the same whistle when the train is receding. Similarly, spectral lines of elements in stars appear bluer (higher frequency) if the star is moving toward the earth and redder (lower frequency) if the star is moving away. Given the observed spectral shift, the formula can be solved for the velocity of approach or recession, which is of enormous importance to astronomers.

1.17. Let $\tau \longmapsto z(\tau) = (t(\tau), \vec{x}(\tau))$ be a particle worldline parametrized by proper time τ. Show that the elapsed proper time between any two events on the worldline is never more than the elapsed coordinate time. That is, for any $\tau_1 < \tau_2$,

(*) $\qquad\qquad t(\tau_2) - t(\tau_1) \geq \tau_2 - \tau_1 .$

This is the origin of the so-called "twin paradox" in which a person who takes a trip will, on returning, have aged less than his twin who stayed home. Of course, this is not a paradox at all, but rather an unfamiliar physical phenomena which, though it may initially seem strange, is now generally accepted. The effect has not been observed in humans (the discrepancy in aging being unmeasurably small for the kinds of trips which people can actually take), but many other physical systems do behave this way. There are now even clocks so accurate that the effect can be observed by carrying them on airplanes. (See the interesting survey article by C. O. Alley in [Meystre and Scully] for detailed descriptions of several such experiments and many references.)

1.18. Show that any affine bijection $T : \mathbf{M} \longmapsto \mathbf{M}$ which preserves the set of null lines is of the form

$$T(x) = \lambda L x + b \quad \text{for all} \quad x \in \mathbf{M} \ ,$$

where b is a fixed vector in $\mathbf{M}$, L is a Lorentz transformation, and λ is a nonzero scalar. (An *affine* bijection T is defined as one of the form $Tx = Cx + b$, where C is an invertible linear transformation and $b \in \mathbf{M}$ is given).

1.19. An experimental aircraft is designed to fly faster than the speed of light, and a test pilot agonizes over whether to accept the dangerous assignment of testing it for the first time. He decides to go ahead, and on the first flight flies 6000 km. in 10^{-2} sec (about twice the speed of light), relative to the usual Lorentz frame used to make measurements on the Earth. Unfortunately, he crashes on landing and is killed. Show that there is some Lorentz frame relative to which the plane crashes before it takes off and the pilot dies before he has decided whether to make the flight. (Assume for simplicity that the Earth is at rest in the given Lorentz frame and that the plane flies in a straight line at constant velocity.) This is one reason that it is usually assumed that ordinary, everyday matter cannot travel faster than the speed of light.

1.20. In the derivation of the famous equation $E = M$ ($= Mc^2$) connecting "mass" and "energy", we started from the Newtonian definition

(1) $\qquad \vec{F} = \dfrac{d\vec{p}}{dt}$

of "force", $\vec{F}$, as the derivative of the three-momentum $\vec{p}$, and we took $\vec{p}$ not as $m\vec{v}$, but rather as

(2) $\qquad \vec{p} := M\vec{v} = m\gamma(\vec{v})\vec{v}$.

Show that if we use these definitions (1) and (2) and define the coordinate acceleration $\vec{A}$ as usual,

$$\vec{A} := \dfrac{d\vec{v}}{dt} \ ,$$

then Newton's law

(3) $\qquad \vec{F} = K\vec{A}$,

where K is a scalar "mass" (which may depend on $\vec{v}$) cannot hold in general. Show that if the coordinate acceleration $\vec{A}$ is orthogonal to the coordinate velocity $\vec{v}$, then K must be given by

(4) $\qquad K = M = \gamma m$,

while if $\vec{A}$ happens to be parallel to $\vec{v}$ (as in one-dimensional motion), then

(5) $\qquad K = \gamma^2 M = \gamma^3 m$.

In the older relativistic literature, (5) is sometimes called the "longitudinal mass", while (4) is the "transverse mass".

These new "masses" are not considered particularly important. One of the points of this exercise is that many Newtonian constructs do not have simple or natural analogs within the framework of special relativity.

1.21. Given an orthonormal basis $\{e_i\}_{i=0}^{3}$ for $\mathbf{M}$, show that every Lorentz transformation L can be written as $L = \pm BR$, where B is a boost and R a spatial rotation, both relative to the Lorentz coordinatization defined by this basis.

1.22. Consider a particle moving on a straight line with worldline $\tau \longmapsto z(\tau) = (t(\tau), x(\tau))$ parametrized by proper time τ with the two irrelevant space dimensions suppressed from the notation. Then, setting $v := dx/dt$, its four-velocity is $u(\tau) := \gamma(v)(1, v)$. Define $w(\tau) := \gamma \cdot (v, 1)$, and observe that w is a spacelike unit vector orthogonal to u. The basis u, w for two-dimensional Minkowski space is geometrically natural for many calculations associated with one-dimensional particle motion.

Recalling that $du/d\tau$ is always orthogonal to u, we may define the scalar proper acceleration A (not to be confused with the quite different $\tilde{A}$ at the end of Section 1.7) by

$$\frac{du}{d\tau} = Aw ,$$

so that A is the scalar acceleration as seen from the rest frame of the particle. Assume that A is constant. For a physical picture, think of a rocket ship accelerated in a straight line in such a way that its occupants always feel constant acceleration (constant "gravity"!).

(a) Write down and solve explicitly the differential equation for the worldline $\tau \longmapsto z(\tau)$. For a neat representation of the solution, recall that

$$\frac{d \sinh \tau}{d\tau} = \cosh \tau \quad \text{and} \quad \frac{d \cosh \tau}{d\tau} = \sinh \tau .$$

(b) Can a material particle ever "outrun" a photon? For a test case, suppose a friend of one of the passengers is located at space coordinate 0 at coordinate time 0 and at that time tries to send a message to the receding spacecraft, which blasted off at coordinate time 0 from space coordinate $b > 0$ with zero velocity and constant positive acceleration (so that it is moving to the right). The message consists of a single goodbye blink of a flashlight. Will the message be received for all initial positions b? If not, for which?

1.23. Classically, a photon is often modelled as a "particle" moving at the speed of light. Though this picture is not entirely appropriate for real photons (cf. the footnote in Section 1.3), it suffices for many purposes, and let us temporarily accept it. The worldline of such a particle is a parametrized null line

(1) $\qquad s \longmapsto z(s) = sn + b , \qquad -\infty < s < \infty ,$

where $n, b \in \mathbf{M}$, $<n, n> = 0$, and $n^0 > 0$.

There is no natural parameter for the worldline to take the place of proper time. The future-pointing null tangent vector n is, of course, only determined up to a positive multiple, corresponding to a linear change in parametrization. Changing from a parameter s to a parameter $\bar{s} := ks$ changes the tangent by the constant factor $k^{-1} = ds/d\bar{s}$. It is rather remarkable and beautiful that this extra freedom can be exploited to include the concept of *frequency* of a photon in the description, as we shall now explain.

First, we *define* a photon to be a parametrized line of the form (1). Different parametrizations of the same worldline will describe different possible attributes (such as energy and frequency) of a photon with this worldline. The *energy-momentum* four-vector of the photon (1) is defined as

$$\text{energy-momentum of photon (1)} := n = \frac{dz}{ds}.$$

If $n = (n^0, \vec{n})$ relative to a given Lorentz coordinatization, the *energy* of the photon relative to that coordinatization is defined as n^0 and the *momentum* as $\vec{n}$. Obviously, the energy and momentum depend on the parametrization as well as the coordinatization. (One doesn't equate the "mass" of a photon with its energy; the analogy between mass and energy doesn't extend this far. Photons are considered to have zero mass, though this doesn't really make sense classically; one has to go into quantum mechanics to see what it means.)

A basic principle of quantum mechanics identifies the energy $E := n^0$ relative to a given Lorentz frame with the *frequency* ν of the photon as observed in that frame via the equation

$$E = h\nu,$$

where h is Planck's constant. The frequency is observed as *color;* higher frequency is more toward the blue end of the spectrum and lower is redder. One also associates a *wavelength*

$$\lambda := \frac{1}{\nu}$$

with the photon; this is just the wavelength which a wave of frequency ν traveling with unit velocity (the speed of light) would have. A natural analog of 1.8(10) is postulated as a law of conservation of energy-momentum in a collision of two entities with initial energy-momenta p_1, p_2 and final energy-momenta p_1', p_2':

$$p_1 + p_2 = p_1' + p_2'.$$

Here p_1 and p_2 can be energy-momenta of either photons or massive particles.

(a) Suppose a photon with frequency ν traveling along the x-axis in a given Lorentz frame hits an elementary particle (in this context this means constant rest mass) such as an electron sitting at rest and is "scattered" into a new photon traveling at an angle θ to the x-axis with frequency ν'. (See diagram.)

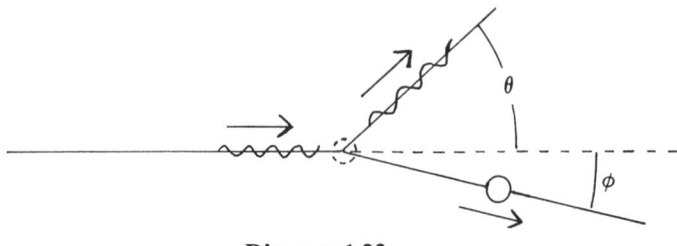

Diagram 1.23

The particle also recoils and moves off at an angle ϕ. Write down a system of equations which can be solved for the frequency ν' of the scattered photon in terms of the scattering angle θ, given the particle mass and the initial frequency ν.

(b) Show that the initial and final frequencies are related to θ by the equation

(2) $$\frac{1}{\nu'} - \frac{1}{\nu} = \lambda' - \lambda = \frac{h}{m}(1 - \cos \theta) \ .$$

(The algebraic solution may not be immediately obvious, but there is a trick by which the equations can be easily solved.) The number h/m is called the *Compton wavelength* of the scattering particle [†], and it is sometimes mentioned as the characteristic length at which classical physics might be expected to break down and quantum effects become important. [‡] Such scattering of light from electrons is actually observed and does behave as (2) predicts.

[†] If units are used in which the velocity of light is c, then the Compton wavelength is h/mc, and some authors define it as $\hbar/mc := h/2\pi mc$.
[‡] Note, however, that the Compton wavelength is not obviously related to any "wavelike" behavior of electrons, such as diffraction.

Chapter 2

Mathematical Tools

In this chapter we review some definitions and results, mainly from multilinear algebra and differential geometry, which we shall need. The reader who is well-versed in differential geometry need only skim it to absorb the notation, while one whose knowledge stems from a half-remembered course taken years ago will probably need to read it more carefully. Few proofs are given, but most of the omitted arguments are trivial or routine; those which are not are clearly flagged. The ability to fill the gaps, or at least sense how the proofs might go, could be taken as a test as to whether the reader has the background for Chapters 3 and 4.

2.1. Multilinear algebra.

All vector spaces considered in this section will be finite-dimensional spaces over the real field **R**. By *inner product space* we mean a vector space equipped with a symmetric, non-degenerate bilinear form, usually denoted $<\cdot,\cdot>$ or $g(\cdot,\cdot)$. (Non-degenerate means that only the zero vector can be orthogonal to all vectors in the space.) The vector space which is the direct sum of n copies of **R** is denoted R^n; when we want to consider it as an inner product space with the usual positive definite inner product, we speak of *Euclidean* R^n.

(a) Dual spaces and tensors.

Definition. A *linear form* f, also called a *1-form*, on a vector space V is a linear map $f: V \longrightarrow R$. That is, f is a real-valued function on V which satisfies

$$f(su + tv) = sf(u) + tf(v) \qquad \text{for all } u, v \in V \text{ and all } s, t \in \mathbf{R}.$$

The set of all linear forms on V is a vector space (under the usual operations of addition and scalar multiplication of functions) which is denoted V^* and is called the *dual space* for V.

Let $e_1, e_2, \cdots, e_n$ be a basis for V. The *dual basis* to $\{e_i\}$ is the basis for V^* consisting of the linear forms $e^1, e^2, \cdots, e^n$ (the superscripts are indices, not exponents) defined by

$$e^j(e_i) := \delta^j{}_i := \begin{cases} 1 & \text{if } i = j \\ 0 & \text{if } i \neq j \end{cases}.$$

Any $f \in V^*$ satisfies

$$f = \sum_{i=1}^{n} f(e_i) e^i ,$$

and from this and the obvious linear independence of the e^i, it follows immediately that $\{e^i\}$ is a basis for V^*.

Now suppose that V has the additional structure of a distinguished inner product (i.e. non-degenerate, symmetric, bilinear function) $<\cdot,\cdot>$. (Since the main application we have in mind is Minkowski space with the Minkowski inner product, of course we do not assume that it is positive definite). Then the given bilinear function induces an isomorphism

$$v \longmapsto v^*$$

from V to V^* defined as follows:

For each $v \in V$, v^* is the linear form defined by

(1) $\qquad v^*(u) = <v, u> \quad$ for all $u \in V$.

Physics texts often refer to an application of this isomorphism as "lowering an index". The terminology arises as follows. Any vector $v \in V$ can, of course, be written

(2) $\qquad v = \sum_{i=1}^{n} v^i e_i$

for a unique list of scalars $v^1, v^2, \cdots, v^n$, called the *components* of v with respect to the given basis. Notice the placement of the indices i. It turns out to be a very useful convention to place the indices so that when a "product" of indexed quantities is summed, as with and e_i in (2), the summed pair always appears with one index in the upper position and the other in the lower, and we shall follow this convention. Also, we shall use without comment the notation u^i for the components of a vector $u \in V$ with respect to a given basis. Under this convention, the components of u^* with respect to the dual basis $\{e^i\}$ should carry a lower index; we follow the standard practice of denoting them as u_i rather than the more logical but more cumbersome u^*_i. Thus

$$u^* = \sum_{1}^{n} u_i e^i \quad ,$$

and in passing from u, denoted by its list of components (u^i), to u^*, denoted by (u_i), one "lowers an index".

Actually, it is more usual to omit the parentheses and denote u as u^i and u^* as u_i. This turns out to be a convenient shorthand in most situations, but it does leave the problem of how to gracefully distinguish between the *vector* $u = u^i$ and a generic component u^i of u, as in the next sentence. Of course, a particular component of u, such as u^2, is not usually numerically equal to its counterpart u_2; the exact relation between the two is that for any fixed j,

(3) $\qquad u_j = \sum_{i=1}^{n} u_i e^i(e_j) = u^*(e_j) = <u, e_j> = \sum_{i=1}^{n} u^i <e_i, e_j>$.

Sums similar to those in (3) appear frequently in computations. In fact, when the same symbol appears as an upper and lower index in a product, as in $u^i <e_i, e_j>$, it is almost always summed, and it is another useful convention to omit the summation sign in all expressions of this type. This is known as the *Einstein summation convention,* and we shall adopt it throughout, using summation signs only for emphasis. Also, in relativity texts the inner product components $<e_i, e_j>$ are almost universally denoted by the slightly simpler expression g_{ij}, and we shall use this also where convenient. Thus in terms of this compactified notation, (3) may be written as

(4) $$u_j = u^i g_{ij} .$$

Another notational device which we find helpful is to distinguish summed indices by writing them in Greek as in

$$u_j = u^\alpha g_{\alpha j} .$$

We shall not do this consistently, but only where special emphasis is desired. Since some authors reserve Greek indices for space indices, we emphasize that all summations are over the full range of indices.

For any form $f \in V^*$ and vector $v \in V$, we have

$$f(v) = f_\alpha e^\alpha(v^\beta e_\beta) = f_\alpha v^\beta e^\alpha(e_\beta) = f_\alpha v^\beta \delta^\alpha{}_\beta$$
$$= f_\alpha v^\alpha .$$

Similarly, the inner product $<u, v>$ of two vectors $u, v \in V$ may be variously written in terms of components as:

(5) $$<u, v> = \sum_{\alpha=1}^{n} \sum_{\beta=1}^{n} u^\alpha v^\beta <e_\alpha, e_\beta> = u^\alpha v^\beta g_{\alpha\beta} = u^*(v)$$
$$= u_\beta v^\beta = v^*(u) = u^\beta v_\beta .$$

Note, in particular, the extremely compact expressions $u_\beta v^\beta$ and $u^\beta v_\beta$ in which the inner product components $g_{\alpha\beta}$ are hidden in the notation for the components of u^* and v^*.

Another common and useful convention is to denote the inverse of the matrix (g_{ij}) as (g^{ij}), so that, by definition,

(6) $$g^{i\alpha} g_{\alpha j} = \delta^i{}_j \qquad \text{for all } i, j .$$

The utility of this notation is shown by the observation, whose proof is left as an easy exercise, that passing from a linear form $f = \sum f_i e^i$ to the corresponding vector $u \in V$ with $u^* = f$ ("raising an index") is accomplished using (g^{ij}) as follows:

(7) $$u^i = g^{i\alpha} f_\alpha .$$

Correspondingly, we define for any form $f \in V^*$,

$$f^i := g^{i\alpha} f_\alpha ,$$

so that raising an index on f produces the components of the vector u satisfying $u^* = f$.

Definition. Let $p, q \geq 0$ be integers, and let V and W be vector spaces. A *W-valued tensor of contravariant rank p and covariant rank q* on V is a multilinear function

$$F: \underbrace{V^* \times V^* \times \cdots \times V^*}_{p \text{ copies}} \times \underbrace{V \times V \times \cdots \times V}_{q \text{ copies}} \longrightarrow W .$$

That is, F is a function of $p+q$ variables, the first p variables from V^* and the last q variables from V, which is linear in each variable separately:

$$F(v_1, \cdots, v_{s-1}, au+bw, v_{s+1}, \cdots, v_{p+q})$$
$$= aF(v_1, \cdots, v_{s-1}, u, v_{s+1}, \cdots, v_{p+q}) + bF(v_1, \cdots, w, \cdots, v_{p+q})$$

for all variables $v_1, \cdots, v_{s-1}, u, w, v_{s+1}, \cdots, v_{p+q}$ and all scalars $a, b \in \mathbf{R}$.

The set of all W-valued tensors of contravariant rank p and covariant rank q is a vector space under the obvious operations which we shall call $L_q^p(V; W)$; elements of this space will be called *W-valued tensors of type p,q*, or simply *tensors of type p,q* when $W = \mathbf{R}$. We shall also abbreviate $L_q^p(V; \mathbf{R})$ as V_q^p.

Readers who are familiar with the general theory of tensor products (helpful, but not a necesary prerequisite) will recognize that V_q^p is naturally isomorphic to the tensor product $V \otimes V \otimes \cdots \otimes V \otimes V^* \otimes V^* \otimes \cdots \otimes V^*$ of p copies of V with q copies of V^*. Those conversant with category theory may wonder, with good reason, why such an object is not called p-covariant and q-contravariant, and the answer is that this branch of mathematics is older than category theory and the historically established terminology happens to be the reverse of what it "should" be. (Of course, there *is* a logic behind the older terminology. See, for example, [Dodson and Poston, p.82].)

By multilinearity, every $F \in V_q^p$ is uniquely determined by the "array"

$$(F^{i_1 i_2 \cdots i_p}{}_{j_1 j_2 \cdots j_q}),$$

where

$$F^{i_1 i_2 \cdots i_p}{}_{j_1 j_2 \cdots j_q} := F(e^{i_1}, e^{i_2}, \cdots, e^{i_p}, e_{j_1}, e_{j_2}, \cdots, e_{j_q}),$$

with each index $i_1, \cdots, i_p, j_1, \cdots, j_q$ running from 1 to n. This array is called the *matrix* of F with respect to the given basis $\{e_i\}$ for V. In many older physics and mathematics texts, such an array (with the parentheses generally omitted) *is* the notation for a tensor. This notation is ugly and typographically difficult, but it does have computational advantages in certain circumstances. We shall use coordinate-free notation most of the time, but shall not hesitate to use component notation when it seems clearer. Also, we shall often omit the parentheses when denoting matrices of tensors, writing, for example, $T^{ij}{}_k$ in place of $(T^{ij}{}_k)$.

Given vectors $v_1, \cdots, v_p \in V$ and linear forms $f_1, \cdots, f_q \in V^*$, define the tensor $v_1 \otimes \cdots \otimes v_p \otimes f_1 \otimes \cdots \otimes f_q \in V_q^p$ as the multilinear map whose value on arbitrary $g_1, \cdots, g_p \in V^*$ and $w_1, \cdots, w_q \in V$ is given by

$$(v_1 \otimes \cdots \otimes v_p \otimes f_1 \otimes \cdots \otimes f_q)(g_1, \cdots, g_p, w_1, \cdots, w_q) := \prod_{i=1}^p g_i(v_i) \prod_{j=1}^q f_j(w_j).$$

Tensors of this form are called *decomposable*. The relation between an arbitrary tensor $F \in V_q^p$ and its matrix with respect to a basis $\{e_i\}$ for V may also be expressed as:

$$F = \sum F^{i_1 \cdots i_p}{}_{j_1 \cdots j_q} e_{i_1} \otimes \cdots \otimes e_{i_p} \otimes e^{j_1} \otimes \cdots \otimes e^{j_q}.$$

where each index i_k and j_m ranges from 1 to n, and it is easy to see that the set of all

$$e_{i_1} \otimes \cdots e_{i_p} \otimes e^{j_1} \otimes \cdots \otimes e^{j_q}$$

is a basis for V_q^p.

(b) Natural isomorphisms.

(i) There is a natural isomorphism between the space $L^0_1(V; V)$ of all linear transformations on V and the space V^1_1 of all real-valued tensors of type 1,1 which assigns to each $T \in L^0_1(V; V)$ the element $\bar{T}$ of V^1_1 defined by

$$\bar{T}(f, v) = f(T(v)) \qquad \text{for all } f \in V^* \text{ and } v \in V.$$

(Note that this isomorphism is independent of the inner product.) If we write the matrix for T as $(T^i{}_j)$ (where i is the row index and j the column index), then the matrices for T and $\bar{T}$ are identical. For this reason, we use the notation $(T^i{}_j)$ for the matrix of a linear transformation instead of the (T_{ij}) more common in abstract linear algebra. In our notation, the action of T on a vector v conforms to the summation convention when expressed in components: $(Tv)^i = T^i{}_\alpha v^\alpha$.

We shall occasionally encounter other natural isomorphisms. For example, given a linear transformation T, the bilinear map

$$f, v \longmapsto f(Tv) \qquad f \in V^*, \quad v \in V$$

just considered may also be viewed as a rule which assigns to each form $f \in V^*$ another form $T^\dagger f \in V^*$ defined by

$$(T^\dagger f)(v) = f(Tv) \qquad \text{for all } v \in V.$$

This linear transformation $T^\dagger : V^* \longrightarrow V^*$ is called the *adjoint* of T. If T has matrix $(T^i{}_j)$ with respect to a basis $\{e_i\}$ for V, then for any $f = \sum f_\alpha e^\alpha \in V^*$,

$$(T^\dagger f)_j = f_\alpha T^\alpha{}_j.$$

In this notation, the matrix of $T^\dagger$ is the same as the matrix of T (instead of its transpose as in the usual notation of abstract linear algebra), but the rule by which it operates on a list on components is different. In effect, when the matrix operates on column vectors, it is T, and on row vectors it is $T^\dagger$. The up-down index conventions keep these natural isomorphisms in plain view.

The *trace* $\text{tr}(T)$ of a linear transformation T with matrix $T^i{}_j$ is, of course, the scalar defined by

(8) $$\text{tr}(T) = \sum_\alpha T^\alpha{}_\alpha.$$

When the operation of taking the trace of a linear transformation is transferred to V^1_1 via the above natural isomorphism, it is called *contraction*.† Thus the contraction of the tensor in V^1_1 with matrix $T^i{}_j$ is the scalar defined by (8). More generally, one can "contract" a tensor in V^p_q, $p, q > 0$, with respect to an upper and lower index to obtain a tensor in V^{p-1}_{q-1}. For example, a tensor T in V^2_1 with matrix $T^{ij}{}_k$ can be contracted to obtain the tensor in V^1_0 with matrix S^j given by

$$S^j = \sum_\alpha T^{\alpha j}{}_\alpha.$$

---

† A definition of contraction which is aesthetically preferable on a higher algebraic level begins by viewing V^1_1 as $V \otimes V^*$ and defining contraction $\kappa(\cdot)$ as the linear map $\kappa : V^1_1 = V \otimes V^* \longrightarrow \mathbf{R}$ corresponding to the bilinear map $v, f \longmapsto f(v)$, $v \in V$, $f \in V^*$, via the universal property of tensor products. That is, κ is the unique linear map with the property that $\kappa(v \otimes f) = f(v)$ for all $v \in V$, $f \in V^*$. This is preferable to (8) because it is manifestly coordinate-independent.

This appears to depend on the choice of basis but is actually basis-independent for the same reason that (8) is. Another, usually different, contraction of T is the tensor U with matrix U^i given by

$$U^i = \sum_\alpha T^{i\alpha}{}_\alpha .$$

One of the virtues of component notation is the ease and clarity with which contractions are indicated. Incidentally, the Einstein summation convention applies to contractions also, though it looks a little strange. For instance, (8) would be written under this convention as

$$tr(T) = T^\alpha{}_\alpha .$$

(ii) The inner product furnishes a natural isomorphism $F \longmapsto \hat{F}$ between the space V_2^0 of bilinear functions on V and the space $L_1^0(V; V)$ of linear transformations on V defined as follows:

(9) For each $F(\cdot, \cdot) \in V_2^0$, $\hat{F}(\cdot) \in L_1^0(V; V)$ is defined by

$$\langle \hat{F}(u), v \rangle = F(u, v) \quad \text{for all } u, v \in V .$$

That there exists a unique $\hat{F}$ satisfying (9) is a standard fact of linear algebra. The matrix $\hat{F}^i{}_j$ of the linear transformation $\hat{F}$ with respect to a given basis is just the matrix F_{ij} of the tensor F with an index raised:

(10) $$\hat{F}^i{}_j = g^{i\alpha} F_{\alpha j} .$$

(iii) Indices are raised and lowered on tensors of arbitrary rank in a similar way; for example, lowering the middle index on a tensor T^{ijk} is defined in terms of components by

$$T^i{}_j{}^k = T^{i\alpha k} g_{\alpha j} .$$

It is easily verified that this is independent of the basis with respect to which the tensor components are defined. Of course, such an index-lowering operation corresponds to a certain natural isomorphism defined in a coordinate-free manner, but we can't say explicitly what the isormorphism is without naming additional spaces of tensors. (Technically, $T^i{}_j{}^k$ is not in V_1^2 because the "form" index is in the middle instead of on the right.) Fortunately, there is no need to pursue this boring matter further because the only index lowerings and raisings which we shall need in this book will be applied to 1-forms and 2-forms.

2.2. Alternating forms.

(a) The wedge product.

An antisymmetric tensor of covariant rank $q \geq 1$ is called a *q-form*. Thus $\alpha \in V_q^0$ is a q-form if and only if for all $v_1, \cdots, v_q \in V$ and all permutations π,

$$\alpha(v_{\pi(1)}, v_{\pi(2)}, \cdots, v_{\pi(q)}) = \text{sgn}(\pi) \alpha(v_1, \cdots, v_q)$$

where $\text{sgn}(\pi)$ denotes the sign of π. It is a standard fact that the set of all q-forms on a vector space of dimension n, $1 \leq q \leq n$, is a vector space of dimension

$$\begin{bmatrix} n \\ q \end{bmatrix} := \frac{n!}{q!(n-q)!} ,$$

which we call $\Lambda_q(V)$. † It is convenient to define $\Lambda_0(V) := \mathbf{R}$, so that a 0-form, by definition, is a real scalar. There are no nonzero q-forms when $q > n$ (Exercise 4).

Any q-form can be built out of 1-forms as follows. Given q 1-forms $\alpha_1, \alpha_2, \cdots, \alpha_q$, define a q-form denoted

$$\alpha_1 \wedge \alpha_2 \wedge \cdots \wedge \alpha_q$$

as the antisymmetrization of $\alpha_1 \otimes \alpha_2 \otimes \cdots \otimes \alpha_q$:

(1) $$\alpha_1 \wedge \alpha_2 \wedge \cdots \wedge \alpha_q := \sum_\pi \text{sgn}(\pi) \alpha_{\pi(1)} \otimes \alpha_{\pi(2)} \otimes \cdots \otimes \alpha_{\pi(q)} ,$$

where the sum is over all permutations π of $\{1, 2, \ldots, q\}$. ‡ It is easy to see that every q-form is a linear combination of q-forms of the type (1). Moreover if $\{e_i\}$ is any basis for V^*, then it is not hard to show that the set of q-forms

(2) $$e^{i_1} \wedge e^{i_2} \wedge \cdots \wedge e^{i_q} \quad \text{with} \quad 1 \leq i_1 < i_2 < \cdots < i_q \leq n$$

constitute a basis for $\Lambda_q(V)$. Thus any q-form can be built from 1-forms as a linear combination (also as a sum) of forms of the type (1).

A q-form which can be written in the form (1) is called *decomposable*. This is not the same as the concept of a decomposable tensor defined in the previous section, but no confusion can arise since no nonzero q-form can be a decomposable tensor when $q \geq 2$, and every 1-form is trivially decomposable in both senses (Exercise 9). In dimension $n \leq 3$, every q-form is decomposable, but this is not true in higher dimensions (Exercise 6).

The "wedge product" $\alpha \wedge \beta$ of a decomposable p-form $\alpha = \alpha_1 \wedge \cdots \wedge \alpha_p$ with a decomposable q-form $\beta = \beta_1 \wedge \cdots \wedge \beta_q$, where the α_i and β_j are 1-forms, is defined by:

$$\alpha \wedge \beta := \alpha_1 \wedge \cdots \wedge \alpha_p \wedge \beta_1 \wedge \cdots \wedge \beta_q ,$$

and the wedge product of an arbitrary p-form α with an arbitrary q-form β is defined by writing each as a sum of decomposable forms and requiring that the wedge product distribute over addition. Naturally, it is necessary to check that the wedge product is well-defined. This requires careful organization if the proof is not to degenerate into a notational nightmare, and

---
† Many authors use $\Lambda_q(V^*)$, which is preferable within a more general algebraic context.
‡ Some authors insert a factor of $1/q!$ in the right side of (1). This must be kept in mind when comparing formulas in various texts.

2.2 Alternating forms

we shall not attempt it here. It can be shown that the wedge product is distributive with respect to addition (i.e. $\alpha \wedge (\beta + \gamma) = \alpha \wedge \beta + \alpha \wedge \gamma$ and $(\beta + \gamma) \wedge \alpha = \beta \wedge \alpha + \gamma \wedge \alpha$), associative (i.e. $(\alpha \wedge \beta) \wedge \gamma = \alpha \wedge (\beta \wedge \gamma)$), and satisfies the law of commutation:

(3) $$\beta \wedge \alpha = (-1)^{pq} \alpha \wedge \beta \qquad \text{for any } p\text{-form } \alpha \text{ and } q\text{-form } \beta.$$

The gaps in the above sketchy description of the wedge product are sufficiently wide that the reader unfamiliar with the theory of alternating forms would probably save time by referring to a proper text rather than filling them himself. However, this is largely unnecessary for our purposes since we shall not use nontrivial properties of the wedge product very often. For most purposes, all we really need is the *notation* (1) of the wedge product of a collection of 1-forms as the antisymmetrization of the corresponding tensor product and the easy fact that the decomposable 1-forms of (2) form a basis for $\Lambda_q(V)$.

(b) The inner product on forms.

Let V be a vector space with non-degenerate inner product $<\cdot,\cdot>$. Let $e_1, \cdots, e_n$ be a basis for V and $e^1, \cdots, e^n$ the dual basis. The inner product on the space $\Lambda_q(V)$ of all q-forms for $q \geq 1$, also denoted $<\cdot,\cdot>$, is defined so as to make the basis (2) consisting of all wedge products

$$e^{i_1} \wedge \cdots \wedge e^{i_q}, \qquad 1 \leq i_1 < i_2 < \cdots < i_q \leq n$$

an orthonormal basis in such a way that the signs of the norms of the basis elements are given by

(4) $$<e^{i_1} \wedge \cdots \wedge e^{i_q}, e^{i_1} \wedge \cdots \wedge e^{i_q}> = \prod_{k=1}^{q} <e_{i_k}, e_{i_k}>.$$

This can be done uniquely by *fiat* for the given basis, and then it can be checked that (4) must hold for all bases if it holds for one (Exercise 7). Notice that the inner product on 1-forms is just the inner product on V transferred to V^* via the index-lowering isomorphism $v \longmapsto v^*$. The inner product of 0-forms (scalars) α, β is defined by $<\alpha, \beta> := \alpha\beta$.

The expression for the inner product of arbitrary q-forms α, β in terms of components is easily seen to be

(5) $$<\alpha, \beta> = \frac{1}{q!} \alpha_{i_1 \cdots i_q} g^{i_1 j_1} \cdots g^{i_q j_q} \beta_{j_1 \cdots j_q},$$

but we shall never need this. It is included only to avoid confusion of this inner product with the usual inner product on *tensors*, which for antisymmetric tensors is just (5) without the factor $1/q!$. The inner product on tensors is defined by requiring that *tensor* (as opposed to wedge) products of an orthonormal basis are to be orthonormal and that the norm of a decomposable tensor is to be the product of the norms of its factors. Since $e^{i_1} \wedge \cdots \wedge e^{i_q}$ is the sum (1) of $q!$ orthonormal decomposable tensors each of norm (4), its norm relative to the tensor inner product is $q!$ times (4). The difference between the two inner products is obviously undesirable, but if the inner product on forms is defined to agree with the inner product on tensors, then annoying factorial factors proliferate throughout the formulae. Fortunately, we shall never use the inner product on tensors, so no confusion can arise.

(c) The Hodge dual.

The inner product on an oriented inner product space V induces a very useful duality operation on alternating forms. If V has dimension n, then this duality operation assigns to each p-form α an $(n-p)$-form denoted $\perp\alpha$ and called the *Hodge dual* of α. (Many authors use $*\alpha$ instead of $\perp\alpha$.) The Hodge duality is defined relative to a given n-form Ω, which is customarily taken to be the so-called "volume n-form", so we shall first explain how the latter is defined.

Recall that the space $\Lambda_n(V)$ of all n-forms on a vector space V of dimension n is one-dimensional, so that any two non-zero n-forms are multiples of each other. Let $\{e_i\}$ be a basis for V and $\{e^i\}$ the dual basis. Then

(6) $$e^1 \wedge e^2 \wedge \cdots \wedge e^n$$

is a nonzero n-form. If $\{f_i\}$ is another basis for V, then

$$e^1 \wedge \cdots \wedge e^n = c\, f^1 \wedge \cdots \wedge f^n$$

for some constant c. The constant c is just the determinant of the linear transformation T satisfying $Te_i = f_i$ for $i = 1, 2, \cdots, n$; in fact, this is essentially the usual abstract definition of the determinant of a linear transformation. If the two given bases are orthonormal, then the transformation T is orthogonal, and so has determinant ± 1. This shows that the n-form (6) is determined up to a sign if the basis $\{e_i\}$ is taken to be orthonormal. Now to say that V is *oriented* means that one of the two equivalence classes of ordered bases has been designated as "positively oriented", where two ordered bases $\{e_i\}$ and $\{f_i\}$ are defined as equivalent if the linear transformation T relating them as above has positive determinant. It follows that the n-form (6) is uniquely specified if it is defined using an orthonormal, positively oriented basis $\{e_i\}$. This n-form is called the *volume n-form* on the oriented inner-product space V.

Definition. † Let Ω denote the volume n-form on an oriented inner-product space V of dimension n, and let $0 \leq p \leq n$. The *Hodge dual* of a given p-form α is defined to be the unique $(n-p)$-form $\perp\alpha$ such that for all p-forms β,

(7) $$\beta \wedge \perp\alpha = \langle \beta, \alpha \rangle \Omega .$$

It is not immediately obvious that such an $(n-p)$-form exists, but it is an easy exercise that if it does exist then it is unique. To establish the existence, first consider the case $p \geq 1$. Since the right side of (7) is linear in α and the left in $\perp\alpha$, it is sufficient to show that $\perp\alpha$ exists for a set of α which spans the space of all p-forms. Thus it is sufficient to define $\perp\alpha$ on the set of p-forms of the type

(8) $$e^{i_1} \wedge e^{i_2} \wedge \cdots \wedge e^{i_p}$$

with $\{e_k\}$ an orthonormal basis for V and $i_1 < i_2 < \cdots < i_p$. Given a collection of p distinct indices $i_1, \cdots, i_p$, denote the other $n-p$ indices from the set $\{1, 2, \cdots, n\}$ as $j_1, j_2, \cdots, j_{n-p}$, and let s denote the sign of the permutation $i_1, \cdots, i_p, j_1, \cdots, j_{n-p}$ of $1, 2, \cdots, n$. Then it is routine to check that a q-form $\perp\alpha$ satisfying (7) is given for a p-form α of the type (8) by

---

† Some authors use a definition of "Hodge dual" which differs from ours by a sign which can depend on the degree of the form, the dimension of the space, and the signature of the inner product.

(9) $$\perp\alpha = \perp(e^{i_1} \wedge \cdots \wedge e^{i_p}) = s <\alpha, \alpha> e^{j_1} \wedge \cdots \wedge e^{j_{n-p}}.$$

In other words, the Hodge dual of the wedge product of a collection of basis vectors from an orthonormal set consists of the wedge product of the remaining basis vectors with the sign $s <\alpha, \alpha>$ attached. For a 0-form α (i.e. a scalar), the definition gives $\perp\alpha := \alpha\Omega$.

Example. If we take V to be Euclidean 3-space with a given orientation and e_1, e_2, e_3 a positively oriented orthonormal basis, then it is easy to check using (9) that for any vectors $\vec{v}, \vec{w}$,

(10) $$\perp(\vec{v}* \wedge \vec{w}*) = (\vec{v} \times \vec{w})*,$$

where $\vec{v} \times \vec{w}$ denotes the ordinary vector cross product in R^3.

Returning to an arbitrary V, if we set $s := \prod_{i=1}^{n} <e_i, e_i>$ with $\{e_i\}$ an orthonormal basis, then the map $\alpha \longmapsto \perp\alpha$ is an isomorphism from the vector space of all p-forms to the space of $(n-p)$-forms with the property that for all p-forms α,

(11) $$\perp(\perp\alpha) = (-1)^{p(n-p)} s\alpha,$$

so that $\perp^{-1} = \pm\perp$, where the sign depends on the degree of the form, the dimension of V, and the number of spacelike (negative norm) directions in V. Two important special cases are:

(12) $\perp(\perp\alpha) = \alpha$ for all p-forms α on Euclidean space R^3, and

(13) $\perp(\perp\alpha) = (-1)^{p+1}\alpha$ for all p-forms α on Minkowski space.

There is a simple geometrical interpretation of the Hodge dual of a *decomposable p-form* α. Such a p-form can be written as

(14) $$\alpha = w_1* \wedge w_2* \wedge \cdots \wedge w_p*$$

for some vectors $w_1, w_2, \cdots, w_p \in V$. We shall show below that the rule by which $\perp\alpha$ assigns a real number to a given $(n-p)$-tuple of vectors $v_1, v_2, \cdots, v_{n-p} \in V$ is:

(15) $$(\perp\alpha)(v_1, v_2, \cdots, v_{n-p}) = \Omega(w_1, w_2, \cdots, w_p, v_1, \cdots, v_{n-p}).$$

That is, $(\perp\alpha)(v_1, v_2, \cdots, v_{n-p})$ may be characterized as the volume of the n-parallelipiped spanned by $w_1, \cdots, w_p, v_1, \cdots, v_{n-p}$.

By linearity in the w_i's, it is sufficient to establish (15) in the case in which the list $w_1, w_2, \cdots, w_p$ is orthonormal. After possible reindexing, it may be embedded in a positively oriented orthonormal basis $w_1, \cdots, w_p, w_{p+1}, \cdots, w_n$ except in the exceptional case $p=1$, $n=1$, which we handle by replacing α by $-\alpha$. Then

$$\alpha = w_1* \wedge \cdots \wedge w_p* = w^1 \wedge \cdots \wedge w^p \prod_{k=1}^{p} <w_k, w_k>$$

$$= <\alpha, \alpha> w^1 \wedge \cdots \wedge w^p,$$

and so from (9),

(16) $$\perp(w_1^* \wedge \cdots \wedge w_p^*) = \perp\alpha = \langle\alpha,\alpha\rangle^2 w^{p+1} \wedge \cdots \wedge w^n$$
$$= w^{p+1} \wedge \cdots \wedge w^n \;.$$

This formula is often handy because all the signs due to the negative norm directions in V cancel.

From (16) we see that for any list of $n-p$ basis vectors $w_{k_{p+1}}, w_{k_{p+2}}, \cdots, w_{k_n}$, the value

(17) $$\perp\alpha(w_{k_{p+1}}, w_{k_{p+2}}, \cdots, w_{k_n})$$

vanishes if the list is not a permutation of $w_{p+1}, w_{p+2}, \cdots, w_n$ and otherwise is the sign of the permutation. Since the expression

(18) $$\Omega(w_1, \cdots, w_p, w_{k_{p+1}}, \cdots, w_{k_n}) \;,$$

has the same property, (17) and (18) are the same, which establishes (15) with each v_i replaced by $w_{k_{p+i}}$. That is, we have established (15) for the case in which the v_i are drawn from the given basis $w_1, \cdots, w_n$, and the general case of (15) then follows immediately from the linearity of both sides in the variables v_i.

From (15) follows easily a formula for the Hodge dual in terms of tensor components: for any p-form α with components with respect to a given basis (not necessarily orthonormal) denoted $\alpha_{i_1 \cdots i_p}$ as usual, we have

(19) $$(\perp\alpha)_{j_1 \cdots j_{n-p}} = \frac{1}{p!} \alpha^{\lambda_1 \cdots \lambda_p} \Omega_{\lambda_1 \cdots \lambda_p j_1 \cdots j_{n-p}} \;,$$

where summation over the indices λ_i is implied. † To see how (19) follows from (15), first note that by the usual formula for transformation of tensor components from one basis to another, if (19) is valid for components with respect to one particular basis, then it is valid for components with respect to any basis, so it is sufficient to verify (19) for components with respect to the previous basis $w_1, \cdots w_p, w_{p+1}, \cdots, w_n$. By linearity, we may also assume that α is given by (14). In this case it is apparent from the definition of α that, $\alpha^{\lambda_1 \cdots \lambda_p}$ vanishes if $\lambda_1, \cdots, \lambda_p$ is not a permutation of $1, \cdots, p$ and otherwise is the sign of the permutation. Hence for any *particular* list of indices $i_1, \cdots, i_p$,

(20) $$\alpha^{i_1 \cdots i_p} \Omega_{i_1 \cdots i_p j_1 \cdots j_{n-p}} \qquad (\textbf{no summation implied})$$

vanishes unless both $i_1, \cdots, i_p$ is a permutation of $1, \cdots, p$ and $j_1, \cdots, j_{n-p}$ is a permutation of $p+1, \cdots, n$, and when it does not vanish, it is the sign of the permutation $j_1, \cdots, j_{n-p}$ of $p+1, \cdots, n$. (The sign of the permutation $i_1, \cdots, i_p$ of $1, \cdots, p$ effectively occurs in both α and Ω and thus cancels.) Equation (15) shows that

$$(\perp\alpha)_{j_1 \cdots j_{n-p}} := (\perp\alpha)(w_{j_1}, \cdots, w_{j_{n-p}})$$
$$= \Omega(w_1, \cdots, w_p, w_{j_1}, \cdots, w_{j_{n-p}}) =: \Omega_{1 \cdots p j_1 \cdots j_p}$$

---

† In literature written in component notation, the component array of the volume n-form on the right is traditionally denoted $\epsilon_{i_1 \cdots i_n} := \Omega_{i_1 \cdots i_n}$ and is called the *Levi-Civita* tensor.

is the sign of the same permutation. In other words, $(\perp\alpha)_{j_1\cdots j_{n-p}}$ is equal to (20) for any particular choice of distinct indices $i_1\cdots i_p$ drawn from the list $1,\cdots,p$, and the right side of (19) is just (20) summed over all $p!$ possible such choices and then divided by $p!$ to compensate for the sum. (Incidentally, notice how awkward the summation convention becomes when one needs to work with something like (20). If arguments like this occured very frequently, one would hesitate to employ it.)

Fortunately, one rarely needs the ugly component formula (19) for the Hodge dual, but we include it because it appears once in a technical proof in Section 2.9. However, the special case of (15) in which $\alpha = w^*$ is a 1-form on Minkowski space $\mathbf{M}$ will occur in a central way in a fundamental calculation of energy-momentum radiation in Section 4.4. For this special case, (15) states that for any four vectors $w, v_1, v_2, v_3 \in \mathbf{M}$,

(21) $\qquad (\perp w^*)(v_1, v_2, v_3) = \Omega(w, v_1, v_2, v_3)$.

The reader who skipped the proof of (15) above and wants to understand Section 4.4 should at least verify (21), which is notationally (though not essentially) simpler than (15).

2.3. Manifolds.

Most of our treatment of electrodynamics does not require the concept of a manifold, the most notable exception being an important computation in Section 4.4. All the ideas can be developed within the framework of special relativity, the only manifolds which need appear are submanifolds of Minkowsksi space, and for most purposes Minkowski space itself is sufficient. Moreover, some of the important results and techniques are special to Minkowski space and do not readily generalize to semi-Riemannian manifolds and general relativity. Nonetheless, I feel that many of the ideas are actually clearer when presented in the context of manifolds. In addition, limiting the exposition to Minkowski space would blur certain distinctions which are important in general relativity and thus limit the usefulness of our treatment for those who wish to go on in that fascinating subject. For this reason, we make free use of the machinery of differential geometry, which may initially seem excessive to the unaccustomed, but whose power and appropriateness should be evident in the end. In the next few section we shall give a brief, informal introduction to the ideas of differential geometry which we shall use. It is not intended to replace a proper treatment of this vast and enormously important subject. For some, it will establish notation and point of view, for others it may bridge the gap between what one should know after a first course and what one actually remembers, and for still others it will point the way to further study in a rich and beautiful field.

To informally introduce the idea of manifolds, let us start with a connected topological space M and an integer n. The following paragraphs explain what it means to say that M is an n-dimensional manifold.

Suppose that for each point $p \in M$ we are given an open neighborhood U of p and a "coordinatization"

$$\phi_U : U \longrightarrow R^n$$

of U. We use the term coordinatization in essentially the same sense in which it was used in the preceding chapter, where the set of all events was coordinatized by a map onto R^4. More precisely, when we say that ϕ_U is a coordinatization, we mean that ϕ_U is a homeomorphism (i.e. bicontinuous bijection) from U onto R^n. † These given coordinatizations, which are assumed to satisfy a compatibility condition given below, are the additional structure which converts the topological space M into a manifold.

---

† Another commonly used definition of coordinatization allows the ϕ_U to map onto open balls in R^n rather than onto R^n itself. Since any open ball can be mapped in a C^∞ and bijective manner onto R^n,

The coordinatization maps ϕ_U allow us to locally identify M with R^n. That is, given any point p in M, some open neighborhood U of p can be identified with R^n via the bijection ϕ_U. If we write

$$\phi_U(p) = (x^1(p), \cdots, x^n(p)),$$

then the functions $x^i(\cdot): U \longrightarrow R$ are called the *coordinate functions* of the given coordinatization of U. (Upper indices are used for the coordinate functions so that the 1-forms which are their differentials will satisfy the index conventions of Section 1.) When we shift attention from U to the larger space M, we sometimes call ϕ_U a *local* coordinatization. The values $x^i(q)$ of the coordinate functions are then called the *local coordinates* of the points $q \in M$ relative to the given local coordinatization ϕ_U.

As an example, consider the unit sphere S^2 in R^3:

$$S^2 = \{ (x, y, z) \in R^3 \mid x^2 + y^2 + z^2 = 1 \}.$$

There is no way to homeomorphically identify S^2 with R^2. (For instance, S^2 is compact while R^2 is not.) However, if any single point is removed from S^2, then the remainder is an open subset of S^2 which is homeomorphic to R^2 via, for example, the usual construction which identifies the complex plane with the Riemann sphere. Thus any point $p \in S^2$ has an open neighborhood (for example, S^2 with the antipodal point of p deleted) which is homeomorphic with R^2.

Nothing we have said so far precludes the possibility that, given $p \in M$, there may be two open sets U and V containing p with corresponding coordinate maps ϕ_U and ϕ_V. Naturally, we want ϕ_U and ϕ_V to be compatible in the sense that the composition of ϕ_V with the inverse of ϕ_U,

$$\phi_V \circ \phi_U^{-1}: \phi_U(U \cap V) \longrightarrow R^n$$

should preserve whatever structure of R^n we are interested in. The definition of C^∞ *manifold* is that all such maps $\phi_V \circ \phi_U^{-1}$ should be C^∞ (i.e. infinitely continuously differentiable). We shall assume that all manifolds under consideration are C^∞ manifolds. The dimension n of the Euclidean space R^n with which the manifold is locally identified is called the *dimension* of the manifold.

For example, we can make S^2 into a C^∞ manifold in the following way. Let U denote S^2 minus the north pole, V denote S^2 minus the south pole, and ϕ_U, ϕ_V the previously mentioned "Riemann sphere" bijections from U and V onto R^2. Then every point of S^2 is in either U or V, and once we check that $\phi_U \circ \phi_V^{-1}$ and $\phi_V \circ \phi_U^{-1}$ are C^∞, we have made S^2 into a C^∞ manifold.

A real-valued function f on M is called a C^∞ *function* if for each coordinatization ϕ_U, the map

$$f \circ \phi_U^{-1}: R^n \longrightarrow R$$

is C^∞. The ring of all C^∞ functions (under the usual operations of addition and multiplication of functions) will be denoted $C^\infty(M)$.

---

the two definitions are equivalent.

2.4. Tangent spaces and vector fields.

For the moment let us consider the unit sphere S^2 as embedded in R^3, though the manifold structure is independent of this circumstance. While it is embedded in R^3, one can consider, at each point $p \in S^2$, vectors in R^3 emanating from p which are tangent to S^2 (Figure 1). Any such vector v can be used to define a real-valued linear map $\tilde{v}$ on $C^\infty(R^3)$ by differentiation in the direction of v:

(1) $$\tilde{v}(f) := \lim_{t \to 0} \frac{f(p+tv) - f(p)}{t} \quad \text{for each } f \in C^\infty(R^3).$$

Since v is tangent to S^2, it is easy to show that $\tilde{v}(f)$ depends only on the values of f on the sphere S^2; that is, if $f = g$ on S^2, then $\tilde{v}(f) = \tilde{v}(g)$.

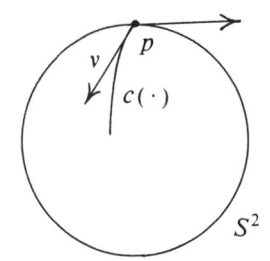

Figure 2-1. Tangent vectors at a point p of the sphere S^2. Differentiating a function f in the direction of v is the same as differentiating f along a suitably parametrized curve $c(\cdot)$ in S^2 tangent to v.

It follows that we may view $\tilde{v}$ as a linear map on $C^\infty(S^2)$ instead of $C^\infty(R^3)$ by extending a given $f \in C^\infty(S^2)$ to $\tilde{f} \in C^\infty(R^3)$, and defining $\tilde{v}(f) := \tilde{v}(\tilde{f})$. From now on we shall view $\tilde{v}$ in this way, and when we want to emphasize the fact that v is a vector emanating from p, we shall sometimes write $\tilde{v}_p$ in place of $\tilde{v}$.

The linear map $f \longmapsto \tilde{v}(f)$ has the "Leibnitz" property:

(2) $$\tilde{v}_p(fg) = \tilde{v}_p(f)g(p) + f(p)\tilde{v}_p(g) \quad \text{for all } f, g \in C^\infty(S^2).$$

By expanding functions in finite Taylor series about p, one easily shows that any real-valued linear map on $C^\infty(S^2)$ with property (2) is of the form $\tilde{v}_p$ for some vector v_p at p. The importance of this observation is that real-valued linear maps on $C^\infty(S^2)$ which satisfy (2) may be identified with tangent vectors, and this suggests the following way of defining the "tangent space" to a general manifold at a point p even when the manifold is not embedded in a larger Euclidean space.

In general, if M is any manifold (which need not be embedded in R^n), a *tangent vector* w at a point $p \in M$ is *defined* to be a real-valued linear map $w \longmapsto w(f)$ on $C^\infty(M)$ such that (2) holds for all $f, g \in C^\infty(M)$ with w in place of $\tilde{v}_p$. The set of all tangent vectors at a given point p is a vector space under the obvious operations called the *tangent space* to M and denoted M_p. This definition may seem rather roundabout, but it neatly captures all the essential properties of "tangent spaces" of manifolds embedded in R^n in an intrinsic manner, without reference to any such embedding.

When M is itself a vector space, equation (1) gives a natural way to identify each tangent space M_p with M. This will be called the *natural identification* of M with M_p.

A *vector field* v on M is defined as a map $p \longmapsto v_p \in M_p$ which assigns to each point $p \in M$ a vector v_p at p. The vector field v is said to be C^∞ if for each C^∞ function f on M, the function $p \longmapsto v_p(f)$ is also a C^∞ function. This last function will also be denoted as $v(f)$. If v is a vector field and g a C^∞ function, then the vector field gv is defined in the obvious way by

(3) $\qquad (gv)_p(f) = g(p)v_p(f) \quad$ for all $f \in C^\infty(M)$,

and the set of all vector fields on M may thus be viewed as a module over $C^\infty(M)$. Of course, we may replace M by an open subset U of itself and speak of C^∞ vector fields and functions defined on U, and similar remarks apply.

Let $p \longmapsto \phi(p) \in R^n$ be a local coordinatization of M with coordinate functions x^j, so that

$$\phi(p) = (x^1(p), \cdots, x^n(p)).$$

For any C^∞ function f on M and any given $p \in M$, the operation of differentiating $f \circ \phi^{-1}$ at $\phi(p)$ with respect to the j'th coordinate is a tangent vector denoted $(\partial/\partial x_j)_p$:

$$\left(\frac{\partial}{\partial x_j}\right)_p(f) := \frac{\partial(f \circ \phi^{-1})}{\partial x_j}$$

$$= \lim_{h \to 0} \frac{f \circ \phi^{-1}(x_1, \cdots, x_j + h, \cdots, x_n) - f \circ \phi^{-1}(x_1, \cdots, x_j, \cdots, x_n)}{h}.$$

We also abbreviate $(\partial/\partial x_j)_p$ as $\partial/\partial x_j$. † The set $\partial/\partial x_1, \cdots, \partial/\partial x_n$ is easily shown to be a basis for M_p which is often convenient for calculations. If we view $\partial/\partial x_j$ as a locally defined vector field (i.e the field $p \longmapsto (\partial/\partial x_j)_p$), then the set $\partial/\partial x_1, \cdots, \partial/\partial x_n$ forms a basis for the module of all locally defined vector fields over the ring of C^∞ functions. That is, every vector field v defined in a sufficiently small neighborhood U of p can be uniquely written as

$$v = \sum_{i=1}^{n} v^i \frac{\partial}{\partial x_i} \qquad \text{with } v^i \in C^\infty(U).$$

If $t \longmapsto c(t) \in M$, $(t \in R)$, is any C^∞ curve in M, define its *derivative* at t, which for any fixed t will be a vector in $M_{c(t)}$ denoted $c'(t)$, by

$$c'(t)(f) := \lim_{h \to 0} \frac{f(c(t+h)) - f(c(t))}{h} \qquad \text{for all } f \in C^\infty(M).$$

Intuitively, $c'(t)$ is the vector tangent to the curve $c(\cdot)$ at the point $c(t)$. (Figure 1.)

The next two paragraphs are included only to make a later definition self-contained and will not play an important role in the sequel. They may be skipped without essential loss.

Consider a C^∞ vector field v on M and a given point $p \in M$. To *integrate* v at p means to find a curve $t \longmapsto c(t)$ such that $c(0) = p$ and

(4) $\qquad c'(t) = v_{c(t)}$

---

† The upper indices on the coordinate functions x^j are shifted to lower indices in $\partial/\partial x_j$ in order that the components v^i of a tangent vector $\sum v^i \partial/\partial x_i$ can carry an upper index and still conform to the summation convention.

for all t in some open interval $(-\delta, \delta)$. Standard results on differential equations imply that this can always be done since when (4) is transferred to R^n via a coordinatization, it becomes a first-order vector differential equation of the form

(5) $$\frac{d\vec{x}}{dt} = \vec{F}(\vec{x}(t)) .$$

The solutions to (4) are called *integral curves* for v. If we consider the solution to (4) as depending on p as a parameter and indicate this dependence by denoting it as $c_p(\cdot)$, then we get a map $t, p \longmapsto c_p(t)$ from $(-\delta, \delta) \times M \longrightarrow M$ which is called the *flow* associated with the vector field. Intuitively, the flow associated with a vector field v on M may be pictured as a collection of curves in M with the property that there is exactly one curve through each point $p \in M$ and the tangent vector to this curve at p is v_p.

Now we review the construct known as the *commutator* of two vector fields. Let v and $\bar{v}$ be vector fields with flows

$$t, p \longmapsto c_p(t) \quad \text{and}$$

$$t, p \longmapsto \bar{c}_p(t), \quad \text{respectively.}$$

Fix $t > 0$, start at a given base point p, and follow the flow of v up to $c_p(t)$. From there, follow the flow of $\bar{v}$ up to $\bar{c}_{c_p(t)}(t)$. If we do this in the opposite order (follow $\bar{v}$ then v), we end up at $c_{\bar{c}_p(t)}(t)$, and of course, there is no reason that these two paths should end at the same point. Intuitively, the *commutator* of $\bar{v}$ with v at p, written $[\bar{v}, v]_p$, is a vector in M_p which measures how large the gap is relative to t. In other words, $[\bar{v}, v]$ measures the extent to which the flows c and $\bar{c}$ fail to commute. The formal definition of $[\bar{v}, v]$ is:

(6) $$[\bar{v}, v]_p(f) = \bar{v}_p(v(f)) - v_p(\bar{v}(f)) \qquad \text{for all } f \in C^\infty(M) .$$

(In (6), $v(f)$ is considered as the function $p \longmapsto v_p(f)$, and similarly for $\bar{v}(f)$.) Note that though neither of the maps $f \longmapsto \bar{v}_p(v(f))$, nor $f \longmapsto v_p(\bar{v}(f))$ satisfy the Leibnitz identity (2), their difference does, and so (6) does define a vector in M_p. Unfortunately, it is not easy to see directly from (6) in exactly what sense $[\bar{v}, v]$ measures the non-commutativity of the flows. For a precise statement along these lines, see [Hicks, p. 15, problem 13].

A mapping $p \longmapsto \omega_p \in M_p^*$, which assigns to each $p \in M$ a 1-form ω_p on M_p is called a *1-form field*, or simply a *1-form*, on M, and is said to be C^∞ if the map $p \longmapsto \omega_p(v_p)$ is C^∞ for all C^∞ vector fields v. Other C^∞ tensor fields of arbitrary degree are defined similarly. We shall assume that all tensor fields are C^∞ unless otherwise specified.

One important example of a 1-form field deserves special mention. With every C^∞ function f on M is associated a 1-form field df, called the *differential* of f, by:

(7) $$(df)_p(v_p) := v_p(f) \qquad \text{for all } p \in M, \; v_p \in M_p .$$

That is, the value of df on a vector v_p is the directional derivative of f in the direction of v_p.

If $\{x^i\}$ are the coordinate functions of a local coordinatization

$$p \longmapsto (x^1(p), \cdots, x^n(p)) ,$$

then at each point p, the set of 1-forms $(dx^1)_p, \cdots, (dx^n)_p$ is a basis for the space M_p^* of 1-forms at p. In fact, it is the dual basis to $\partial/\partial x_1, \cdots, \partial/\partial x_n$ and is very useful for calculations. In terms of this basis, the differential of any function f may be written,

$$df = \sum_{i=1}^{n} \frac{\partial f}{\partial x_i} dx^i .$$

Since $\{dx^i\}$ is a basis for M_p^*, a general p-form on M, $1 \le p \le n$, can be locally written as a sum of decomposable p-forms of the type

(8) $\qquad f\, dx^{i_1} \wedge dx^{i_2} \wedge \cdots \wedge dx^{i_p} \qquad$ with $f \in C^\infty(M)$,

and this often simplifies computations. Since a 0-form on a vector space is by convention a scalar, a 0-form field on a manifold is simply a function.

2.5. Covariant derivatives.

Let us return to the example of the sphere S^2 embedded as the unit sphere in R^3. Given any vector v at a point $p \in S^2$ and any vector field w on R^3, we can define the derivative of w in the direction of v at p, denoted $D_v w$, by the usual formula of vector calculus:

$$D_v w := \lim_{t \to 0} \frac{w_{p+tv} - w_p}{t} .$$

Although this works for any vector field $p \longmapsto w_p \in R^3$, we shall only be interested in the case in which $w_p \in (S^2)_p$ for all p; i.e. when $p \longmapsto w_p$ is a vector field of *tangent* vectors on S^2. If the components of w are $w = (w_1, w_2, w_3)$, then $D_v w = (v(w_1), v(w_2), v(w_3))$. In other words, this is just componentwise differentiation. As noted in the previous section, $v(w^i)$ depends only on the values of w^i on S^2, and hence $D_v w$ depends only on the values w_q of w for q on S^2. It follows that we may view the map $w \longmapsto D_{v_p} w$ as a vector-valued map on C^∞ tangent vector fields w defined near p, and this is a natural way to differentiate tangent vector fields on S^2 when S^2 is embedded in R^3.

One might think that there would be some way to intrinsically characterize such differentiation without reference to the embedding analogous to the way that tangent spaces were intrinsically defined, but unfortunately, there is none. "Vector differentiation" is a structure which must be added to the basic manifold structure. The minimal properties which one wants to demand of such a differentiation on a manifold M are:

(1) $\qquad$ Given a point $p \in M$ and a C^∞ vector field w defined near p (i.e. defined in some neighborhood of p), the map $v \longmapsto D_v w$, $v \in M_p$, should be a linear map from M_p to M_p, and

(2) $\qquad$ For any v in M_p, any such vector field w, and C^∞ function f defined near p,

$$D_v(fw) = v(f) D_v w + f D_v w .$$

Any such D is called a *covariant derivative*, or *connection* on M.

There are many possible covariant derivatives, and in the present generality there is no obvious condition to add to (1) and (2) to single out a "natural" one. (However, we shall later specialize to semi-Riemannian manifolds, where there is such a condition.) One desirable property of a covariant derivative is that

(3) $$D_v w - D_w v = [v, w]$$

for all vector fields v, w, since componentwise differentiation of vector fields on R^n satisfies this. A covariant derivative for which (3) holds is said to be *torsion-free*. (There seems to be no good motivation for the use of the term "torsion" in this context.) From now on, we shall consider only torsion-free covariant derivatives.

The addition of condition (3) still does not uniquely characterize D. To see this, choose a local coordinatization $\{x^i\}_{i=1}^n$, and for each i and j, expand

(4) $$D_{\frac{\partial}{\partial x_i}} \frac{\partial}{\partial x_j} = \sum_{k=1}^n \Gamma_{ij}^k \frac{\partial}{\partial x_k} ,$$

where this defines the functions $p \longmapsto \Gamma_{ij}^k(p)$ on the coordinate neighborhood. These function are called *connection coefficients,* or sometimes, *Christoffel symbols.* Since $[\partial/\partial x_i, \partial/\partial x_j] = 0$, any covariant derivative which satisfies (3) will also satisfy $\Gamma_{ij}^k = \Gamma_{ji}^k$, and conversely. Thus locally specifying an arbitrary torsion-free connection is equivalent to specifying connection coefficients which are symmetric under interchange of the two lower indices.

A useful theorem [Hicks, p. 131] states that given any point $p \in \mathbf{M}$, it is always possible to choose a local coordinatization of some neighborhood of p such that $\Gamma_{ij}^k(p) = 0$ for all i, j, k. (However, it is not usually possible to make $\Gamma_{ij}^k(\cdot)$ vanish throughout the entire coordinate neighborhood.) Note that this also shows that the array $(\Gamma_{ij}^k(p))$ does not transform as the components of a tensor would under change of coordinates because if the components of a tensor vanish in one coordinate system, then they must vanish in all systems.

A covariant derivative D tells us how to differentiate *vector* fields. Given this, there is also a natural way to differentiate *1-form* fields, defined as follows. Given a 1-form field α defined near $p \in M$ and a vector $v \in M_p$, define the covariant derivative of α in the direction v, denoted $D_v \alpha$ by requiring that for any vector field w, the following Leibnitz-type identity should hold:

(5) $$(D_v \alpha)(w) = v(\alpha(w)) - \alpha(D_v w) .$$

(The $\alpha(w)$ on the right denotes, of course, the function $p \longmapsto \alpha_p(w_p)$.) At first glance, it may appear that the definition is incomplete because for any vector $v_p \in M_p$, $D_{v_p}\alpha \in M_p^*$ is supposed to be a form *at* p, so to define it we should only have to specify $(D_{v_p}\alpha)(w_p)$ for $w_p \in M_p$, not for all vector *fields* w defined in a neighborhood of p. However, it is easily seen that both sides of (5) are linear over the ring $C^\infty(M)$ in the sense that replacing w by fw with $f \in C^\infty(M)$ only multiplies (5) by f. Although f gets differentiated by the v's on the right side, the resulting derivatives in the two terms cancel. This can be used to show that (5) actually only depends on the value w_p of w at p (Exercise 14). Thus given a vector $w_p \in M_p$, we extend it to a locally defined vector field w in any old way and then define $(D_{v_p}\alpha)(w_p) := (D_{v_p}\alpha)(w)$ by (5).

If D is given by (4), then the statement of (5) in terms of the connection coefficients and the basis $\{\partial/x_i\}$ is:

(6) $$D_{\frac{\partial}{\partial x_i}}(dx^j) = -\sum_k \Gamma_{ik}^j dx^k ,$$

and for an arbitrary 1-form $\alpha = \sum \alpha_j dx^j$,

(7) $$D_{\frac{\partial}{\partial x_i}} \alpha = \sum_k \left(\frac{\partial \alpha_k}{\partial x_i} - \sum_j \alpha_j \Gamma_{ik}^j \right) dx^k .$$

For reference, the corresponding formula for a vector field $w = \sum w^j \frac{\partial}{\partial x_j}$ is

(8) $$D_{\frac{\partial}{\partial x_i}} w = \sum_k \left(\frac{\partial w^k}{\partial x_i} + \sum_j w^j \Gamma_{ij}^k \right) \frac{\partial}{\partial x_k} .$$

Notice how the index conventions make these formulas easy to remember: just sum whichever indices of Γ_{ij}^k are in the appropriate positions.

Finally, D is extended to tensor fields of arbitrary higher rank as follows. Given any point $p \in M$, vector $v \in M_p$ and tensor field T of type s,q defined near p, $D_v T \in (M_p)_q^s$ is to be a tensor at p of the same rank s,q. The map $T \longmapsto D_v(T)$ is defined by requiring firstly that for each v in M_p, $D_v(\cdot)$ be linear over $\mathbf{R}$; i.e. for any two tensor fields S, T of the same rank and any $v \in M_p$,

(9)(i) $\qquad D_v(aS+bT) = aD_v S + bD_v T \qquad$ for all $a, b \in \mathbf{R}$,

and secondly that for any vector fields $v_1, \cdots, v_s$ and 1-form fields $\alpha_1, \cdots, \alpha_q$, the following "derivation" property hold:

(9)(ii) $$D_v(v_1 \otimes \cdots \otimes v_s \otimes \alpha_1 \otimes \cdots \otimes \alpha_q) = (D_v v_1) \otimes v_2 \otimes \cdots \otimes \alpha_q$$
$$+ v_1 \otimes D_v v_2 \otimes \cdots \otimes \alpha_q + \cdots + v_1 \otimes \cdots \otimes \alpha_{q-1} \otimes D_v \alpha_q .$$

Of course, if $q = 0$ or $s = 0$, the corresponding v's or α's are omitted from (ii). It is clear that (i) and (ii) uniquely determine $D_v(T)$ for any tensor field T with at least one index given that $D_v(T)$ is already defined when T is a vector field and when T is a 1-form field. To check that $D_v(T)$ is *well-defined* is either a simple exercise on a fairly high algebraic level (use the universal property of tensor products) or a routine but notationally complicated computation on the level of indicial tensor analysis. It will also be convenient to define $D_v(\alpha)$ for 0-form fields α (i.e. C^∞ functions) by

$$D_v \alpha := v(\alpha) \qquad \text{for all } v \in M_p .$$

Note that (i) and (ii) imply the analog of (2) for tensor fields of arbitrary rank; for any tensor field α and any C^∞ function f,

$$D_v(f\alpha) = v(f)\alpha + fD_v \alpha .$$

Let T be a given C^∞ tensor field of type s,q, so T has components $T^{i_1 \cdots i_s}{}_{j_1 \cdots j_q}$. Given $p \in M$, the map $v_p \longmapsto D_{v_p} T$ from M_p to the space $(M_p)_q^s$ of tensors of type s,q on M_p is linear, and may be viewed in an obvious way as an element of $(M_p)_{q+1}^s$, which we denote as $(DT)_p$ or simply DT. Thinking of DT as a function $p \longmapsto (DT)_p$ of $p \in M$ defines DT as a new tensor field of type $s,q+1$. More explicitly, for any $v_1, \cdots, v_q, v_{q+1}$ in M_p and $\beta_1, \cdots, \beta_s$ in M_p^*,

(10) $$(DT)(\beta_1, \cdots, \beta_s, v_1, \cdots, v_q, v_{q+1})$$
$$:= (D_{v_{q+1}} T)(\beta_1, \cdots, \beta_s, v_1, \cdots, v_q) .$$

That is, **the covariant derivative may be viewed as the map** D **from tensor fields of type** s,q

to tensor fields of type $s, q+1$ defined by (10).

In particular, we may apply D to a q-form α. The result, $D\alpha$, is a tensor of type $0, q+1$, but is not necessarily a $(q+1)$-form because it is not necessarily antisymmetric. Its antisymmetrization is called the *differential* of α and is denoted $d\alpha$; explicitly, the value of $d\alpha$ on any vectors $v_1, v_2, \cdots, v_{q+1}$ is given by

$$(11) \qquad d\alpha(v_1, \cdots, v_{q+1}) := \sum_{i=1}^{q+1} (-1)^{i+1} (D_{v_i}\alpha)(v_1, \cdots, \hat{v}_i, \cdots, v_{q+1}),$$

where $\hat{v}_i$ means that v_i is omitted. When α is a 0-form, $d\alpha$ was previously defined by 2.4(7).)

This definition of $d\alpha$ appears to depend on the connection D but is actually independent of D so long as D is torsion-free. In fact, we shall see that d can be computed in local coordinates on decomposable q-forms $f dx^{i_1} \wedge \cdots \wedge dx^{i_q}$ via the formula:

$$(12) \qquad d(f dx^{i_1} \wedge \cdots \wedge dx^{i_q}) = df \wedge dx^{i_1} \wedge \cdots \wedge dx^{i_q}$$

for all $f \in C^\infty(M)$. Since any q-form can be written as a sum of decomposable q-forms, (12) completely determines d and shows that it is independent of D. Actually, (12) is the usual definition of the differential operator d. Its advantages are that it is computationally convenient and obviously independent of D; its disadvantages are that it appears to depend on the choice of local coordinate system (though of course it doesn't in the end) and lacks intuitive content. Each is more convenient than the other in some applications, and so it is good to know both.

To see how (12) follows from the first definition, choose a local coordinate system $\{x^i\}$, and consider a decomposable q-form $f dx^{i_1} \wedge \cdots \wedge dx^{i_q}$ with $f \in C^\infty(M)$. Since $dx^{i_1} \wedge \cdots \wedge dx^{i_q}$ is an antisymmetrization of $dx^{i_1} \otimes \cdots \otimes dx^{i_q}$, it follows from (5) and (9) that for any vector v,

$$(13) \qquad D_v(f dx^{i_1} \wedge \cdots \wedge dx^{i_q}) = v(f) dx^{i_1} \wedge \cdots \wedge dx^{i_q}$$

$$+ f \sum_{s=1}^{q} dx^{i_1} \wedge \cdots \wedge D_v(dx^{i_s}) \wedge \cdots \wedge dx^{i_q}.$$

Recalling that $v(f) = df(v)$, one verifies (Exercise 16) that the antisymmetrization operation (11) applied to the first term gives $df \wedge dx^{i_1} \wedge \cdots \wedge dx^{i_q}$, so we need to show that the antisymmetrization of the rest vanishes. From (6),

$$D(dx^{i_s}) = -\sum_{j,k} \Gamma^{i_s}_{kj} dx^j \otimes dx^k,$$

which is a *symmetric* tensor when D is torsion-free, and given this observation, it is again straightforward to verify that the antisymmetrization of the connection-dependent terms does vanish. (If one already knows that (12) is coordinate-independent, the second verification can be avoided by using a coordinate system in which all the connection coefficients vanish at the point at which the equivalence is to be proved.)

We state as a proposition a slightly different way of computing the differential which will be useful later:

Proposition.

Let $e_1, e_2, \cdots, e_n$ be a collection of vector fields defined on an open subset U of an n-dimensional manifold M with the property that at each point $p \in U$, the collection $(e_1)_p, \cdots, (e_n)_p$ forms a basis for the tangent space M_p. Let $e^1, \cdots, e^n$ be the dual basis of form fields; i.e. $e^i(e_j) = \delta^i{}_j$. Let D be any (torsion-free) covariant derivative on M. Then for any q-form $\alpha \in C^\infty(U)$,

$$(14) \qquad d\alpha = \sum_{i=1}^n e^i \wedge D_{e_i}\alpha \ .$$

Proof. The reader may verify (Exercise 16) that for any 1-form λ, q-form β, and vectors $v_1, v_2, \cdots, v_{q+1}$,

$$(15) \qquad (\lambda \wedge \beta)(v_1, \cdots, v_{q+1}) = \sum_{k=1}^{q+1} (-1)^{k+1} \lambda(v_k) \beta(v_1, \cdots, \hat{v}_k, \cdots, v_{q+1}) \ .$$

Applying this with $\lambda = e^i$ and $\beta = D_{e_i}\alpha$ yields for any collection $e_{j_1}, \cdots, e_{j_{q+1}}$ of basis vectors:

$$\sum_{i=1}^n (e^i \wedge D_{e_i}\alpha)(e_{j_1}, \cdots, e_{j_{q+1}}) = \sum_{i=1}^n \sum_{k=1}^{q+1} (-1)^{k+1} e^i(e_{j_k})(D_{e_i}\alpha)(e_{j_1}, \cdots, \hat{e}_{j_k}, \cdots, e_{j_{q+1}})$$

$$= \sum_{k=1}^{q+1} (-1)^{k+1} (D_{e_{j_k}}\alpha)(e_{j_1}, \cdots, \hat{e}_{j_k}, \cdots, e_{j_{q+1}}) = d\alpha(e_{j_1}, \cdots, e_{j_{q+1}}) \ .$$

Of course, equality on all collections $e_{j_1}, \cdots, e_{j_{q+1}}$ of $q+1$ basis vectors implies by linearity equality on all collections of $q+1$ vectors.

An important property of the differential operator d is that for any p-form α and q-form β,

$$(16) \qquad d(\alpha \wedge \beta) = (d\alpha) \wedge \beta + (-1)^p \alpha \wedge d\beta \ .$$

The proof is a routine application of exterior algebra.

Another very important property of d is that its square is zero:

$$(17) \qquad d(d\alpha) = 0 \qquad \text{for any } p\text{-form } \alpha \ .$$

This follows for 0-forms (i.e C^∞ functions) from an easy calculation (essentially just a restatement of the fact that partial derivatives commute), and the general case follows immediately from the 0-form case and (15) applied to decomposable p-forms.

2.6. Stokes' Theorem.

Stokes' Theorem is essentially a generalization of the Fundamental Theorem of Calculus to manifolds formulated in terms of the differential operator d on form fields. Unfortunately, a proper treatment of this result in reasonable generality requires many fussy definitions and considerable care despite the fact that the basic ideas are quite intuitive. For this reason, we shall only describe this result informally. Good accounts of the technical details can be found in [Lang], [Warner], [Spivak], and many other texts. Our outline follows [Lang] most closely.

Consider the closed unit ball

(1) $$B = \{ (x, y, z) : x^2+y^2+z^2 \leq 1 \}$$

in R^3. This is not a manifold (at least not in any obvious way), since points on the unit sphere

$$S^2 = \{ (x, y, z) : x^2+y^2+z^2 = 1 \}$$

have no neighborhood which is obviously homeomorphic to R^3. However, the open ball

$$B_0 = \{ (x, y, z) : x^2+y^2+z^2 < 1 \}$$

is a manifold in a natural way, and the closed ball B consists of the union of the open ball B_0 with its topological boundary S^2 in R^3. The closed ball B is the prototypical example of what is called a *manifold with boundary*. The definition of "manifold with boundary" is essentially the same as the definition of "manifold" except that the local coordinatization maps are allowed to map onto half-spaces

(2) $$\{ (x_1, \cdots, x_n) \in R^n \mid x_1 \leq 0 \}$$

in R^n as well as onto R^n itself. The "boundary" of a manifold with boundary is defined as the union of the inverse images of the hyperplanes $x_1 = 0$ under those coordinatization maps which map onto half-spaces (2). Since a hyperplane in R^n can be identified with R^{n-1}, the boundary of an n-dimensional manifold with boundary should be itself a manifold of dimension $n-1$, under the manifold structure induced by these coordinate maps, and the formal definition of manifold with boundary is set up so that this does follow. We shall denote the boundary of a manifold M with boundary as ∂M. The "boundary" as defined above of a manifold with boundary should not be confused with other naturally ocurring topological boundaries. For example, the closed ball B, like any topological space, has no boundary when considered as a subspace of itself. Another simple example is a closed interval embedded as a line segment in R^3. As a subset of the topological space R^3, the line segment is its own boundary, whereas if it is considered in the natural way as a "manifold with boundary", its "boundary" consists of its endpoints.

A manifold M is said to be *oriented* if its manifold structure is defined by a collection of coordinate maps ϕ_U such that for any two such maps ϕ_U, ϕ_V, the Jacobian determinant of

$$\phi_U \circ \phi_V^{-1} : \phi_V(U \cap V) \longrightarrow R_n$$

is always positive when $U \cap V$ is nonempty. This condition allows us to consistently define a *positively oriented basis* for the tangent space at a point p as one which is mapped by the derivative of the coordinate maps onto a positively oriented basis for R^n, where the standard basis $(1, 0, \cdots, 0), (0, 1, 0, \cdots, 0), \cdots, (0, \cdots, 0, 1)$ for R^n is taken as positively oriented. The definition of an oriented manifold with boundary is similar and is formulated so that the $(n-1)$-dimensional boundary ∂M of an oriented n-dimensional manifold M with boundary is itself oriented by an orientation induced in a natural way by the original orientation on M. We shall elaborate on this last point after stating Stokes' Theorem; to say more about it here would only clutter the discussion and obscure the main ideas.

Let M be an n-dimensional oriented manifold with or without boundary, and let α be an n-form on M. We can intuitively conceive of "integrating" α over M as follows. Think of tangent vectors v_p at a point $p \in M$ as actually lying in the manifold, though of course, they don't, and think of partitioning the manifold into a large number of small pieces, each piece being approximately an n-cube spanned by a positively oriented basis $v_1, \cdots, v_n$ of tangent vectors. (See Figure 4, in which a two-dimensional surface in R^3 is divided into a large number of small pieces, each piece approximately a rectangle.) In terms of this picture, the integral of α over M would naturally be conceived as an appropriate limit, as the number of n-cubes gets large and the "size" of each cube gets small, of the sum of the numbers

(3) $\qquad \alpha(v_1, v_2, \cdots, v_n)$

obtained by applying α to the bases spanning the cubes. (The reason it is important that M be oriented is that if the bases were written in a different order, it could change the sign of (3).)

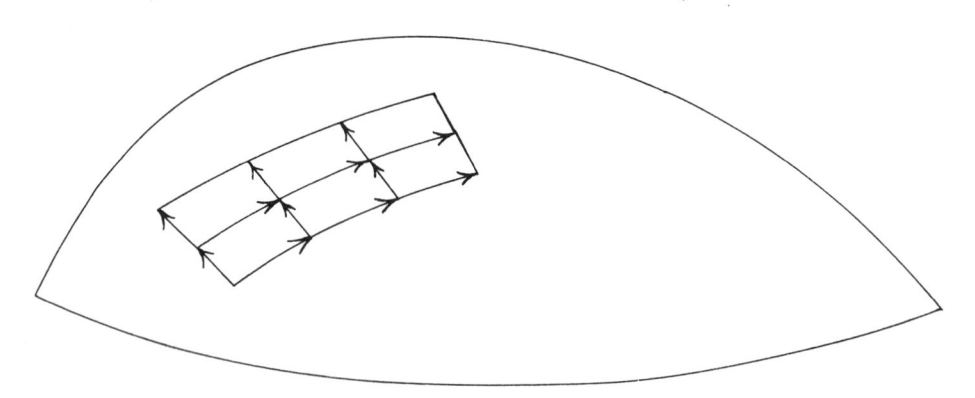

Figure 2-2. A surface is divided into a large number of small regions, each approximately a rectangle spanned by two tangent vectors.

There is a rigorous definition of the integral of an n-form α with compact support over an n-dimensional oriented manifold which captures this intuitive picture. In outline, it goes as follows. Consider a local coordinatization

$$p \longmapsto \phi_U(p) = (x_1(p), \cdots, x_n(p)),$$

of an open set U and write the restriction of α to U as

$$\alpha = f\,dx_1 \wedge \cdots \wedge dx_n \qquad \text{with } f \in C^\infty(M).$$

Define

(4) $\qquad \displaystyle\int_U \alpha = \int_{R^n} f \circ \phi_U^{-1}(x)\, d^n x \quad ,$

where the integral on the right is an ordinary Riemann integral, and the transformation properties of n-forms imply that it is independent of the coordinatization. (If the manifold were not oriented, a change in coordinatization could change the sign.) Finally,

$$\int_M \alpha$$

is defined by a suitable patching together of integrals (4) over coordinate patches U. The patching is the only part of this definition which is not straightforward; see [Lang] for the technical details.

With this integral at our disposal we can state

Stokes' Theorem.

If ω is any C^∞ $(n-1)$-form on an n-dimensional, compact, oriented manifold M with boundary, then

(5) $$\int_M d\omega = \int_{\partial M} \omega \; .$$

We have stated (5) for a manifold M with boundary, but we shall often need it for M an n-cube such as

$$M = \{ (x_1, \cdots, x_n) \in R^n \mid 0 \leq x_i \leq 1 \text{ for all } i \},$$

which is technically not a manifold (or at least not obviously one) because of its "sharp edges". It should be clear that the obvious analog of (5) for n-cubes can be obtained by approximating such a cube by a suitable manifold ("rounding the edges"). Alternatively, (5) can be proved directly for n-cubes, which is even easier than the proof for manifolds, since the "patching" argument is no longer necessary. In fact, if ω is a decomposable $(n-1)$-form

$$\omega = f \, dx_1 \wedge dx_2 \wedge \cdots \wedge dx_{n-1},$$

and both sides of (5) are calculated for an n-cube M using Riemann integrals and the obvious definitions, their equality is seen to follow from the Fundamental Theorem of Calculus and the rule (to be given below) relating the orientations on M and ∂M. This exercise is strongly recommended to the reader who is unfamiliar with this form of Stokes' Theorem. To avoid encumbering the exposition, it will sometimes be convenient to pretend that n-cubes and similar simple geometric figures with "sharp edges" are manifolds, and we leave to the reader who worries about such things the task of either rounding the edges or adding the necessary qualifiers to the exposition.

The convention by which an orientation for ∂M is induced from that for M is as follows. It is enough to settle this when M is the half-space (2) and ∂M the hyperplane $x_1 = 0$, for the orientations on general M and ∂M are induced from these by the corresponding coordinatization maps. The convention for the half space is that an ordered basis $e_2, \cdots, e_n$ for the hyperplane is positively oriented if and only if the augmented ordered basis $(1, 0, \cdots, 0), e_2, \cdots, e_n$ is a positively oriented basis for R^n (i.e related to the standard basis by a linear transformation with positive determinant). Note that $(1, 0, \cdots, 0)$ may be intuitively viewed as the "outward normal" to the hyperplane, and this suggests another way to think about the relations between the orientations: prefixing an "outward normal" to a positively oriented basis for the hyperplane produces a positively oriented basis for R^n. (See Figure 3. Actually, what is important is not that n is normal but that it is outward, as will be discussed below.)

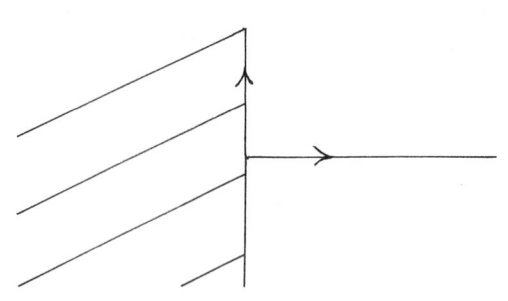

Figure 2-3. The orientation of the vertical axis in R^2, considered as the boundary of the left half-space, is "up".

This latter criterion is particularly useful when the n-dimensional manifold M is concretely realized as a subset of R^n with its manifold structure and orientation induced in an obvious way from R^n. (Most of the manifolds which we shall consider will be obtained in this way; a subset of R^n, such as the unit ball (1), is written down, and it is left to the reader to understand in what obvious way the its manifold structure is induced from R^n.) In this case, vectors in the tangent space $(\partial M)_p$ to the boundary at a point $p \in R^n$ may be identified in an obvious way with vectors in R^n, and it makes sense to prefix a vector n in R^n to a basis $e_2, \cdots, e_n$ for $(\partial M)_p$ to obtain a basis for R^n. The conventions are such that prefixing an "outward" vector n to a positively oriented basis for $(\partial M)_p$ produces a positively oriented basis for R^n, and this is usually the easiest way to remember the relation between the orientations.

In Euclidean space with a positive definite metric it is customary to state the above in terms of an "outward normal" n, but in spaces with indefinite metrics it is important to realize that the vector n does not have to be normal to ∂M; any "outward" vector will do. Consider, for example, a subset M of Minkowski space **M** bounded by   light cones, as in Figure 4. Here any normal vector is tangent to the cone. In this case, there is no outward normal, but there is still an obvious intuitive sense in which any vector emanating from a point of ∂M and not tangent to ∂M points either "into" or "out of" M. When we make such a set into a manifold in the "natural" way, we *define* the orientation on the manifold so that a positively oriented basis for the tangent space at any point is a positively oriented basis for **M**, and then the orientation on the boundary is defined so that prefixing an outward vector to a positively oriented basis for a tangent space of the boundary produces a positively oriented basis for **M**.

As a simple illustration of Stokes' Theorem, consider a 2-form ω on R^3, say

$$\omega = E_3 \, dx \wedge dy + E_2 \, dz \wedge dx + E_1 \, dy \wedge dz.$$

Then

$$d\omega = \left(\frac{\partial E_3}{\partial z} + \frac{\partial E_2}{\partial y} + \frac{\partial E_1}{\partial x}\right) dx \wedge dy \wedge dz = (\nabla \circ \vec{E}) \, dx \wedge dy \wedge dz,$$

where $\vec{E} := (E_1, E_2, E_3)$. Take M to be a compact, three-dimensional submanifold with boundary of R^3. Then

$$\int_{\partial M} \omega$$

is easily seen to be what would be written in the traditional notation of vector calculus as the

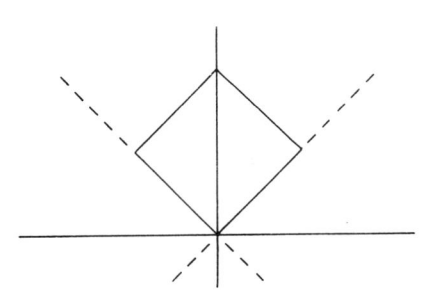

Figure 2-4. Any normal to the boundary of the diamond-shaped region in Minkowski space is a null vector tangent to the boundary, so there is neither an outward normal nor a unit normal. This region is not quite a C^∞ submanifold with boundary due to its sharp "corners", but similar C^∞ examples can be obtained by rounding the corners.

surface integral

$$\int_{\partial M} \vec{E} \cdot \vec{n} \, dS ,$$

where dS denotes surface area, and $\vec{n}$ is the outward normal on ∂M, so in this case, Stokes' Theorem is just Gauss' Divergence Theorem:

$$\int_{\partial M} \vec{E} \cdot \vec{n} \, dS = \int_M \vec{\nabla} \cdot \vec{E} \, dV ,$$

where dV is the usual three-dimensional volume element on M. Also, the classical "Stokes' Theorem" of three-dimensional vector analysis:

$$\int_C \vec{v} \cdot d\vec{c} = \int_S (\vec{\nabla} \times \vec{v}) \cdot \vec{n} \, dS ,$$

which relates the line integral of a vector field $\vec{v}$ around a closed curve C with the integral of the normal component of its curl over a two-dimensional surface S bounded by C, is easily seen to be (5) with $\omega := \vec{v}^*$.

2.7. The metric tensor.

A *semi-Riemannian* manifold is a manifold with the additional structure of a C^∞ non-degenerate real inner product $<\cdot,\cdot>$ on the tangent spaces. More explicitly, the inner product is a C^∞ assignment $p \longmapsto <\cdot,\cdot>_p$ of an inner product $<\cdot,\cdot>_p$ on M_p to each point p, i.e. a C^∞ tensor of type 0,2, with the additional property that $<\cdot,\cdot>_p$ is non-degenerate at each point p. When the condition that the inner product at each p be nondegenerate is replaced by the condition that it be positive definite, the manifold is called a *Riemannian* manifold. A four-dimensional semi-Riemannian manifold whose inner product has signature +,-,-,- at each point is called *Lorentzian*. The inner product is called the *metric tensor*.

The starting point for general relativity is the assumption that spacetime is a Lorentzian manifold. Very roughly speaking, a small piece of such a manifold "looks like" Minkowski space up to first order in the size of the piece, just as a tangent plane to a two-dimensional surface approximates the surface to first order. When translated into physics, this means that most experiments (many would say "all", but I am naturally cautious) done in a laboratory of linear (and time) dimension s will agree with the predictions of special relativity up to order s. The differences which appear in order s^2 are described by Newtonian physics as the effects of a "gravitational field", but in general relativity no such additional structure is necessary.

Much of the more elementary part of electrodynamics can be done as easily in a Lorentzian manifold as in Minkowski space, and the exposition is written with this in mind. Special properties of Minkowski space, such as the vector space structure, are used only where absolutely necessary (as in Chapter 4) or where they produce major simplifications (as in Section 3.7). The places where our arguments may fail in a general Lorentzian manifold are clearly marked.

Traditionally, the metric of general relativity is called $g(\cdot,\cdot)$, or simply g, so its components $g(e_i, e_j)$ with respect to a given basis $\{e_i\}$ are g_{ij}. The inner product notation $<u,v>$ is more compact than $g(u,v)$ or $g_{ij}u^i v^j$ and usually seems more natural, but when one wants to refer to the inner product abstractly, it is easier to write g in place of $<\cdot,\cdot>$, and when the tensor components are needed to convert vectors to forms, g_{ij} is simpler than $<e_i, e_j>$. The best solution seems to be to use both notations, each where it is most convenient.

It is a theorem, the usual proof of which is a short but tricky computation, that there is a unique torsion-free connection D on a semi-Riemannian manifold with the property that for any C^∞ vector fields u, v, w,

(1) $\qquad w(<u,v>) \;=\; <D_w u, v> + <u, D_w v>$.

This connection is called the *Riemannian connection*. (Recall that vectors are considered as differential operators on $C^\infty(M)$, so the left side of (1) is shorthand for the map $p \longmapsto w_p(<u,v>)$, where $<u,v>$ is again shorthand for the map $p \longmapsto <u,v>_p$.) Since (1) is a form of Leibnitz identity which is satisfied by classical vector fields on R^n, this is clearly a natural connection to use. Henceforth, we shall always work in a semi-Riemannian manifold M, and D will always denote the Riemannian connection.

Another nice property of the Riemannian connection is that it commutes with raising and lowering of indices. We shall need this property only in the following simple form: for any vector $v \in M_p$ and vector field u,

(2) $\qquad D_v(u^*) \;=\; (D_v u)^*$.

To see this, note that by the definition 2.5(5) of the covariant derivative of a 1-form, for any vector $w \in M_p$,

$$(D_v(u^*))(w) := v(u^*(w)) - u^*(D_v(w))$$
$$= v(<u, w>) - <u, D_v(w)> = <D_v u, w> = (D_v u)^*(w).$$

2.8 The covariant divergence.

(a) The primary definition.

Let T be a tensor with at least one contravariant index; say T is of type $r+1, s$ with $r, s \geq 0$. We have already pointed out that DT may be regarded as the tensor of type $r+1, s+1$ defined at a point $p \in M$ by 2.5(10):

$$(DT)(\omega_1, \cdots, \omega_{r+1}, v_1, \cdots, v_{s+1}) := (D_{v_{s+1}} T)(\omega_1, \cdots, \omega_{r+1}, v_1, \cdots, v_s)$$

for all $v_i \in M_p$, $\omega_i \in M_p^*$.

If we contract the ω_1 slot with the v_{s+1} slot, we obtain a tensor of type r, s called the *covariant divergence* of T and denoted $\delta(T)$, or simply δT. Explicitly, δT is defined as follows, where $e_1, \cdots, e_n$ is any given basis and $e^1, \cdots, e^n$ the dual basis:

(1)
$$(\delta T)_p(\omega_1, \cdots, \omega_r, v_1, \cdots, v_s)$$
$$:= \sum_{k=1}^{n} (D_{e_k} T)(e^k, \omega_1, \cdots, \omega_r, v_1, \cdots, v_s).$$

In terms of indices relative to a local coordinatization $\{x^i\}$,

$$(\delta T)^{i_1 \cdots i_r}{}_{j_1 \cdots j_s} = \sum_{k=1}^{n} (D_{\frac{\partial}{\partial x_k}} T)^{k i_1 \cdots i_r}{}_{j_1 \cdots j_s}.$$

Many authors denote

$$(D_{\frac{\partial}{\partial x_k}} T)^{i_1 \cdots i_{r+1}}{}_{j_1 \cdots j_s} \quad \text{as} \quad T^{i_1 \cdots i_{r+1}}{}_{j_1 \cdots j_s | k},$$

the stroke denoting covariant differentiation. In this notation,

(2)
$$(\delta T)^{i_1 \cdots i_r}{}_{j_1 \cdots j_s} = T^{k i_1 \cdots i_r}{}_{j_1 \cdots j_s | k},$$

where summation over k is implied.

At this point some remarks on notation may help the beginner. The indicial notation (2) is more compact than (1), but can easily mislead one into thinking that it is somehow the indiviual tensor components which are being differentiated in (2). † Moreover, it is easy to be seduced into writing expressions such as

---
† If the tensor components are with respect to a constant basis in Minkowski space, then (2) *is* equivalent to differentiation of the individual components. This case is easier to think about, but the reader who intends to study general relativity will find it profitable to learn the more general definition from the start.

$$T^i{}_{|j|k} \quad , \text{ or worse, } \quad T^{ij}{}_{|i|j}$$

which are actually ambiguous. (Which differentiation is performed first? Covariant differentiations do not commute in general.) However, there are situations in which indicial notation does seem clearer than more abstract notations such as (1). For example, half of Maxwell's equations will turn out to be expressible in the form

(3) $\qquad \delta F = \rho u \; ,$

where $F = (F^{ij})$ is the contravariant form of the electromagnetic field tensor, and ρu is the charge current four-vector. In the above indicial notation with the summation convention, this reads

(4) $\qquad F^{ij}{}_{|i} = \rho u^j \; .$

Though slightly more complicated than (3), it gives more information since it effectively defines δF, and reminds the reader which index is contracted by the operator δ. The analog of (4) in abstract notation is

(5) $\qquad \sum_{i=0}^{3} (D_{\frac{\partial}{\partial x_i}} F)(dx^i, \alpha) \;=\; \rho \alpha(u) \qquad \text{for all } \alpha \in M_p{}^* \; ,$

which most will probably find less attractive than (4). This is one of a fairly small class of situations in which indicial notation seems preferable to the coordinate-free notation of differential geometry, and we shall not hesitate to use the former when it seems clearer.

(b) The covariant divergence and codifferential of alternating forms.

Let α be a q-form field on an oriented semi-Riemannian manifold M, so that the indicial expression for α is

$$\alpha_{j_1 j_2 \cdots j_q} \; .$$

We cannot directly take the covariant divergence of α because it does not have a contravariant index to contract. However, we can always raise the first index to obtain

(6) $\qquad \alpha^{j_1}{}_{j_2 \cdots j_q} \; ,$

and then take the covariant divergence of this to obtain the $(q-1)$-form

(7) $\qquad \sum_k \alpha^k{}_{j_2 \cdots j_q | k} \quad .$

This will be called the *covariant divergence* of α and denoted as $\delta(\alpha)$ or simply $\delta\alpha$.

Let n be the dimension of M and b the number of negative-norm elements in an orthonormal basis for a tangent space M_x. The following very useful formula states that, up to a sign, δ is the transform under Hodge duality of the differential operator "d" on forms: for any q-form α,

(8) $\qquad \delta\alpha \;=\; (-1)^{b+(q+1)n} \perp d \perp \alpha \; .$

An alternate form of (8) from which b and n are eliminated is obtained by taking the dual of both sides and using 2.2(11):

(9) $$\perp \delta\alpha = (-1)^{q+1} d\perp\alpha .$$

When M is a vector space, the proof is a straightforward computation, and the case of general M can be reduced to the vector space case by the device of doing the calculation in a coordinate system $\{x^i\}$ in which both all the connection coefficients vanish and $\{\partial/\partial x_i\}$ is orthonormal at the point at which (8) is evaluated. For details, see Exercise 20. We shall need (8) only for the case of Minkowski space, where it reads

(10) $$\delta\alpha = -\perp d\perp\alpha .$$

The main idea is most easily seen by working out a special case. Consider the case $\alpha = f dx^1$ in Minkowski space with Lorentz coordinates. Then $\perp\alpha$ will be $\pm f\, dx^0 \wedge dx^2 \wedge dx^3$, and the sign, which turns out to be "+", can be read off from the defining relation

$$\alpha \wedge \perp\alpha = <\alpha, \alpha> dx^0 \wedge dx^1 \wedge dx^2 \wedge dx^3 = -f^2 dx^0 \wedge dx^1 \wedge dx^2 \wedge dx^3.$$

Thus

$$\perp\alpha = f dx^0 \wedge dx^2 \wedge dx^3 ,$$

$$d\perp\alpha = \frac{\partial f}{\partial x^1} f dx^1 \wedge dx^0 \wedge dx^2 \wedge dx^3 ,$$

and

$$\perp d \perp\alpha = \frac{\partial f}{\partial x^1} .$$

On the other hand, raising an index on α gives $-f\dfrac{\partial}{dx^1}$, the minus sign appearing because $\dfrac{\partial}{\partial x^1}$ has negative norm, and so

$$\delta\alpha = -\frac{\partial f}{\partial x^1} .$$

Thinking about this example should make it clear that for any given q-form α on Minkowski space, (10) holds up to a sign.

There is a sense in which the operator $-\delta$ on form fields is the "adjoint" of the differential operator d (see Exercise 23). Thus within the context of the theory of differential forms, $-\delta$ is more important than δ and is sometimes called the *codifferential* operator. Some authors denote the codifferential operator by δ instead of our $-\delta$.

(c) Gauss's Theorem.

If $w = w^i$ is a vector field on Minkowski space with components written with respect to a constant basis, then one easily computes that δw is the scalar function

(11) $$\delta w = w^i{}_{,i} = \sum_{i=0}^{3} \frac{\partial w^i}{\partial x_i} ,$$

which looks like an ordinary vector divergence. (However, (11) does not hold in general semi-Riemannian manifolds or even in Minkowski space if the tensor components are with respect to a non-constant basis; cf. Exercise 17.) Hence one looks for some form of Gauss's theorem involving δw, and there is one which holds for any oriented, semi-Riemannian manifold with boundary M even though (11) does not.

Let M have dimension m, and let Ω be the volume m-form. Recall from 2.2(15) that the Hodge dual $\perp w^*$ may be characterized as the $(m-1)$-form obtained by inserting w into the first slot of Ω:

$$(12) \qquad (\perp w^*)(v_1, v_2, \cdots, v_{m-1}) = \Omega(w, v_1, v_2, \cdots, v_{m-1}) .$$

Suppose that we integrate $\perp w^*$ over the boundary ∂M of M. Intuitively, such integration consists of approximating ∂M by a large number of small parallelipipeds spanned by tangent vectors $v_1, v_2, \cdots, v_{m-1}$ as in Figure 2 in Section 6, adding up all the numbers

$$(13) \qquad (\perp w^*)(v_1, v_2, \cdots, v_{m-1}) = \Omega(w, v_1, v_2, \cdots, v_{m-1}) ,$$

and taking a limit over better and better approximations. Of course, this picture cannot be taken literally (since, for example, tangent vectors do not lie in the manifold), but it provides useful insights.

Fix attention on a particular parallelipiped spanned by tangent vectors $v_1, v_2, \cdots, v_{m-1}$ at $x \in \partial M$. (The two-dimensional case is drawn in Figure 5.) Suppose that there exists an outward unit normal vector n to ∂M at x. This means that n is a unit vector in the tangent space M_x which is orthogonal to all vectors in ∂M_x and such that prefixing n to a positively oriented basis for ∂M_x produces a positively oriented basis for M_x. In this case, we can write

$$(14) \qquad w = \frac{<w, n>}{<n, n>} n \quad + \quad \text{linear combination of the } v_i ,$$

and since the linear combination of the v_i will not contribute to the right side of (13) because Ω is alternating, we have

$$(15) \qquad \perp w^*(v_1, \cdots, v_{m-1}) = \frac{<w, n>}{<n, n>} \Omega(n, v_1, \cdots, v_{m-1}) .$$

Now $<w, n>/<n, n>$ is the component of w in the direction of n, and since n is an outward unit vector orthogonal to $v_1, \cdots, v_{m-1}$,

$$\Omega(n, v_1, \cdots, v_{m-1})$$

is just the $(m-1)$-dimensional volume of the parallelipiped spanned by $v_1, \cdots, v_{m-1}$ (cf. Exercise 2.4 and Figure 5).

Summing over all such parallelipipeds, we see that in the special case in which M has an outward unit normal everywhere,

$$\int_{\partial M} \perp w^*$$

is intuitively the integral of the normal component of w over ∂M. For instance, if M is a three-dimensional submanifold of R^3, this integral could be written in traditional calculus notation as

$$\int_{\partial M} \vec{w} \cdot \vec{n} \, dS ,$$

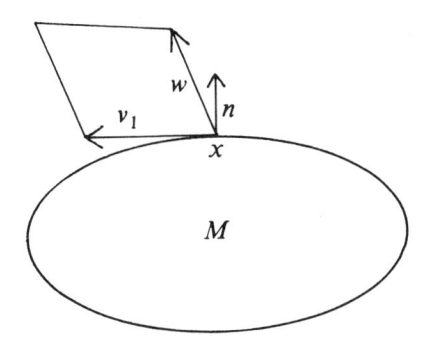

Figure 2-5. A two-dimensional semi-Riemannian manifold with boundary. For any tangent vector v_1 at p, $\perp w^*(v_1)$ is the "volume" (signed area) of the parallelipiped spanned by w, v_1. If there is a unit normal vector n at p, this is also $(<n,w>/<n,n>)\Omega(n, v_1)$.

with dS area measure on ∂M.

By Stokes' Theorem,

$$(16) \qquad \int_{\partial M} \perp w^* = \int_M d\perp w^* .$$

Recalling from (9) that $d\perp w^* = \perp \delta w^* = \perp \delta(w) = \delta(w)\Omega$, we rewrite (16) as

$$(17) \qquad \int_{\partial M} \perp w^* = \int_M \delta(w)\Omega ,$$

which is like the familiar three-dimensional Gauss Theorem in that it equates the integral of the normal component of a vector field over the boundary of a region to the integral of its divergence over the interior.

The interpretation of the left side as the integral of the normal component of w over the boundary required that an outward unit normal exist everywhere, but (17) is of course independent of any such interpretation. In spaces with an indefinite metric, there can be points of ∂M on which there is neither a unit normal nor an outward normal. For instance, if M is a submanifold of Minkowski space this will occur whenever ∂M is tangent to a light cone. At such a point, any normal vector lies in the cone (Exercise 1.3), and so is a null vector tangent to ∂M. In such a case, there is no outward unit normal or other distinguished normal vector, and it simply doesn't make sense to speak of the normal component of w. Although the left side of (17) can be interpreted as the integral of the normal component of w when that interpretation makes sense, this is not necessarily the best way to think about (17). A more accurate geometrical picture views the left side of (17) as the integral over ∂M of the $(n-1)$-form whose value on an $(n-1)$-tuple of tangent vectors $v_1, v_2, \cdots, v_{n-1}$ is the signed volume of the n-dimensional parallelipiped spanned by $w, v_1, \cdots, v_{n-1}$. The latter idea is sometimes expressed (not all too clearly) in component notation by writing the left side of (17) as

$$\int_{\partial M} w^i d\sigma_i ,$$

where $w^i d\sigma_i$ (taken as a whole) is supposed to denote this $(n-1)$-form, with $d\sigma_i := \perp e_i^*$ representing the $(n-1)$-form obtained as in (12) by inserting the i'th vector e_i from a basis into the first slot of Ω.

(d) Conservation of charge.

One important application of (17) will occur with $w = \rho u$, where u is the four-velocity and ρ the proper charge density of a charged fluid. To have a simple physical picture in mind, think of a "charged fluid" as a continuum of infinitesimal charged particles which move coherently in the sense that nearby particles have nearly the same velocity. (Our considerations will apply to more complicated fluid models as well, but all the essential ideas are present in this simple one.) The map which assigns to each spacetime point p the four-velocity u_p of the particle at p is a unit vector field on spacetime M. The physical definition of the "proper charge density" ρ is as follows when M is Minkowski space $\mathbf{M}$; similar ideas apply in general spacetimes. Given a point $p \in \mathbf{M}$, choose a Lorentz frame in which the particle at p is at rest; i.e. the components of u_p relative to this coordinatization are $(1, 0, 0, 0)$. Any such coordinatization is called a *rest frame* for the particle, and $\rho(p)$ is defined as the charge density in a rest frame. Intuitively, we may conceive of measuring $\rho(p)$ by counting the number of particles in a small spatial volume (where now the particles are conceived as very small but not truly infinitesimal) and dividing by the volume. Since any two rest frames differ by a spatial rotation, $\rho(p)$ is a well-defined physical quantity.

If e_1, e_2, e_3 is an orthonormal set of spacelike unit vectors at a point p which are orthogonal to u at p and such that u, e_1, e_2, e_3 is positively oriented, then by (12),

$$(18) \qquad (\perp \rho u^*)(e_1, e_2, e_3) = \rho \Omega(u, e_1, e_2, e_3) = \rho .$$

Thus the left side of (18) may be thought of as the charge in the unit volume spanned by e_1, e_2, e_3. To be pedantically precise, (18) gives

$$\lim_{t \to 0} t^{-3}[\text{charge in the volume spanned by } te_1, te_2, te_3] .$$

Even if e_1, e_2, e_3 are not orthogonal to u, we can consider them as part of a positively oriented orthonormal set e_0, e_1, e_2, e_3, which defines a coordinatization of the tangent space at the point in question relative to which $u = \gamma(\vec{v})(1, \vec{v})$ for some $\vec{v}$ in R^3, and so, again from (12),

$$(19) \qquad (\perp \rho u^*)(e_1, e_2, e_3) = \rho \gamma(\vec{v}) .$$

From the spacetime diagram in Figure 6, it is apparent that a collection of ρ stationary particles per unit volume in the coordinate system defined by the new orthonormal basis $e'_0 = u$, $e'_1 = \gamma(\vec{v})(|\vec{v}|, \vec{v}/|\vec{v}|)$, $e'_2 = e_2$, $e'_3 = e_3$ will be seen in the e_0, e_1, e_2, e_3 system as a flow of $\rho\gamma(\vec{v})$ particles per unit volume with velocity $\vec{v}$. Thus in all cases, (19) may be viewed as the charge in the unit spatial volume spanned by e_1, e_2, e_3. It follows that if S is an oriented spacelike hyperplane (or, more generally, an oriented spacelike three-dimensional submanifold), then

$$(20) \qquad \int_S \perp(\rho u^*) \qquad \text{may be interpreted as the \textbf{total charge} in } S$$

up to a sign which depends on the orientation of S.

Now consider a four-dimensional volume

$$(21) \qquad V := \{(x^0, x^1, x^2, x^3) \mid t_1 \leq x^0 \leq t_2, \ -a \leq x^i \leq a \ \ i = 1, 2, 3 \}$$

which is large enough so that all the charged particles in the universe (or, more modestly, in the system under consideration) are contained in the three-dimensional cube

2.8 The covariant divergence

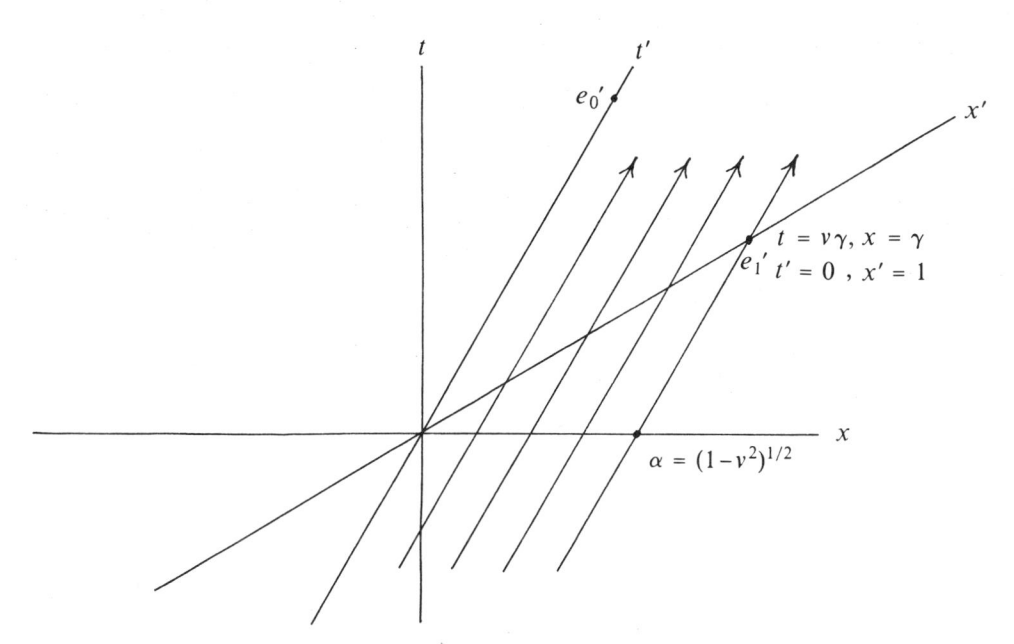

Figure 2-6. Worldlines of particles at rest in the t'-x' frame. The worldline of the particle at rest at $x' = 1$ intersects the x-axis at α, and elementary geometry gives $\alpha = (1-v^2)^{1/2}$. If there are ρ particles stationary between $x' = 0$ and $x' = 1$ in the primed system, then their worldlines cross the x-axis between 0 and α, and hence the particle density as measured in the t-x system is $\rho/\alpha = \rho\gamma$ particles per unit length.

$$C := \{ (x^1, x^2, x^3) \mid -a \leq x_i \leq a, \ i=1, 2, 3 \}$$

for all times x^0 with $t_1 \leq x^0 \leq t_2$.

Then ρ vanishes on the faces of ∂V with spacelike normals, and the only nonzero contributions to

$$\int_{\partial V} \perp(\rho u^*)$$

come from the integrals over the faces $x^0 = t_1$ and $x^0 = t_2$. Moreover, the orientation induced on these two faces by a given orientation on V is opposite in the sense that positive ρ will give a positive contribution on one face and a negative one on the other. The orientation induced from Minkowski space is such that positive ρ gives a positive contribution on $x_0 = t_2$. Then we see from (17) and (20) that

(22) $$\int_V \delta(\rho u)\Omega = \int_{\partial V} \perp \rho u^* =$$

= (total charge in all space at time t_2) - (total charge in all space at time t_1),

and so conservation of charge requires that

(23) $$\int_V \delta(\rho u)\Omega = 0 \quad \text{for all such regions } V.$$

Even if ρ does not vanish on the "sides" of V, defined as the sets

$$S_{j\pm} := \{ (x^0, x^1, x^2, x^3) \mid t_1 \leq x_0 \leq t_2, \ x_j = \pm a, \ |x_i| \leq a \ \ i \neq 0, j \},$$

the quantity

$$\int_{S_{j\pm}} \lrcorner \rho u^*$$

can be seen to represent the amount of charge which flows out of the three-dimensional cube C through the two-dimensional rectangle

$$\{ (x^1, x^2, x^3) \mid x_j = \pm a, \ |x^i| \leq a, \ i \neq j \}$$

during the time interval $t_1 \leq x_0 \leq t_2$.

Thus the integral of $\lrcorner \rho u^*$ over all the sides $S_{j\pm}$ represents the amount of charge which flowed out of C during this time interval, and so

(total charge in C at time t_2) - (total charge in C at time t_1)

 + (charge which flowed out of C between times t_1 and t_2) =

$$= \int_{\partial V} \lrcorner \rho u^* = \int_V \delta(\rho u)\, \Omega\, .$$

Charge will be conserved if and only if both sides vanish, and since V can now be an arbitrary four-dimensional rectangle which we may shrink to a point, it follows that the law of conservation of charge can be stated in differential form as

(24) $\delta(\rho u) = 0$.

In component notation in Minkowski space, this reads

(25) $\dfrac{\partial(\rho u^i)}{\partial x_i} = 0$,

where summation over i is understood. In general spacetimes, the ordinary derivatives in (25) must be replaced by covariant derivatives:

(26) $0 = \delta(\rho u) = (\rho u)^i{}_{|i} = \dfrac{\partial(\rho u^i)}{\partial x_i} + \rho u^j\, \Gamma^i_{ij}\, .$

In Minkowski space one can see in an elementary way that (25) expresses conservation of charge by integrating the left side over the 4-cube C. Direct applications of the fundamental theorem of calculus reduce the integral to integrals over the various three-dimensional boundaries of C, and examination of each of these yields the stated physical interpretation.

(e) A warning.

Before leaving the topic of the four-dimensional Gauss theorem in semi-Riemannian manifolds, we warn the reader that attempts to generalize (17) to tensors of higher order are fraught with peril. The issue arises in practice because (17) is used in special relativity to formulate a law of conservation of energy-momentum which one would like to extend to general relativity. Specifically, we shall later define an "energy-momentum" tensor $T = (T^{ij})$ of contravariant rank 2 whose covariant divergence vanishes. Recall the definition (2) of covariant divergence in this context:

$$(27) \qquad (\delta T)^j = \sum_{i=0}^{3} T^{ij}{}_{\backslash i} \; ,$$

so our hypothesis is:

$$(28) \qquad \sum_{i=0}^{3} T^{ij}{}_{\backslash i} = 0 \; .$$

(We do not use the summation convention here because for pedagogical reasons we want to make everything in this section completely explicit.) Now fix j and consider the vector field

$$\sum_{i=0}^{3} T^{ij} \frac{\partial}{\partial x_i} \; ,$$

where $\{x^i(\cdot)\}$ is the coordinate system with respect to which the tensor components are defined. Let us call this vector field

$$T^{\bullet j} \; ,$$

so that the i'th component of $T^{\bullet j}$ is T^{ij}. We can apply Gauss's Theorem (17) to convert the integral of $T^{\bullet j}$ over the boundary δV of the four-rectangle V to an integral of a divergence over V:

$$(29) \qquad \int_{\delta V} \bot T^{\bullet j} = \int_V \delta(T^{\bullet j}) \, \Omega \; .$$

Now by definition of covariant divergence,

$$(30) \qquad \delta(T^{\bullet j}) = \sum_{i=0}^{3} T^{ij}{}_{\backslash i} \; ,$$

which looks identical to (27). It would seem that one could conclude from (28) that

$$(31) \qquad \delta(T^{\bullet j}) = 0 \; ,$$

from which it would follow that both sides of (29) vanish. If this conclusion were correct, then the left side of (29) would give four "quantities", one for each j, which would be "conserved" in the same way that we concluded conservation of charge from the equation $\delta(\rho u) = 0$. The details are not important, for we went astray earlier; (28) and (30) do not imply (31) ! The difficulty is that the operations of passing from T to $T^{\bullet j}$, j fixed, and of taking the covariant divergence do not commute in general:

$$(32) \qquad (\delta T)^j \text{ does not necessarily equal } \delta(T^{\bullet j}) \; .$$

The best way to see what is happening is to write out the two covariant divergences in (32) in gory detail using definition (1):

since

$$T = \sum_{i,j=0}^{3} T^{ij} \frac{\partial}{\partial x_i} \otimes \frac{\partial}{\partial x_j} = \sum_{j=0}^{3} (\sum_{i=0}^{3} T^{ij} \frac{\partial}{\partial x_i}) \otimes \frac{\partial}{\partial x_j} = \sum_{j=0}^{3} T^{\bullet j} \otimes \frac{\partial}{\partial x_j} ,$$

we have for any k,

(33) $\quad \delta(T)^k := \delta(T)(dx^k) := \sum_{\alpha=0}^{3} (D_{\frac{\partial}{\partial x_\alpha}} T)(dx^\alpha, dx^k)$

$$= \sum_{\alpha,\beta} ((D_{\frac{\partial}{\partial x_\alpha}} T^{\bullet \beta}) \otimes \frac{\partial}{\partial x_\beta})(dx^\alpha, dx^k) + \sum_{\alpha,\beta} (T^{\bullet \beta} \otimes D_{\frac{\partial}{\partial x_\alpha}} \frac{\partial}{\partial x_\beta})(dx^\alpha, dx^k)$$

$$= \sum_{\alpha} (D_{\frac{\partial}{\partial x_\alpha}} T^{\bullet k})(dx^\alpha) + \sum_{\alpha,\beta} T^{\alpha\beta}(D_{\frac{\partial}{\partial x_\alpha}} \frac{\partial}{\partial x_\beta})(dx^k)$$

$$= \delta(T^{\bullet k}) + \sum_{\alpha,\beta} T^{\alpha\beta} \Gamma^k_{\alpha\beta} .$$

This example illustrates very graphically the dangers of indicial notation such as $T^{ij}{}_{\|i}$. Nonetheless, we shall use it on occasion because it is simple and convenient for some calculations.

Notice that the above calculation does show that the two divergences in (32) *are* equal if

(34) $\quad \sum_{\alpha,\beta} T^{\alpha\beta} \Gamma^k_{\alpha\beta}$

happens to vanish. If we use Lorentz coordinates in Minkowski space, then all the connection coefficients $\Gamma^k_{\alpha\beta}$ vanish identically, and a vector-valued form of (17) can be obtained by the above argument. Another special case in which (34) vanishes is when $T^{\alpha\beta}$ is antisymmetric (assuming as always that D is torsion-free, so that Γ is symmetric in its two lower indices).

2.9. The equations $d\beta = \gamma$ and $\delta\beta = \alpha$.

This section establishes a fundamental result which states that under appropriate analytical hypotheses, the differential and covariant divergence of a p-form β may be arbitrarily specified subject only to the obvious consistency conditions stemming from the relations $d^2\beta = 0$ and $\delta^2\beta = 0$. The precise statement of the theorem, which is cluttered with technical hypotheses and rarely used conclusions, is given at the end of the section. For the purposes of introduction and explanation, it is more enlightening to state it as an algebraic scholium with technical hypotheses suppressed:

Scholium.

Let $p \geq 1$, and let γ be a given $(p+1)$-form satisfying $d\gamma = 0$ and α a given $(p-1)$-form satisfying $\delta(\alpha) = 0$ in a neighborhood of a point q in a oriented semi-Riemannian manifold M. Under appropriate technical hypotheses, there exists a p-form β, defined in a neighborhood of q, with

(1)   (i) $d\beta = \gamma$ and

(ii) $\delta\beta = \alpha$.

Physicists routinely use several special cases of this result. For instance, if β is a 1-form on R^3, which may be identified with a vector field by raising an index, then it translates into the statement that one can always find a vector field with a given divergence and given curl subject only to the necessary condition that the divergence of a curl must vanish. † Another common application is when one is given a two-form F describing the electromagnetic field with $dF = 0$ and seeks a "potential" one-form A such that $dA = F$; the result not only assures that this is possible locally, but also allows us to arbitrarily specify δA, the latter specification being what physicists call a "choice of gauge". Also, a global version of the scholium for compact Riemannian manifolds is one of the main results of the branch of differential geometry known as *Hodge theory* [Warner, Chapter 6].

The scholium is something of a "folk theorem", but I am aware of no rigorous proof (or even precise statement) in reasonable generality in the literature. However, the special case $p = 2$, $\gamma = 0$ in four-dimensional spacetimes is proved in [Sachs and Wu, p. 98, Thm. 3.11.1], and several of the ideas in the proof below are implicit in that proof. Though we shall need the scholium only for four-dimensional spacetimes (and usually only for Minkowski space),* the general result seems of sufficient mathematical interest to justify examining the case of arbitrary semi-Riemannian manifolds, particularly since the proof (to the extent that it goes through in this generality) is not essentially harder than in four-dimensional spacetime.

However, the proof *is* somewhat less complicated when M is an inner product space, and to make it as transparent as possible, we shall do this case first, indicating later the modifications necessary for general semi-Riemannian manifolds. With this in mind, we shall set up the notation so that the transition to the general case will go smoothly. We take M to be of dimension $n+1 \geq 2$ (the case $n = 0$ being a simple exercise) and when M is an inner product space will realize it as $M = R^1 \oplus R^n$. This decomposition is purely algebraic; we do not assume that R^n carries the Euclidean inner product. However, we do assume that the R^1 factor is orthogonal to the R^n factor and that $(1, 0, \cdots, 0)$ has

---

† However, we shall not prove this for $M = R^3$ in the generality for which one would like to have it; e.g. for C^∞ forms α, γ. Fortunately, we need it only for M a spacetime, where the proof is particularly easy.

* A simple proof for Minkowski space is sketched in Appendix 2.

positive norm in M (see below). The general point of M will be described relative to this decomposition by coordinates named

$$(t, x_1, x_2, \cdots, x_n) \in R^1 \oplus R^n \ .$$

In other words, we distinguish one positive-norm direction and call it "time", t. Of course, this is not possible if the inner product on M happens to be negative definite, but that case can be handled by simply reversing the sign of the inner product. Since that simply reverses the sign of the covariant divergence operation $\delta(\cdot)$ on forms, it is clear that if the result is true for positive definite inner products, then it is true in the negative definite case as well.

In the generalization of the proof to arbitrary semi-Riemannian manifolds we shall need local coordinate systems with some of the special properties of the "time-space" linear coordinates above. Specifically, we again assume M to have dimension $n+1 \geq 2$, and we need at each point of M a local coordinate system $t(\cdot), x^1(\cdot), \cdots, x^n(\cdot)$ with the property that:

(2) (i) $\qquad \left\langle \dfrac{\partial}{\partial t}, \dfrac{\partial}{\partial x_j} \right\rangle = 0 \qquad$ for $j = 1, 2, \cdots, n$.

For one version of our theorem we shall need to assume in addition that these coordinates satisfy

(ii) The restriction of the metric to the subspace of any tangent space spanned by $\dfrac{\partial}{\partial x_1}, \cdots, \dfrac{\partial}{\partial x_n}$ is negative definite and $\dfrac{\partial}{\partial t}$ is timelike (i.e. has positive norm).

The reader who only wants to follow the proof for an inner product space M can skip the next indented paragraph. This is recommended on first reading for all but experts in differential geometry. In fact, since the ideas of the proof (as opposed to the result itself) will not be important in the rest of the book, the entire proof may be skipped if desired.

> Condition (2)(i) can be satisfied in a neighborhood of any given point m in a semi-Riemannian manifold, and (2)(ii) can be assured in addition if the restriction of the metric to the tangent space at m has exactly one timelike vector in an orthonormal basis. This can be done as follows. Begin by choosing a local coordinate system $\bar{t}, x^1, \cdots, x^n$ for a neighborhood V of m. If there is exactly one timelike direction, this coordinate system may be chosen so that the restriction of the metric to the time-zero submanifold
>
> $$S_0 := \{ p \in V \mid \bar{t}(p) = 0 \}$$
>
> is negative definite. This can be accomplished by applying an appropriate linear transformation to any local coordinate system and restricting to a smaller coordinate neighborhood, if necessary. Through each point p in S_0, construct a geodesic $t \longmapsto c_p(t)$, parametrized by arc length t, with $c_p(0) = p$ and $c_p'(0)$ a unit vector orthogonal to S_0. The union of these geodesics can be shown to fill an open subset U of M containing S_0. Moreover, U can be taken small enough so that each point $q \in U$ lies on exactly one such geodesic. If
>
> $$q = c_p(t) \ ,$$
>
> then we assign coordinates
>
> $$t, x^1(p), x^2(p), \cdots, x^n(p)$$
>
> to q. In other words, we are coordinatizing a generic point q of U by the "spatial" coordinates $\{x^i(p)\}$ of the point $p \in S_0$ where the geodesic on which q lies cuts S_0

together

with the arc length t along the geodesic from p to q. (See Figure 7.)

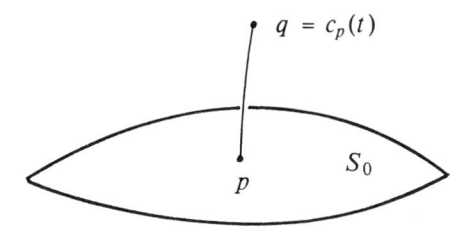

Figure 2-7.

These are known as *Gaussian* coordinates, and, though it is not obvious, it is not hard to show that they have property (2)(i). [Synge, p. 35] Of course, U can also be taken small enough so that (2)(ii) holds.

Now we return to the main proof, which will apply to an arbitrary semi-Riemanian M until further notice. We work exclusively in a particular coordinate system named $t, x^1, \cdots, x^n$ which will not change throughout the proof. Eventually we shall assume that this system satisfies (2)(i) and later shall impose (2)(ii) as well, but the conditions (2) are not necessary at the outset. For any given time t_0, let

$$S_{t_0} := \{ (t, x_1, \cdots, x_n) \mid t = t_0 \}$$

denote the submanifold of constant time t_0. We shall also use "S" as a generic symbol for any constant-time submanifold S_{t_0}; thus "S" means "S_{t_0} for some t_0 which we do not care to name".

For $p \geq 1$, any p-form can be written as a linear combination of wedge products of the dx_i and (possibly) dt. If this decomposition contains no dt's, we shall call the form *spatial*. Thus a spatial p-form σ is one which can be written as

$$(3) \qquad \sigma = \sum_{1 \leq i_1 < \cdots < i_p \leq n} f_{i_1 i_2 \cdots i_p} dx^{i_1} \wedge \cdots \wedge dx^{i_p} .$$

We also make the convention that any 0-form (i.e. a function) is spatial.

If we write an arbitrary p-form β with $p \geq 1$ as a sum of wedge products of the dx^i and dt and collect terms containing dt, we see that β may be uniquely written in the form

$$(4) \qquad \beta = dt \wedge \beta_T + \beta_S ,$$

with β_S is a spatial p-form and β_T a spatial $(p-1)$-form. We shall call β_S the *space* part of β, and β_T the *time* part. For a 0-form β, we define $\beta_S := \beta$, and β_T is undefined.

Note that β_S and β_T may be viewed as forms on any constant-time submanifold S. We shall denote the differential operation within this submanifold as d_S; for instance, if σ is given by (3), then

$$(5) \qquad d_S \sigma := \sum_{1 \leq i_1 < \cdots < i_p \leq n} \sum_{j=1}^{n} \frac{\partial f_{i_1 \cdots i_p}}{\partial x_j} dx^j \wedge dx^{i_1} \wedge \cdots \wedge dx^{i_p} .$$

For any spatial p-form σ, define

(6) $$\frac{\partial \sigma}{\partial t} := \sum_{1 \le i_1 < \cdots < i_p \le n} \frac{\partial f_{i_1 \cdots i_p}}{\partial t} dx^{i_1} \wedge \cdots \wedge dx^{i_p},$$

and note that $\partial\sigma/\partial t$ is the unique spatial p-form satisfying

(7) $$d\sigma = d_S\sigma + dt \wedge \frac{\partial \sigma}{\partial t}.$$

From (7) it follows that $\partial\sigma/\partial t$ is independent of the choice of spatial coordinates. †

Taking the differential of (4) and using (7) with $\sigma := \beta_S$ gives

$$d\beta = -dt \wedge d\beta_T + d\beta_S = -dt \wedge d_S\beta_T + d_S\beta_S + dt \wedge \frac{\partial \beta_S}{\partial t}$$

$$= dt \wedge \left(\frac{\partial \beta_S}{\partial t} - d_S\beta_T\right) + d_S\beta_S.$$

This shows in particular that the space part $(d\beta)_S$ of $d\beta$ is given by

(8) $$(d\beta)_S = d_S\beta_S,$$

and the time part $(d\beta)_T$ is

(9) $$(d\beta)_T = \frac{\partial \beta_S}{\partial t} - d_S\beta_T.$$

Of course, (8) holds also for 0-forms β.

Now let us consider just half of the problem to be solved: the equation

(10) $$d\beta = \gamma.$$

Here we seek β given a $(p+1)$-form γ satisfying $d\gamma = 0$. If we decompose γ as

$$\gamma = dt \wedge \gamma_T + \gamma_S,$$

then (8) and (9) show that (10) is equivalent to

(11)   (i) $d_S\beta_S = \gamma_S$, and

      (ii) $\dfrac{\partial \beta_S}{\partial t} = d_S(\beta_T) + \gamma_T$.

Given γ and β_T, (ii) can be integrated for β_S, with β_S arbitrarily specified on the time-

---

† Actually, (6) may be more elegantly invariantly characterized as the Lie derivative $L_{\partial/\partial t}\sigma$ of σ along the vector field $\partial/\partial t$, as may be seen from the famous formula

$$L_{\frac{\partial}{\partial t}}\sigma = d\left(\frac{\partial}{\partial t} \lrcorner \sigma\right) + \left(\frac{\partial}{\partial t}\right) \lrcorner d\sigma,$$

where $\lrcorner$ denotes contraction, but we do not need this.

zero submanifold S_0 as initial condition. We emphasize that this is both trivial and rigorous (little more than the Fundamental Theorem of Calculus applied to the components of β_S); no high-powered existence theorems for partial differential equations are necessary. Note also that β_T can be chosen arbitrarily, since it doesn't appear in (i).

Now let us look at (i). For each fixed time t, we may view (i) as an equation on the constant-time submanifold S_t. This is encouraging because S_t is of dimension one less than that of the original manifold M, and so we may hope to apply induction. Let us suppose for the moment that we can solve (i) on the time-zero submanifold S_0 and that we use that solution as the initial condition for integrating (ii). In that case, we obtain a p-form β which satisfies (ii) on all of M and (i) at time 0. What we clearly need in order to solve (11), and shall now show, is that (i) at time 0 and (ii) for all time imply (i) for all time; i.e. imply (i) on S_t for all t.

We are assuming throughout that γ satisfies the consistency condition

$$d\gamma = 0 \ ,$$

which, by (11) with β replaced by γ, is equivalent to

(12)　(i)　$d_S \gamma_S = 0$　and

　　(ii)　$\dfrac{\partial \gamma_S}{\partial t} = d_S \gamma_T \ .$

Assuming this, it follows from (11)(ii) that

$$\frac{\partial \gamma_S}{\partial t} = d_S \gamma_T = d_S\left(\frac{\partial \beta_S}{\partial t} - d_S \beta_T\right) = d_S \frac{\partial \beta_S}{\partial t} = \frac{\partial (d_S \beta_S)}{\partial t} \ ,$$

where we have used the easily checked fact that the operator $\partial/\partial t$ on forms commutes with d_S. This makes it clear that (11)(i) at time 0 and (11)(ii) for all time imply (11)(i) for all time.

Thus we have just reduced (10) on the $(n+1)$-dimensional manifold M to the same problem on the n-dimensional manifold S_0, and so have solved (10) by induction. (Of course, the problem becomes trivial when the dimension of M descends to p; then there are no nonzero $(p+1)$-forms.) Essentially the same argument works also in a general manifold, which doesn't even have to be semi-Riemannian. There is, however, one important difference between the latter case and the case in which M is an inner product space; since the coordinates on a general manifold need not be globally defined, only local solutions are obtained by integrating (11)(ii).

The fact that any $(p+1)$-form γ with vanishing differential is locally the differential of some p-form is, of course, one version of the classical Poincaré Lemma † and the reader may be wondering why we have gone to so much trouble to prove this standard fact. The reason is that though our proof of the theorem is just a generalization of the ideas just presented, the theorem itself does not seem to follow directly from the Poincaré Lemma.

The above analysis proves the following Proposition, which in turn implies the Poincaré Lemma.

---

† Some authors call the trivial converse that $d^2 = 0$ the Poincaré Lemma.

Proposition.
   Let M be an $(n+1)$-dimensional C^∞ manifold, let q be a given point of M, and let $t(\cdot), x^1(\cdot), \cdots, x^n(\cdot)$ be a coordinate system for a neighborhood U of q. Let
$$S_0 = \{ u \in U \mid t(u) = 0 \} .$$
Let γ be a given C^∞ spatial $(p+1)$-form, $0 \leq p \leq n-1$, satisfying $d\gamma = 0$ in U, σ a given C^∞ spatial $(p-1)$-form on U, and η a given spatial C^∞ p-form satisfying $d_S \eta = \gamma_S$ on S_0. Then there exists a C^∞ p-form β on U such that

(i) $\qquad d\beta = \gamma \quad$ on U,

(ii) $\qquad \beta_T = \sigma \quad$ on U,

and for any $q \in S_0$,

(iii) $\qquad (\beta_S)_q = \eta_q$.

In other words, we can always solve $d\beta = \gamma$ on U while arbitrarily specifying β_T for all times and β_S at time 0, the latter specification subject to the necessary consistency condition $d_S \beta_S = \gamma_S$ at time 0.

As a corollary, we obtain one version of the classical Poincaré Lemma. The resulting proof is similar to [Sternberg, p.121, Thm. 4].

Lemma 1. (Poincaré Lemma.)
   Let $p \geq 0$ and let γ be a $(p+1)$-form defined in a coordinate neighborhood U in an $(n+1)$-dimensional C^∞ manifold M, $n \geq 0$ (so that U is identified with R^n under the coordinate map) and satisfying $d\gamma = 0$ in U. Then there exists a C^∞ p-form β defined on U such that $d\beta = \gamma$.

Proof. By induction on the dimension of M. If $n+1 = \dim M \leq p$, then γ necessarily vanishes (Exercise 3) and we may choose $\beta := 0$. For arbitrary n, we may apply induction to solve $d_S \beta_S = \gamma_S$ on the submanifold S_0 (in the notation of the Proposition) and then apply the Proposition to obtain $d\beta = \gamma$ on U.

Let V be a vector space of dimension n. Let p, $0 \leq p \leq n-1$, and a 1-form η be given, and define a linear transformation
$$W_\eta : \Lambda_p(V) \longrightarrow \Lambda_{p+1}(V)$$
whose action is "wedging with η" as follows: for any $\alpha \in \Lambda_p(M)$,
$$W_\eta \alpha := \eta \wedge \alpha .$$
Given any vector $u \in V$, define a linear transformation
$$C_u : \Lambda_{p+1}(V) \longrightarrow \Lambda_p(V)$$
of "contraction" with u: for any $\beta \in \Lambda_{p+1}(V)$ and any $v_1, v_2, \cdots, v_p \in V$,

$$(C_u\beta)(v_1, \cdots, v_p) := \beta(u, v_1, \cdots, v_p) .$$

Given any basis $e_1, e_2, \cdots, e_n$ and dual basis $e^1, e^2, \cdots, e^n$, consider the linear transformations W_{e^i} and C_{e_i}.

Lemma 2.

Let V be a vector space of dimension n with basis e_i, and let p be given, $1 \leq p \leq n-1$. Let $[W_{e^i}]$ and $[C_{e_i}]$ denote the matrices of W_{e^i} and C_{e_i}, respectively, with respect to the basis for $\Lambda_p(V)$ consisting of all

$$e^{i_1} \wedge e^{i_2} \wedge \cdots \wedge e^{i_p} , \qquad 1 \leq i_1 \leq i_2 \leq \cdots \leq i_p \leq n$$

and the analogous basis for $\Lambda_{p+1}(V)$. Then the transpose of $[W_{e^i}]$ is $[C_{e^i}]$, and the following $m \times m$ matrix, where

$$m = \binom{n}{p} + \binom{n}{p+1} ,$$

is symmetric:

$$\begin{bmatrix} 0 & [W_{e^i}] \\ [C_{e_i}] & 0 \end{bmatrix} .$$

(Of course, the zeros in the matrix stand for zero matrices of the appropriate sizes.)

Proof. If $i, j_1, \cdots, j_p$ are distinct indices, then

$$C_{e_i}(e^i \wedge e^{j_1} \wedge \cdots \wedge e^{j_p}) = e^{j_1} \wedge \cdots \wedge e^{j_p} \qquad \text{(no summation)},$$

This shows that $[C_{e_i}]$ is the transpose of $[W_{e^i}]$.

It is obvious that when $\{e_i\}$ is an orthonormal basis for an inner product space, contraction with e_i is, up to a sign, the transform under Hodge duality of wedging with e^i. The sign, which depends on n, p, and the norm of e_i will be important later. The following lemma computes this sign for the case for which we need it. The "V" of the lemma will later be the "space" part of an inner product space satisfying (2)(ii), so that the metric on V will be negative definite.

Lemma 3.

Let V be an oriented inner product space of dimension n with negative definite inner product, and let $\{e_i\}$ be an orthonormal basis. Then for any p-form λ on V, $1 \leq p \leq n$, and any i,

(13) $\qquad \perp W_{e^i} \perp \lambda = (-1)^{np+1} C_{e_i} \lambda .$

More generally, for any $u \in V$ and any p-form λ,

(14) $\qquad \perp W_{u*} \perp \lambda = (-1)^{np} C_u \lambda .$

Proof. The last statement is an immediate consequence of the first (and conversely): just embed u as the first element $u = e_1$ of an orthonormal basis and note that $e^1 = -u^*$. We state them separately because (14) is basis-independent but (13) is used in that form. To establish (13), note that by linearity, it is sufficient to consider decomposable λ of the form

(15) $$\lambda = e^{j_1} \wedge e^{j_2} \wedge \cdots \wedge e^{j_p}$$

with $j_1, \cdots, j_p$ distinct. Let $k_1, k_2, \cdots, k_{n-p}$ be the complement of the set $j_1, \cdots, j_p$, arranged so that $j_1, j_2, \cdots, j_p, k_1, \cdots, k_{n-p}$ is an even permutation of $1, 2, \cdots, n$. Then, assuming that $e_1, \cdots, e_n$ is positively oriented (which is clearly no loss of generality), we have

$$\perp(e^{j_1} \wedge \cdots \wedge e^{j_p}) = (-1)^p e^{k_1} \wedge \cdots \wedge e^{k_{n-p}} .$$

If i is not among the set $j_1, \cdots, j_p$, then (13) is obvious (both sides vanish), so assume i is one of the j_a's.

There is no real loss of generality in assuming $i = j_1$. To check this, note that if the set $j_1, j_2, \cdots, j_p$ has one element or more than two elements, then there is always an even permutation of $j_1, \cdots, j_p$ which brings any given j_b to the front. (Move it successively toward the front in $b-1$ steps and then interchange two other elements, if necessary, to make the permutation even.) Thus in this case we can make j_1 equal i by permuting the one-forms e^{j_a} in (15) and reindexing the basis. In the remaining case, $p = 2$,

$$\lambda = e^{j_1} \wedge e^{j_2} = -e^{j_2} \wedge e^{j_1} ,$$

and we can simply prove (13) with λ replaced by $-\lambda$.

When $i = j_1$, we have

(16) $$\perp W_{e^i} \perp (e^{j_1} \wedge \cdots \wedge e^{j_p}) = (-1)^p \perp (e^{j_1} \wedge e^{k_1} \wedge \cdots \wedge e^{k_{n-p}})$$

$$= (-1)^p (-1)^{n-p+1} (-1)^{(n-p)(p-1)} e^{j_2} \wedge \cdots \wedge e^{j_p}$$

$$= (-1)^{np+1} e^{j_2} \wedge \cdots \wedge e^{j_p} = (-1)^{np+1} C_{e_i}(e^{j_1} \wedge \cdots \wedge e^{j_p}) .$$

To understand the signs, note that the factor $(-1)^{n-p+1}$ is the norm of $e^{j_1} \wedge e^{k_1} \wedge \cdots \wedge e^{k_{n-p}}$ and the factor $(-1)^{(n-p)(p-1)}$ the number of interchanges necessary to change

$$e^{j_1} \wedge e^{k_1} \wedge \cdots \wedge e^{k_{n-p}} \wedge e^{j_2} \wedge \cdots \wedge e^{j_p}$$

into the volume 4-form

$$\Omega = e^{j_1} \wedge e^{j_2} \cdots \wedge e^{j_p} \wedge e^{k_1} \wedge \cdots \wedge e^{k_{n-p}}$$

by the obvious method of moving each of the $p-1$ j_a's successively to the left through $n-p$ k_b's. These signs arise from considering the relation

$$\alpha \wedge \perp \alpha = \langle \alpha, \alpha \rangle \Omega$$

with $\alpha := e^{j_1} \wedge e^{k_1} \wedge \cdots \wedge e^{k_{n-p}}$, since it is obvious, *a priori*, that $\perp \alpha = \pm e^{j_2} \wedge \cdots \wedge e^{j_p}$. The simplification of the sign in the final line of (16) comes from remembering that for any p, $p(p-1) \equiv 0 \pmod 2$, and so $(n-p)(p-1) \equiv np - n \pmod 2$.

For any spatial p-form σ, let $\perp_S \sigma$ denote its Hodge dual relative to the spatial submanifold S; i.e. $\perp_S$ is the Hodge dual relative to the restriction of the metric on M to S. In order for this to be well-defined, we must specify an orientation on S. Assuming M to be oriented, we define the orientation on S by the condition that $\frac{\partial}{\partial x_1}, \cdots, \frac{\partial}{\partial x_n}$ is positively oriented in S if and only if $\frac{\partial}{\partial t}, \frac{\partial}{\partial x_1}, \cdots, \frac{\partial}{\partial x_n}$ is positively oriented in M.

Lemma 4.

Let M be an oriented semi-Riemannian manifold with a coordinate system $t, x^1, \cdots, x^n$ satisfying (2)(i) (i.e. "time" is orthogonal to "space"). For any p-form β on M with $1 \leq p \leq n$,

(17)   (i) $(\perp \beta)_S = \langle dt, dt \rangle \perp_S \beta_T$

       (ii) $(\perp \beta)_T = (-1)^p \perp_S \beta_S$.

Also, (ii) holds when β is a 0-form.

Proof. Let $g_S(\cdot, \cdot) = \langle \cdot, \cdot \rangle_S$ denote the "restriction" of the metric $g(\cdot, \cdot) = \langle \cdot, \cdot \rangle$ to a constant-time submanifold S. In other words, g_S is the metric induced on S by g in the obvious way: for any two vectors u, v tangent to S at a point, $g_S(u, v) := g(u, v)$. First we observe that if $\mu = \mu_S$ and $\eta = \eta_S$ are spatial p-forms, then

$$\langle \mu, \eta \rangle_S = \langle \mu, \eta \rangle .$$

(This is not as quite as obvious as it looks and in fact is not generally true if "time" is not orthogonal to "space".) We use the component expression for the inner product on p-forms

$$\langle \mu, \eta \rangle = \frac{1}{p!} \mu_{i_1 i_2 \cdots i_p} g^{i_1 j_1} g^{i_2 j_2} \cdots g^{i_p j_p} \eta_{j_1 j_2 \cdots j_p} ,$$

where the components are of course with respect to the given coordinate system in which "time" is orthogonal to "space". (Recall that the matrix (g^{ij}) is the inverse of the matrix (g_{ij}).) Of course, $\langle \mu, \eta \rangle_S$ is the same expression with g replaced by g_S. Obviously, $(g_S)_{ij} = g_{ij}$, but in general $g_S{}^{ij}$ need not equal g^{ij}. However, since "time" is orthogonal to "space", the matrix (g_{ij}) is of the special form

$$(g_{ij}) = \begin{bmatrix} \langle \frac{\partial}{\partial t}, \frac{\partial}{\partial t} \rangle & 0 \\ 0 & [g_S] \end{bmatrix} ,$$

where $[g_S]$ denotes the matrix for g_S, and the inverse of such a matrix is of the same form with the inverse of $[g_S]$ in the lower right; in other words, $g^{ij} = g_S{}^{ij}$ for $1 \leq i, j \leq n$.

Let Ω denote the volume $(n+1)$-form on M and Ω_S the volume n-form on "space" S, and note that the orientations were chosen so that

$$\Omega = dt \wedge \Omega_S .$$

For any spatial $(p-1)$-form $\alpha = \alpha_S$, observe that

$$dt \wedge \alpha \wedge \perp \beta = dt \wedge \alpha \wedge [dt \wedge (\perp \beta)_T + (\perp \beta)_S] = dt \wedge \alpha \wedge (\perp \beta)_S ,$$

and so
$$dt \wedge \alpha \wedge (\perp\beta)_S = (dt \wedge \alpha) \wedge \perp\beta = \langle dt \wedge \alpha, \beta\rangle \Omega$$
$$= \langle dt \wedge \alpha, dt \wedge \beta_T + \beta_S\rangle dt \wedge \Omega_S$$
$$= \langle dt, dt\rangle\langle \alpha, \beta_T\rangle dt \wedge \Omega_S = \langle dt, dt\rangle \, dt \wedge \langle \alpha, \beta_T\rangle_S \Omega_S$$
$$= \langle dt, dt\rangle \, dt \wedge (\alpha \wedge \perp_S \beta_T) \quad,$$

where the defining relation 2.2(7) for the Hodge dual was used several times. Since $\alpha \wedge (\perp\beta)_S$ and $\alpha \wedge \perp_S \beta_T$ are purely spatial, we can cancel the dt's and obtain

$$\alpha \wedge [(\perp\beta)_S - \langle dt, dt\rangle \perp_S \beta_T] = 0 \quad.$$

It is an easy exercise (Exercise 8) to show that this can hold for all spatial $p-1$-forms α only if

$$(\perp\beta)_S = \langle dt, dt\rangle \perp_S \beta_T \quad.$$

Now we prove (ii). For any spatial p-form $\alpha = \alpha_S$, $\alpha \wedge (\perp\beta)_S$ is an $(n+1)$-form on the space S of dimension n and so must vanish (Exercise 3). Hence

$$\alpha \wedge (\perp\beta) = \alpha \wedge [dt \wedge (\perp\beta)_T + (\perp\beta)_S]$$
$$= \alpha \wedge dt \wedge (\perp\beta)_T = (-1)^p dt \wedge \alpha \wedge (\perp\beta)_T \quad.$$

It follows that

$$(-1)^p dt \wedge \alpha \wedge (\perp\beta)_T = \alpha \wedge \perp\beta = \langle \alpha, \beta\rangle \Omega$$
$$= \langle \alpha, \beta_S\rangle_S \, dt \wedge \Omega_S = dt \wedge \alpha \wedge \perp_S \beta_S \quad.$$

Since this is true for all spatial p-forms α, we must have

$$(\perp\beta)_T = (-1)^p \perp_S \beta_S \quad.$$

Returning to our explanation of the scholium, let us now try to solve

(19) $\qquad \delta\beta = \alpha$

assuming that $\delta\alpha = 0$. **We assume until further notice the condition (2)(ii) that $\partial/\partial t$ is time-like (i.e positive norm) and that the metric on "space" S is negative definite.** The proof does go through to a certain extent without this hypothesis, but many complicated signs appear which turn out to be irrelevant to our approach, so for simplicity we assume it. At the end the true meaning of this assumption will become clear. Recall from Exercise 20 that

$$\delta\beta = s \perp d \perp \beta \quad,$$

where s is a sign, and from 2.2(11) that for any p-form σ,

$$\perp\perp\sigma = s'\sigma$$

where s' is another sign. Hence (19) is equivalent to

(20) $\qquad d_\perp \beta = ss' \perp \alpha$,

and similarly, the consistency condition $\delta\alpha = 0$ is the same as

$$d_\perp \alpha = 0 \ .$$

Replacing α by $-\alpha$ if necessary, we may assume that $ss' = 1$ in (20) and thus avoid carrying these inessential signs.

With this simplification, we are now trying to solve simultaneously

(21) $\qquad d\beta = \gamma \qquad$ and

(22) $\qquad d_\perp \beta = \perp\alpha$

assuming the consistency conditions

(23) $\qquad d\gamma = 0 \qquad$ and

(24) $\qquad d_\perp \alpha = 0$.

By the analysis which led to the Proposition, (21) and (22) are respectively equivalent to:

(25) (i) $\qquad d_S \beta_S = \gamma_S \qquad$ on S_0 ,

(ii) $\qquad \dfrac{\partial \beta_S}{\partial t} = d_S(\beta_T) + \gamma_T$,

and

(26) (i) $\qquad d_S((\perp\beta)_S) = (\perp\alpha)_S \qquad$ on S_0 ,

(ii) $\qquad \dfrac{\partial(\perp\beta)_S}{\partial t} = d_S((\perp\beta)_T) + (\perp\alpha)_T$.

Lemma 4 shows that when $p \geq 2$, so that α has degree at least one, we may replace (26) by

(27) (i) $\qquad d_S(\perp_S \beta_T) = \perp_S \alpha_T \qquad$ on S_0

(ii) $\qquad \dfrac{\partial(\perp_S \beta_T)}{\partial t} = (-1)^p d_S(\perp_S \beta_S) + (-1)^{p-1} \perp_S \alpha_S$.

If $p = 1$, so that α is a 0-form, (26)(i) is automatic (both sides vanish), and in order not to have to do this case separately, we declare (27)(i) (which is meaningless as written because α_T is undefined) as true by definition.

At this point the proof simplifies considerably when M is an inner product space, and we assume this until further notice. We shall use linear coordinates $t, x^1, \cdots, x^n$ with respect to an orthonormal basis. The orthonormality guarantees that the operations of taking $\perp_S$ and $\partial/\partial t$ commute: for any spatial p-form σ ,

(28) $\qquad \dfrac{\partial(\perp_S \sigma)}{\partial t} = \perp_S \dfrac{\partial \sigma}{\partial t}$.

To see this quickly, recall that when the dx^i are orthonormal, $\perp_S(dx^{i_1} \wedge \cdots \wedge dx^{i_p}) = \pm(dx^{j_1} \wedge \cdots \wedge dx^{j_{n-p}})$, where $i_1, \cdots, i_p, j_1, \cdots, j_{n-p}$ is a permutation of $1, \cdots, n$. In a general semi-Riemannian manifold (28) no longer holds; we shall see later that what does hold is

$$\text{(29)} \qquad \frac{\partial \perp_S \beta_T}{\partial t} = \perp_S \frac{\partial \beta_T}{\partial t} + L(\beta_T) \; ,$$

where L is a linear transformation (varying from point to point in the manifold) from p-forms to $(n-p)$-forms.

Substituting (28) with $\sigma := \beta_T$ in (27), and recalling from 2.2(11) that when S has dimension n and negative definite metric,

$$\perp_S \perp_S \sigma = (-1)^{p(n-p)}(-1)^n \sigma$$

for p-forms σ, we may replace (27) by

$$\text{(30)} \quad \text{(i)} \qquad d_S(\perp_S \beta_T) = \perp_S \alpha_T$$

$$\quad \text{(ii)} \qquad \frac{\partial \beta_T}{\partial t} = (-1)^{np+1} \perp_S d_S \perp_S \beta_S + (-1)^{p-1}\alpha_S \; .$$

The sign $(-1)^{np+1}$ of the first term on the right of (ii) is the critical one.

The decomposition

$$\beta = \beta_S + dt \wedge \beta_T$$

gives an obvious way to uniquely identify a form field $z \longmapsto \beta_z$, $z \in M$ with a pair $z \longmapsto (\beta_S)_z$, $z \longmapsto (\beta_T)_z$ of spatial form fields on M. Moreover, spatial form fields may be identified in an obvious way with vector-valued functions from $R^1 \times R^n$ to $\Lambda_p(R^n)$: if $\{e_i\}$ is the orthonormal basis with respect to which the coordinates $\{x^i\}$ are defined, then

$$\text{(31)} \qquad \sum_{i_1 < i_2 < \cdots < i_p} f_{i_1 \cdots i_p} dx^{i_1} \wedge dx^{i_2} \wedge \cdots \wedge dx^{i_p}$$

is identified with

$$\text{(32)} \qquad \sum_{i_1 < i_2 < \cdots < i_p} f_{i_1 \cdots i_p} e^{i_1} \wedge e^{i_2} \wedge \cdots \wedge e^{i_p} \; .$$

In this way, a form field β on M is identified with a function from $R^1 \times R^n$ to the fixed vector space $\Lambda_p(R^n) \oplus \Lambda_{p-1}(R^n)$.

The intuitive idea here is that we are identifying β with the symbolic "vector" function

$$z \longmapsto \begin{bmatrix} (\beta_S)_z \\ (\beta_T)_z \end{bmatrix}, \qquad z = (t, x^1, \cdots, x^n) \in R^1 \times R^n \; .$$

Viewed in this way, the equations (25)(ii) and (27)(ii) may be combined as the single symbolic "vector" equation

$$\text{(33)} \qquad \frac{\partial}{\partial t} \begin{bmatrix} \beta_S \\ \beta_T \end{bmatrix} = \begin{bmatrix} 0 & d_S \\ (-1)^{np+1} \perp_S d_S \perp_S & 0 \end{bmatrix} \begin{bmatrix} \beta_S \\ \beta_T \end{bmatrix} + \begin{bmatrix} \gamma_T \\ (-1)^{p-1}\alpha_S \end{bmatrix} \; .$$

2.9 Equations $d\beta = \gamma$ and $\delta\beta = \alpha$

The matrix

$$\begin{bmatrix} 0 & d_S \\ (-1)^{np+1} \perp_S d_S \perp_S & 0 \end{bmatrix}$$

is a matrix of first-order partial derivative operators, which may be symbolically written as

$$\begin{bmatrix} 0 & d_S \\ (-1)^{np+1} \perp_S d_S \perp_S & 0 \end{bmatrix} = \sum_{i=1}^{n} A^i \frac{\partial}{\partial x_i} ,$$

where the A^i are linear transformations on the vector space

$$Y := \Lambda_p(R^n) \oplus \Lambda_{p-1}(R^n) \quad \text{of dimension} \quad m := \binom{n}{p} + \binom{n}{p-1}.$$

Thus (33) may be more compactly written as a partial differential equation for a vector-valued function $u : R^1 \times R^n \longrightarrow Y$ (or equivalently, as a system of m PDE's for the scalar components of u with respect to some convenient basis):

$$(34) \qquad \frac{\partial u}{\partial t} = \sum_{i=1}^{n} A^i \frac{\partial u}{\partial x_i} + f .$$

where f is a given vector-valued function corresponding to the last term on the right of (33). The equation (34) has been extensively studied, and there is a good existence theorem for the case in which there exists a basis for Y with respect to which the matrices of the A^i, $1 \leq i \leq n$ are symmetric. In fact, the theorem applies to the more general equation

$$(35) \qquad \frac{\partial u}{\partial t} = \sum_{i=1}^{n} A^i \frac{\partial u}{\partial x_i} + Bu + f ,$$

where u is viewed as a column vector with respect to some given basis, $t, x \longmapsto f$ is a given column vector function of $t, x = t, x^1, x^2, \cdots, x^n \in R^1 \times R^n$, the $A^i = A^i(t, x)$ are symmetric matrices which can depend on t and x and $B = B(t, x)$ is likewise a matrix-valued function which *doesn't have to be symmetric.* (This last fact will be important later.)

For our system (33), the condition 2(ii) implies that the matrices of the A^i's with respect to the bases of Lemma 2 (consisting of wedge products of the e_i's) are symmetric. To see this, note that the differential d_S may be symbolically written as

$$d_S = \sum_{i=1}^{n} W_{dx^i} \frac{\partial}{\partial x_i} ,$$

and

$$(36) \qquad \perp_S d_S \perp_S = \sum_{i=1}^{n} \perp_S W_{dx^i} \frac{\partial}{\partial x_i} \perp_S = \sum_{i=1}^{n} \perp_S W_{dx^i} \perp_S \frac{\partial}{\partial x_i}$$

$$= \sum_{i=1}^{n} (-1)^{np+1} C_{\frac{\partial}{\partial x_i}} \frac{\partial}{\partial x_i} ,$$

where we have used Lemma 3 and the fact that in the special case in which the dx^i are everywhere orthonormal, $\perp_S$ commutes with $\partial/\partial x_i$ (for the same reason that it commutes with $\partial/\partial t$). † Under the identification of (31) and (32), the dx^i's get changed into e^i's.

---
† There is a distasteful notational ambiguity on the right side of (36), in which the symbol $\partial/\partial x_i$ is used both in the sense of a tangent vector (as a subscript on C) and as a partial differential operator on form

Hence the matrix of A^i with respect to the bases of wedge products of e^i's of Lemma 2 is

$$A_i = \begin{bmatrix} 0 & [W_{e^i}] \\ [C_{e_i}] & 0 \end{bmatrix},$$

which is symmetric by that lemma.

From the analysis of (35) in [John, Chapter 5, Section 3; especially, problem 2, p.181], one concludes the existence of local C^∞ solutions u to (35) given C^∞ A^i, B, and f.‡
Moreover, the C^∞ initial values $u(0, x^1, \cdots, x^n)$ may be arbitrarily specified. In terms of our original problem, this says that insofar as obtaining a solution to (25)(ii) and (27)(ii) in concerned, the initial values of β_S and β_T may be arbitrarily specified. Of course, the full system (25) and (27) requires that the initial values satisfy (25)(i) and (27)(i), but solutions of these can be obtained from the Poincaré Lemma. That is, we solve

(25)   (i) $d_S \beta_S = \gamma_S$

for β_S on S_0 (the necessary consistency condition $d_S \gamma_S = 0$ is automatic from $d\gamma = 0$), and use the resulting β_S as initial condition for (33). Similarly, we solve (27)(i) for $\perp_S \beta_T$, and hence for β_T, on S_0, the consistency condition again being automatic, and use the resulting β_T as initial condition for (33). This completes the proof of the scholium for M an inner product space with exactly one timelike dimension. The "technical hypothesis" mentioned there is that the given data α, γ are C^∞, and the conclusion is that there exists a local C^∞ solution β in a neighborhood of any given point of M.

Now we indicate the modifications necessary to adapt the proof to a general semi-Riemannian manifold with a local coordinate system $t, x^1, \cdots, x^n$ satisfying (2)(i) and (ii). Again, we shall rewrite (25) and (27) as a partial differential equation of the form (35) with symmetric A^i. The only difference with the previous case will be that B will now be nonzero. This arises from the fact that the partial differential operators $\partial/\partial t$ and $\partial/\partial x^i$ do not necessarily commute with $\perp_S$ in the more general setting. To see what the commutation relation is, we recall the formula 2.2(19) for the Hodge dual in terms of components:

$$(\perp_S \beta_T)_{j_1 \cdots j_{n-p+1}} = \frac{1}{(p-1)!} \Omega_S{}^{\lambda_1 \cdots \lambda_{p-1}}{}_{j_1 \cdots j_{n-p+1}} (\beta_T)_{\lambda_1 \cdots \lambda_{p-1}},$$

where Ω_S is the volume n-form on "space" S at q. Differentiating with respect to t gives (29) with

$$(L \beta_T)_{j_1 \cdots j_{n-p+1}} := \frac{1}{(p-1)!} (\beta_T)_{\lambda_1 \cdots \lambda_{p-1}} \frac{\partial}{\partial t} \Omega_S{}^{\lambda_1 \cdots \lambda_{p-1}}{}_{j_1 \cdots j_{n-p+1}}.$$

This adds a term of the general form Bu to the right of (34). Other such terms come from the $\perp_S d_S \perp_S$ in the matrix of (33).

Unfortunately, there is a further complication. If the dx^i are not orthonormal, the step in (36) which used Lemma 3 to conclude that

$$\perp_S W_{dx^i} \perp_S = (-1)^{np+1} C_{\frac{\partial}{\partial x_i}},$$

is no longer valid, and we can no longer be sure that the matrices of the A^i with respect to the bases of Lemma 2 are symmetric. (They could be made symmetric at any given point by

---
fields, but as the sense is clear from the context, it should cause no confusion once the reader is alerted to it.
‡ See, however, Remark (1) at the end of this section.

choosing a spatial coordinate system in which the dx^i are orthonormal at that point, but this can not in general be accomplished over an entire coordinate neighborhood.) However, we can get around this by changing the identification of p-forms with vector-valued functions as follows. Choose a collection $\bar{e}_1, \cdots, \bar{e}_n$ of C^∞ vector fields defined on some coordinate neighborhood U of the point at which we want to establish the theorem and everywhere orthonormal in that neighborhood. Such a basis can be obtained, for example, by applying the Gram-Schmidt process to a coordinate basis $\{\partial/\partial x_i\}$. Let $\bar{e}^1, \cdots, \bar{e}^n$ denote the corresponding dual basis: $\bar{e}^i(\bar{e}_j) = \delta^i{}_j$. As before, let $e^1, \cdots, e^n$ denote an orthonormal dual basis for R^n. Identify spatial p-form fields μ on U with functions from $R^1 \times R^n$ to $\Lambda^p(R^n)$ by identifying

$$\mu = \sum_{i_1 < \cdots < i_p} f_{i_1 \cdots i_p} \bar{e}^{i_1} \wedge \cdots \wedge \bar{e}^{i_p} , \qquad f_{i_1 \cdots i_p} \in C^\infty(U)$$

with

$$\sum_{i_1 < \cdots < i_p} \tilde{f}_{i_1 \cdots i_p} e^{i_1} \wedge \cdots \wedge e^{i_p} , \qquad f_{i_1 \cdots i_p} \in C^\infty(U) ,$$

where $(t, x^1, \cdots, x^n) \longmapsto \tilde{f}_{i_1 \cdots i_p}(t, x^1, \cdots, x^n)$ is the scalar function on $R^1 \times R^n$ obtained from $f_{i_1 \cdots i_p}$ by identifying U with $R^1 \times R^n$ via the coordinatization map. Now use 2.5(14) to compute d_S of a spatial p-form μ as

$$d_S \mu = \sum_{i=1}^n \bar{e}^i \wedge D_{\bar{e}_i} \mu ,$$

where D is the Riemannian covariant derivative. If $\mu = f \bar{e}^{j_1} \wedge \cdots \wedge \bar{e}^{j_p}$, then

(38) $$\sum_i \bar{e}^i \wedge D_{\bar{e}_i} \mu = \sum_i \bar{e}_i(f) \, \bar{e}^i \wedge \bar{e}^{j_1} \wedge \cdots \wedge \bar{e}^{j_p}$$

$$+ f \sum_i \bar{e}^i \wedge D_{\bar{e}_i}(\bar{e}^{j_1} \wedge \cdots \wedge \bar{e}^{j_p}).$$

Now

$$\sum_i \bar{e}_i(f) \, \bar{e}^i \wedge \bar{e}^{i_1} \wedge \cdots \wedge \bar{e}^{i_p} = d_S f \wedge \bar{e}^{i_1} \wedge \cdots \wedge \bar{e}^{i_p} ,$$

which shows that the first term on the right of (38) is just what was previously obtained for the special case $M = R^{n+1}$ and $\bar{e}^i = dx^i$. Since the $\bar{e}^i$ are orthonormal, the same is true when using (36) to compute $\perp_S d_S \perp_S$, which means that the part of the matrix A^i due to the first term of (38) is the same as in the previous case of an inner product space and so is symmetric. On the other hand, the second term of (38) contributes nothing to A^i (because the differentiation operator does not hit the coefficient function f); rather, it contributes to the zeroth-order linear term Bu. Hence we obtain an equation of the same form (34) if we use this identification, and this completes the proof.

Let us summarize what we have done. Assume $p \geq 1$, the case $p = 0$ being covered by the Poincaré Lemma. First we algebraically reduced the original problem

$$d\beta = \gamma \qquad \text{and} \qquad \delta\beta = \alpha ,$$

to a first order vector partial differential equation of the form (35). The intuitive identification is

$$\beta \longleftrightarrow \begin{bmatrix} \beta_S \\ \beta_T \end{bmatrix} \longleftrightarrow u = \begin{bmatrix} u^1 \\ \cdots \\ u^m \end{bmatrix},$$

where the components of u are just the components of β reindexed. Inspection of the matrices of the A^i with respect to an appropriate basis (wedge products of the $\bar{e}^k$) shows that each decomposes into a direct sum of 2×2 matrices of the form

(39)   (i) $\begin{bmatrix} 0 & 1 \\ 1 & 0 \end{bmatrix}$, or

         (ii) $\begin{bmatrix} 0 & 1 \\ -1 & 0 \end{bmatrix}$.

This is because

$$A^i(0 \oplus \bar{e}^{j_1} \wedge \cdots \wedge \bar{e}^{j_{p-1}}) = \pm \bar{e}^i \wedge \bar{e}^{j_1} \wedge \cdots \wedge \bar{e}^{j_{p-1}} \oplus 0 \quad \text{and}$$

$$A^i(\bar{e}^i \wedge \bar{e}^{j_1} \wedge \cdots \wedge \bar{e}^{j_{p-1}} \oplus 0) = \pm 0 \oplus \bar{e}^{j_1} \wedge \cdots \wedge \bar{e}^{j_{p-1}},$$

where the signs are not necessarily the same. To apply the existence theorem mentioned above, we need to check that only summands of the form (i) can occur, and this was done by a direct computation for the case in which M has exactly one timelike dimension. †

Thus the whole problem is reduced to the study of (35) where the A^i are known matrices with constant entries. Intuitively, we expect to be able to integrate (35) forward in time given arbitrary initial data $u(0, \cdot)$. If the initial data and f happen to be real-analytic, we can conclude this rigorously from the Cauchy-Kowalevski Theorem, and in this case the solution u is also real-analytic. Although in the above analysis we assumed the condition (2)(ii) that there was exactly one timelike direction, this assumption was only used to compute the signs which led to the conclusion that the A^i are symmetric. If we are willing to use the Cauchy-Kowalewski Theorem the symmetry of the A^i is irrelevant, and the analysis goes through exactly as before regardless of the signature of the metric. When the A^i are symmetric, which we showed occurs in spacetimes of arbitrary dimension, we can obtain C^∞ solutions from C^∞ initial data via the fundamental existence theorem for (35). We summarize with the following theorem.

---

† If M has more than one timelike dimension, in general both (39)(i) and (ii) can occur. If M is Riemannian, only (ii) occurs, and there is a sense in which the associated system is elliptic rather than hyperbolic. In view of the Hodge theory, there would presumably be hope of extending the theorem to this case, but we do not address this problem.

Theorem.

Let M be an oriented C^∞ semi-Riemannian manifold, q a given point of M, and $p \geq 1$. Let α be a given C^∞ $(p-1)$-form and γ a given C^∞ $(p+1)$-form defined in a neighborhood of q and satisfying

$$d\gamma = 0 \quad \text{and} \quad d\perp\alpha = 0$$

in that neighborhood. Then there exists a C^1 (see Remark (1) below) p-form β defined on some (possibly smaller) neighborhood U of q and satisfying

(40) $\qquad d\beta = \gamma \quad \text{and} \quad d\perp\beta = \perp\alpha$

on U provided that either of the following two conditions hold:

(i) the metric on the tangent space M_q is of type $+, -, -, \cdots, -$; in other words, there is precisely one timelike direction; or

(ii) the components of α and γ relative to some local coordinatization are real analytic.

In case (ii), the solution β can be chosen to be real analytic.

Moreover, suppose that we choose a hypersurface S_0 containing q whose normal has positive norm at q. Then we can arbitrarily specify β on $S_0 \cap U$ subject only to the necessary consistency conditions

(41) $\qquad d_S\beta_S = \gamma_S \quad \text{and} \quad d_S(\perp_S\beta_T) = \perp_S\alpha_T \quad \text{on } S_0 \cap U.$

(When $p = 1$, the condition $d_S(\perp_S\beta_T) = \perp_S\alpha_T$ is to be interpreted as automatically true.)

More precisely, if η and σ are, respectively, a p-form and a $(p-1)$-form (assumed C^∞ in case (i) and real analytic in case (ii)), defined on the submanifold S_0 and satisfying

$$(d_S\eta)_x = (\gamma_S)_x \quad \text{and} \quad (d_S(\perp_S\sigma))_x = (\perp_S\alpha_T)_x \quad \text{for all } x \in S_0 \cap U,$$

then there exists a solution β of (40) (C^1 in case (i), real analytic in case (ii)) satisfying

$$(\beta_S)_x = \eta_x \quad \text{and} \quad (\beta_T)_x = \sigma_x \quad \text{for all } x \in S_0.$$

Remarks.

(1) The rather unnatural conclusion that β is C^1 instead of C^∞ is because the most accessible reference for the existence theorem for (35) which I have been able to find [John, Chapter 5] concludes C^1 solutions from C^∞ given data. A special case of a problem in that reference (p. 181, problem 2) implies that actually, C^∞ solutions result from C^∞ data, but it seems inappropriate to use a rather complicated problem as a reference for a fundamental result. The difference between C^1 and C^∞ is of no importance to the present work, and this is not the place to examine such questions.

(2) The condition $\delta\beta = \alpha$ of the scholium was replaced by $d\perp\beta = \perp\alpha$ to avoid having to carry the complicated sign relationship between $\perp d\perp$ and δ throughout the whole proof. In addition, the consistency condition $d_S(\perp_S\beta_T) = \perp_S\alpha_T$ acquires a complicated sign when expressed in terms of the spatial covariant divergence operation on S.

(3) Of course, the Poincaré Lemma establishes the theorem for $p = 0$ as well; in this case hypothesis (i) is unnecessary, and a C^∞ solution β is always obtained.

References.

The best single reference for this chapter is probably [Abraham, Marsden, and Ratiu]. Most graduate level differential geometry texts also contain much of this material, though specific topics which are important to us, such as the Hodge dual, may be absent. The point of view of [Hicks] is close to ours. † The text of [Flanders] on differential forms is written for beginners with applications in mind. [Misner, Thorne, and Wheeler] includes several well-chosen illustrations of various geometrical interpretations of differential forms and is worth looking at just for the pictures.

Exercises 2

The first few exercises are meant to accustom the reader to the notation of indicial tensor analysis. Although not used very much in this book, its mastery is essential for the serious student, as it was nearly universal in the physics and relativity literature up until the 1970's and is still extensively used.

2.1. Let $f \in V^*$ be a given form. Show that the vector $u \in V$ whose components are

$$u^i := g^{i\alpha} f_\alpha$$

satisfies $u^* = f$.

2.2. Let $\{e_i\}_{i=1}^n$ and $\{\bar{e}_i\}_{i=1}^n$ be two bases for a vector space V, and denote the components of a vector $v \in V$ with respect to the first by v^i and with respect to the second by $\bar{v}^i$. Thus

$$\bar{v}^\alpha \bar{e}_\alpha = v^\alpha e_\alpha .$$

(a) We know from linear algebra that for some matrix $A^i{}_j$,

(1) $\qquad \bar{v}^i = A^i{}_\alpha v^\alpha .$

Also, each $\bar{e}_i$ is a linear combination of the e_α:

(2) $\qquad \bar{e}_i = B^\alpha{}_i e_\alpha .$

Calculate the relation between the matrices A and B.

(b) Let $\{e^i\}$ and $\{\bar{e}^i\}$ be the dual bases for $\{e_i\}$ and $\{\bar{e}_i\}$, respectively. Let f_i be the components of a form with respect to $\{e^i\}$ and $\bar{f}_i$ with respect to $\{\bar{e}^i\}$. Calculate the relations between the e^i and $\bar{e}^i$ and between the f_i and $\bar{f}_i$.

---

† This is not surprising, since I originally learned differential geometry in a course of the late Noel Hicks and would like to take this opportunity to express my appreciation for his humane approach.

2.3. Show that if V is a vector space of dimension n and $q > n$, then any q-form on V is identically zero.

2.4 (a) Let V be a vector space of dimension n. Show that the space $\Lambda_n(V)$ of all n-forms on V is one-dimensional.

(b) Let $e_1, e_2, \cdots, e_n$ be a basis for V and $e^1, \cdots, e^n$ the dual basis. Show that if

$$v_i = \sum_{j=1}^{n} a_i{}^j e_j \quad , \quad i = 1, 2, \cdots, n$$

is any collection of n vectors, then

$$(e^1 \wedge e^2 \wedge \cdots \wedge e^n)(v_1, v_2, \cdots, v_n) = \det(a_i{}^j) .$$

(Since the matrix $(a_i{}^j)$ does not function as a linear transformation or tensor in this context, the previous index conventions do not apply, but we adhere to them anyway. The index positioning in $(a_i{}^j)$ looks peculiar, but if we wrote (a_{ij}) instead, some might wonder why.)

2.5. Show that no nonzero decomposable q-form can be a decomposable tensor when $q \geq 2$.

2.6. Let Ω denote the volume n-form on an oriented n-dimensional inner product space V. Let $\{e_i\}_{i=1}^{n}$ be a positively oriented basis for V (not necessarily orthonormal) and $\{e^i\}$ the dual basis for V^*. Let $G = (g_{ij})$ denote the matrix of the inner product with respect to this basis: $G_{ij} := \langle e_i, e_j \rangle$. Show that

(*) $$\Omega = |\det(G)|^{1/2} e^1 \wedge \cdots \wedge e^n .$$

Hint: Consider how (*) transforms under a change of basis.

2.7. This problem carries out the details in the definition of the inner product on forms of Section 2.2(b). Let V be a vector space with inner product, and choose an orthonormal basis $e_1, \cdots, e_n$ for V. This basis will be fixed throughout the discussion. Define the inner product on q-forms as explained in the text, so that the set of q-forms

$$e^{i_1} \wedge \cdots \wedge e^{i_q} \quad , \quad 1 \leq i_1 < i_2 < \cdots < i_q \leq n \quad ,$$

is orthonormal with the signs of the norms given by

$$\langle e^{i_1} \wedge \cdots \wedge e^{i_q}, e^{i_1} \wedge \cdots \wedge e^{i_q} \rangle = \prod_{k=1}^{q} \langle e_{i_k}, e_{i_k} \rangle .$$

Show that for any 1-forms $\alpha_1, \cdots, \alpha_q, \beta_1, \cdots, \beta_q \in V^*$

(*) $$\langle \alpha_1 \wedge \alpha_2 \wedge \cdots \wedge \alpha_p, \beta_1 \wedge \cdots \wedge \beta_p \rangle = \det(\langle \alpha_i, \beta_j \rangle)$$

(i.e. the determinant of the matrix whose i, j entry is $\langle \alpha_i, \beta_j \rangle$).

This establishes a coordinate-free characterization of the inner product on q-forms and thus shows that it is independent of the basis used in the original definition. Hint:

For an expeditious proof, use the universal property of $\Lambda_q(V)$.

2.8. (a) Show that an $(n-p)$-form σ satisfying $\beta \wedge \sigma = 0$ for all p-forms β must vanish.

(b) Show that given any p-form α, there is only one $(n-p)$-form γ satisfying 2.2(7):

(*) $\qquad \beta \wedge \gamma = <\beta, \alpha> \Omega \qquad$ for all p-forms β ;

i.e. show that the Hodge dual of α is unique.

2.9. (a) Show that every $n-1$-form on a vector space V of dimension n is decomposable. (Hint: If brute force doesn't work, try defining an inner product on V and consider the Hodge dual of the given form.)

(b) Conclude that every q-form on a space of dimension ≤ 3 is decomposable.

(c) Show that every 2-form on a vector space of dimension 4 can be written as a sum of two decomposable 2-forms.

(d) Give an example of an indecomposable q-form on a vector space of dimension 4.

2.10. Let e^0, e^1, e^2, e^3 be a positively oriented orthonormal dual basis for Minkowski space M (so that e^0 has norm 1 and the other e^i have norm -1). Mentally compute the Hodge duals:

(a) $\quad \bot e^0 \qquad$ (b) $\quad \bot e^1 \qquad$ (c) $\quad \bot e^2$

(d) $\quad \bot(e^0 \wedge e^3) \qquad$ (e) $\quad \bot(e^1 \wedge e^2) \qquad$ (f) $\quad \bot(e^1 \wedge e^2 \wedge e^3)$

(g) $\quad \bot(e^0 \wedge e^2 \wedge e^3) \qquad$ (h) $\quad \bot(e^0 \wedge e^1 \wedge e^2 \wedge e^3)$

This may seem a bit elementary, but it is a good way to familiarize oneself with the Hodge dual. With a little practice one should be able to do these instantly.

2.11. Show that for any 1-forms $\alpha, \beta, \gamma,$ on an oriented vector space of dimension n with positive definite inner product,

(*) $\qquad \bot(\alpha \wedge \bot(\beta \wedge \gamma)) = (-1)^n (<\alpha, \beta> \gamma - <\alpha, \gamma> \beta)$.

Observe that in Euclidean R^3, after identifying vectors with forms via the map $\vec{v} \longmapsto \vec{v}*$, this becomes (cf. 2.2(10)) the familiar vector identity

$$\vec{u} \times (\vec{v} \times \vec{w}) = (\vec{u} \cdot \vec{w}) \vec{v} - (\vec{u} \cdot \vec{v}) \vec{w} .$$

2.12. Let V be an oriented vector space with inner product, $\{e_i\}_{i=1}^n$ an orthonormal basis, and set

$$s := \prod_{i=1}^n \langle e_i, e_i \rangle \;.$$

Show that for any p-form α,

$$\bot\bot \alpha = (-1)^{p(n-p)} s\, \alpha \;.$$

2.13. Show that any isometry on an inner product space (with nondegenerate but not necessarily positive definite inner product) has determinant ± 1.

2.14. Show that the covariant derivative $D_v \alpha$ of a 1-form α in the direction of a vector v is well-defined by 2.5(5). In other words, show that

    (a) Any vector $w_p \in M_p$ can be extended to a C^∞ vector field w defined in a neighborhood of p, and

    (b) If u and w are vector fields with $u_p = w_p$, then for any $v_p \in M_p$,

(*) $\qquad v_p(\alpha(u)) - \alpha(D_{v_p} u) = v_p(\alpha(w)) - \alpha(D_{v_p} w) \;.$

Verify (b) as follows:

    (i) Check that if f is any C^∞ function and w any vector field, then

$$v_p(\alpha(fw)) - \alpha(D_v(fw)) = f(p) \cdot [v_p(\alpha(w)) - \alpha(D_v w)] \;,$$

the point being that the left side depends only on the value of f at p.

    (ii) Show that (i) implies (b).

2.15. Prove 2.5(6):

$$D_{\frac{\partial}{\partial x_i}} (dx^j) = -\sum_k \Gamma^j_{ik} dx^k \;.$$

2.16. Show that for any 1-form λ, q-form β, and vectors $v_1, v_2, \cdots, v_{q+1}$,

2.5(15) $\qquad (\lambda \wedge \beta)(v_1, \cdots, v_{q+1}) = \sum_{k=1}^{p+1} (-1)^{k+1} \lambda(v_k) \beta(v_1, \cdots, \hat{v}_k, \cdots, v_{q+1}) \;.$

2.17. Let v be a vector field on a manifold M with a given covariant derivative V, and let $\{x^i\}_{i=1}^n$ be a coordinate system. The analog of 2.8(6) in this situation will read

$$\delta v = v^i{}_{|i} = \sum_{i=1}^n \frac{\partial v^i}{\partial x_i} + \text{something} \; .$$

Write down an expression for the "something" in terms of the connection coefficients Γ_{ij}^k.

2.18. Let $T \in V_q^p$ be an arbitrary tensor. Show that the definition 2.5(9) of $D_v T$ implies that for any form fields $\alpha_1, \cdots, \alpha_p$ and vector fields $v_1, \cdots, v_q$,

$$(D_v T)(\alpha_1, \cdots, \alpha_p) = v(T(\alpha_1, \cdots, \alpha_p, v_1, \cdots, v_q)) - T(D_v \alpha_1, \alpha_2, \cdots, \alpha_p, v_1, \cdots v_q)$$
$$- T(\alpha_1, D_v \alpha_2, \cdots, \alpha_p, v_1, \cdots, v_p) - \cdots - T(\alpha_1, \cdots, \alpha_p, v_1, \cdots, D_v v_p)$$

2.19. Let D be the Riemannian connection on a semi-Riemannian manifold M with metric $g(\cdot, \cdot)$.

(a) Show that $Dg = 0$.

(b) Let $p \in M$ be a given point, and let $\{x^i\}$ be a coordinate system for a neighborhood of p such that the connection coefficients for this coordinate system vanish at p: $\Gamma_{ij}^k(p) = 0$ for all i, j, k. (Incidentally, given p, there always exists such a coordinate system; for example, a so-called *geodesic* coordinate system has this property [Hicks, p. 131].) Show that

$$\frac{\partial g_{ij}}{\partial x_k}(p) = 0 \; .$$

2.20. Let M be an n-dimensional inner product space whose metric has exactly $b \geq 0$ negative norm elements in any orthonormal basis.

(a) Show that for any p-form α on M,

(*) $\qquad \delta(\alpha) = (-1)^{b+(p+1)n} \perp d \perp \alpha$.

(b) Generalize (a) to the case in which M is a semi-Riemannian manifold whose metric at any point has exactly b negative norm directions.

2.21. (a) Show that if v is a vector field on a spacetime M and f any function, then

$$\delta(fv) = f\delta(v) + v(f) \; .$$

(b) Show that for any two vector fields v and w,

$$\delta(v \otimes w) = \delta(v)w + D_v w \; .$$

(c) Show that

$$\delta(v^* \wedge w^*) = \delta(v)w^* - \delta(w)v^* + [v, w]^* \; .$$

2.22. Consider the two-dimensional region V in two-dimensional Minkowski space $\mathbf{M}^2$ bounded by the timelike surface

$$T := \{ (t, 0) \mid 0 \leq t \leq 1 \} ,$$

the spacelike surface

$$S := \{ (1, x) \mid 0 \leq x \leq 1 \} ,$$

and the light cone

$$L := \{ (x, x) \mid 0 \leq x \leq 1 \} .$$

Give $\mathbf{M}^2$ and V the orientation in which $e_0 := (1, 0)$, $e_1 := (0, 1)$ is a positively oriented basis. (Remember that we are following the relativists' convention of drawing the time axis vertical and that $(1, 0)$ refers to $t = 1, x = 0$, which makes this "vertical-horizontal" orientation the opposite of the usual horizontal-vertical one on R^2.)

(a) Sketch the orientations on T, S, and L according to the rules of Section 2.6; i.e. at any given point of the boundary, which way does a positively oriented basis vector for the tangent space go?

(b) Let $w = w^0 \dfrac{\partial}{\partial t} + w^1 \dfrac{\partial}{\partial x}$ be a given vector field on V. Compute

$$\int_L \bot w^* , \quad \int_T \bot w^* , \quad \text{and} \quad \int_S \bot w^* .$$

(c) Let $\Omega = dt \wedge dx$ denote the volume 2-form. Compute

$$\int_V \delta(w) \, \Omega$$

and thereby directly verify the Gauss-Stokes Theorem

$$\int_V \delta(w) \, \Omega = \int_{\partial V} \bot w^*$$

for this special case.

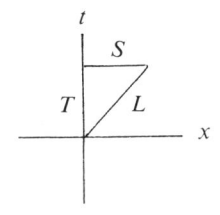

2.23. Let M be a compact, connected semi-Riemannian manifold of dimension n, let $0 \leq p \leq n$, and denote by Λ_p the (infinite-dimensional) vector space of all C^∞ p-form fields on M. Define an inner product $\alpha, \beta \longmapsto (\alpha, \beta)$ on Λ_p by:

$$(\alpha, \beta) := \int_M \langle\alpha, \beta\rangle \, \Omega \, ,$$

where $\langle\alpha, \beta\rangle$ denotes the function $q \longmapsto \langle\alpha_q, \beta_q\rangle$ on M.

Consider the differential d as a linear transformation from Λ_p to Λ^{p+1}. Show that its adjoint $d^\dagger : \Lambda_{p+1} \longrightarrow \Lambda_p$ is given by

$$d^\dagger = -\delta \, .$$

That is, show that for any C^∞ p-form α and $(p+1)$-form β,

(*) $$\int_M \langle d\alpha, \beta\rangle \, \Omega = -\int_M \langle\alpha, \delta\beta\rangle \, \Omega \, .$$

Note that (*) is also algebraically correct for compactly supported p-forms on a noncompact manifold M.

2.24. Let $\vec{v}$ be a vector field on Euclidean R^3, and $\vec{v}*$ the corresponding 1-form field.

(a) Show that

$$d(\vec{v}*) = \perp(\nabla \times \vec{v})* \, .$$

Thus when vectors are identified with 1-forms via $*$ and 1-forms with 2-forms via $\perp$, the differential operation d on 1-forms corresponds to the curl operation on vector fields.

(b) Show that under these identifications, the algebraic Scholium of Section 2.9 for the case $m = R^3$ and $p = 1$ translates into the statement that given any scalar function α and vector field $\vec{w}$ on R^3 satisfying $\nabla \cdot \vec{w} = 0$ in a neighborhood of a point $p \in R^3$, there exists a vector field $\vec{v}$ defined in a (possibly smaller) neighborhood of p satisfying

(i) $\qquad \nabla \times \vec{v} = \vec{w} \qquad$ and

(ii) $\qquad \nabla \cdot \vec{v} = \alpha \, .$

(Note, however, that the Theorem of that section rigorously establishes the Scholium only for the case in which α and $\vec{w}$ are real-analytic.)

Chapter 3

The Electrodynamics of Infinitesimal Charges

3.1. Introduction.

For simplicity of exposition, we shall take spacetime to be Minkowski space **M** except where otherwise specified. However, everything that we do in Sections 1 through 6 extends almost immediately to general spacetimes.

In classical, Newtonian mechanics, the motion of a single particle is usually determined by a second-order differential equation:

(1) $$\frac{d^2\vec{x}}{dt^2} = \vec{G}(\vec{x}(t), \frac{d\vec{x}}{dt}, t),$$

where $\vec{x}(t)$ denotes the position of the particle at time t. The function $\vec{G}$ in general differs from particle to particle, and we may take account of this in the notation by including parameters $a_1, \cdots, a_m$ which label various characteristics of the particle, such as its mass and charge:

(2) $$\frac{d^2\vec{x}}{dt^2} = \vec{G}(a_1, \cdots, a_m; \vec{x}(t), \frac{d\vec{x}}{dt}, t).$$

Thus particles with the same parameter labels will be dynamically indistinguishable.

In this formulation, the function $\vec{G}$ is regarded as given, *a priori*. In most cases, however, $\vec{G}$ arises from interactions of the given particle with others, and the $\vec{G}$ which acts on a particular particle depends on the motion of the other particles, which in turn is influenced by the original particle. Thus we really need to solve a system:

(3) $$\frac{d^2\vec{x}_i}{dt} = \vec{G}_i(\vec{x}_1(t), \cdots, \vec{x}_n(t), \frac{d\vec{x}_1}{dt}, \cdots, \frac{d\vec{x}_n}{dt}, t), \qquad 1 \leq i \leq n,$$

where $\vec{x}_i(t)$ denotes the position of the i'th particle at time t, and we have omitted the particle parameters for brevity. System (3) is recognized as being not essentially different from (2), since it may be rewritten as the single equation

(4) $$\frac{d^2x}{dt^2} = G(x(t), \frac{dx}{dt}, t)$$

of the form (2) over R^{3n} instead of R^3 via the substitutions

$$x(t) := (\vec{x}_1(t), \cdots, \vec{x}_n(t)) \in R^{3n},$$

$$G := (\vec{G}_1, \cdots, \vec{G}_n).$$

One problem in adapting this approach to a theory of relativistic interacting particles is that interactions which propagate faster than the speed of light lead to perplexing (though not necessarily fatal) conceptual difficulties (cf. Exercise 1.19). Relativistically, it would seem more natural to write the force on particle 1 at time t as a function of the positions and velocities of the other particles on, or perhaps within, the backward light cone with vertex at $(t, \vec{x}(t))$ rather than as a function of the positions and velocities of the others at the same coordinate time t as in (3). However, it is not easy to formulate a conceptually acceptable and mathematically tractable set of equations which embodies this concept.

The general approach used in electrodynamics is to go back to an analog of equation (2) for the motion of a single particle with worldline $\tau \longmapsto z(\tau) \in \mathbf{M}$ parametrized by proper time τ :

$$(5) \qquad \frac{d^2z}{d\tau^2} = L(m, q; z(\tau), \frac{dz}{d\tau}) ,$$

where m is the rest mass of the particle and q is called the *charge*. We view (5) as *defining* q. That is, given an explicit L on the right side of (5), q is defined to be the constant for which (5) describes the motion of the particle. In practice there is a unique such q. No explicit time parameter has been included in (5) because the coordinate time is already included in $z(\tau) = (t(\tau), \vec{x}(\tau))$. As before, the function L, which describes the "electromagnetic field", is thought of as being determined by the other charged particles in the universe, but one does not attempt to write an analog of the system (3). Rather, the particles determine the field via the so-called "retarded" solution of Maxwell's equations (which will be described later), and the field determines the motion of the particles via (5). The problem of solving Maxwell's equations for the field *given* the particle motions is mathematically tractable and well understood, as is the even easier problem of solving (4) for the particle worldline $z(\tau)$ given the function L. However, subtle and intricate difficulties arise when one attempts to solve these problems simultaneously, and the focus of this monograph is an understanding of the relations between them. First, however, we need to understand each separately. In Sections 2, 3, and 4, we state the laws of motion of "infinitesimal" charged particles in an "external field"; that is, we consider the field as given, *a priori*. In Sections 5 and 6 we state Maxwell's equations, and in Section 7 the retarded solution is introduced.

3.2. The Lorentz force law.

To present the Lorentz force law honestly, we first have to face up to some features of electrodynamics which make it quite different from classical Newtonian mechanics. The correct equation to describe the motion of point particles is a matter of some controversy, but the most mentioned candidate, the Lorentz-Dirac equation, is of the general form:

$$(1) \qquad \frac{d^2z}{d\tau^2} = \frac{q}{m} G_1(z(\tau), \frac{dz}{d\tau}) + \frac{q^2}{m} G_2(z(\tau), \frac{dz}{d\tau}, \frac{d^2z}{d\tau^2}, \frac{d^3z}{d\tau^3}) ,$$

with m the rest mass and q the charge of the particle. This is not of the form 3.1(5) due to the dependence of G_2 on the second and third (!) derivatives of z, but the same general remarks apply. However, this modification of 3.1(5) is not postulated at the outset, but rather, is the culmination of a long chain of arguments, some more compelling than others, beginning with the simpler equation:

$$(2) \qquad \frac{d^2z}{d\tau^2} = \frac{q}{m} G_1(z(\tau), \frac{dz}{d\tau}) .$$

(The correct response on reading the above sentence is puzzlement. How can (2) possibly imply (1)?) The right side of (2) may be viewed as the limit as $m \to 0$ of the right side of (1) in which the charge/mass ratio $\frac{q}{m}$ is held constant. This suggests the interpretation of (2) as the equation of motion of an "infinitesimal" particle with a given charge/mass ratio. This in turn suggests that (2) may properly pertain to a "charged fluid", conceptually composed of a continuum of infinitesimal charged particles. This is, in fact, the way in which the logic of electrodynamics is organized. Rather than deducing the laws of fluid motion from the laws of motion of point particles as Newtonian theory does, electrodynamics takes the opposite route and obtains the dynamics of point particles from those of charged fluids.

This raises several potentially embarassing issues. What is a charged fluid? How can we hope to formulate correct laws of motion for charged fluids in the absence of the laws of motion for point particles?

In answer to the first question, it is easy enough to define mathematical objects corresponding to the common notion of a fluid with charge. The simplest physical picture is that of a continuous distribution of infinitesimal point charges which move coherently in the sense that nearby charges have nearly the same velocity. Associated with such a system is a unit timelike vector field u and a *proper charge density* function ρ on spacetime M. At any point p of M, u_p represents the four-velocity of the infinitesimal particle at p, and $\rho(p)$ is the charge per unit volume at p *as measured in the rest frame* (cf. Section 1.7) of the particle at p. Such a system is sometimes called a *charged dust,* to distinguish it from a more general "charged fluid" to which one adds a pressure function and perhaps additional structure as well. We shall never need to consider a charged fluid more general than that described above, and we shall use the terms "charged fluid" and "charged dust" interchangeably.

Ideally, one would like to formulate and verify the laws of motion of charged fluids by observing their behavior. Even for Newtonian, non-charged fluids, this is very difficult, but for charged fluids it is even harder because it is not even clear that they exist! The charged dust model which we shall develop and more general charged fluid models are sometimes used to attempt to analyze real physical phenomena such as dilute electron streams in outer space and plasmas [†], but the order of approximation of the models to such systems is not known, and it is safe to say that one cannot verify in detail the charged fluid model of electrodynamics by observing real charged fluids.

Nevertheless, the charged fluid model probably is the best physical crutch to aid in understanding the logical structure of electrodynamics. This may seem strange in view of the fact that massive charged particles clearly exist and can easily be tracked in cloud and bubble chambers as they pass through electromagnetic fields. Why can't we simply obtain (1) by direct observation and postulate it as a starting point of the theory? The reason is that the second term G_2 is too small to observe directly. That is, real charged particles seem to obey the equation (2) for infinitesimal particles insofar as their motion can be directly observed. On the other hand, we shall see in the next chapter that equation (2) for massive particles appears to lead to the unpleasant conclusion that energy (as traditionally defined) is not necessarily conserved, and one way to avoid this consequence is to replace (2) by (1) (though (1) brings new troubles of its own). To summarize, (2) is deeply embedded into the logic of the subject, but the only easy way to reconcile it with the final structure appears to be to interpret it as an equation for infinitesimal particles.

There are other reasons for considering the charged fluid model as an integral part of the theory, one of the most important of which is that the otherwise unmotivated definition of energy-momentum of the electromagnetic field itself is adopted precisely because it enables a proof of conservation of energy-momentum in the fluid model. Having thus acquired status, the definition is exported to the theory of massive point particles and eventually motivates the replacement of (2) by (1). Another reason is that Maxwell's equations are most naturally stated within the framework of the fluid model.

[†] A *plasma,* which is usually defined as a gas consisting of equal numbers of free electrons and ions, is macroscopically neutral and so is not a charged fluid in the sense in which we use the term. However, plasmas are sometimes modelled as two interpenetrating charged fluids.

Having motivated the necessity of considering (2) as the equation of motion only for infinitesimal charged particles, we now state the Lorentz force law, which specifies the form of the function G_1 in (2). First, anticipating later needs, we change notation and write $\hat{F}_z(u)$ in place of $G_1(z, u)$; for the moment the hat is simply a decoration.

Consider a charged particle with worldline $\tau \longmapsto z(\tau) \in \mathbf{M}$ parametrized by proper time τ. As usual, the particle's four-velocity will be denoted $u(\tau) := dz/d\tau$. In this notation, the basic equation of motion for infinitesimal charged particles is more compactly written as:

$$(3) \qquad m\frac{du}{d\tau} = q\,\hat{F}_{z(\tau)}(u) = q\,\hat{F}_z(u) ,$$

In its abstract form, the Lorentz force law states that for fixed $x \in \mathbf{M}$ the function $\hat{F}_x(\cdot)$ is *linear*. Explicitly, for fixed x, the map $u \longmapsto \hat{F}_x(u)$ satisfies

$$(4) \qquad \hat{F}_x(sw + s'w') = s\hat{F}_x(w) + s'\hat{F}_x(w') \qquad \text{for all } w, w' \in \mathbf{M}_x \text{ and all } s, s' \in \mathbf{R}.$$

Equation (3) under the assumption that $\hat{F}$ satisfies (4) will be called the *Lorentz equation*, or *Lorentz force law*.

The fundamental assumption (4) is so simple that it is easy to overlook its remarkable character. Note that in the dynamical equation (3), the only vectors u which appear as the argument of $\hat{F}_z(u)$ are four-velocities, which of course are timelike unit vectors in $\mathbf{M}_z$. Addition of two such four-velocities u does not produce a third four-velocity, and so the linearity condition (4) would seem to have no obvious physical interpretation. In addition, Maxwell's equations are formulated in terms of the the 2-covariant tensor associated with the linear map $u \longmapsto \hat{F}_z(u)$, and if $\hat{F}_z$ were nonlinear, this tensor would not even exist.

We have been working in Minkowski space $\mathbf{M}$ for ease of exposition, but the same general considerations apply in any spacetime. To adapt the Lorentz equation (3) to an arbitrary spacetime, all we need is an appropriate substitute for $du/d\tau$ (which does not make sense in an arbitrary spacetime because the conventional difference quotient of which it is a limit requires subtracting vectors which reside in different tangent spaces). The reader who is well versed in differential geometry will immediately recognize the appropriate substitute as $D_u u$. We shall now explain this in detail for the less knowlegeable reader, for it will sometimes be helpful to replace $du/d\tau$ by $D_u u$ even in Minkowski space.

Consider a particle worldline $\tau \longmapsto z(\tau)$, where τ is proper time, and let $u := z'(\tau)$. We think of u as a vector field $x \longmapsto u_x$ defined on the worldline; that is, defined only for $x = z(\tau)$ for some τ. Then, by definition of $z'(\tau)$, for any function f defined along the worldline,

$$u_{z(\tau)}(f) = \lim_{\Delta\tau \to 0} \frac{f(z(\tau + \Delta\tau)) - f(z(\tau))}{\Delta\tau} .$$

If we could replace the function f by a vector field v, then this equation would be a natural definition of $dv/d\tau$. On the other hand, differentiating a vector field along a curve should carry the same meaning as differentiating it in the direction of the tangent vector to the curve, and the latter is defined by $D_{z'(\tau)} v$. Hence in any manifold equipped with a covariant derivative D, a natural definition of the derivative $dv/d\tau$ of a vector field v in the direction of a curve $\tau \longmapsto z(\tau)$ is

$$(5) \qquad \frac{dv}{d\tau} := D_{z'(\tau)} v = D_u v .$$

Notice that if we write $v = v^j \dfrac{\partial}{\partial x_j}$ (summation implied) with respect to some coordinate system $\{x_j\}$, then

(6) $$\frac{dv}{d\tau} := D_u v = u^i \frac{\partial v^j}{\partial x_i} \frac{\partial}{\partial x_j} + u^i v^j \Gamma_{ij}{}^k \frac{\partial}{\partial x_k}$$
$$= u(v^j)\frac{\partial}{\partial x_j} + u^i v^j \Gamma_{ij}{}^k \frac{\partial}{\partial x_k} .$$

The right side depends only on the values of v on the curve $\tau \longmapsto z(\tau)$ (since $u(v^j)$ are the only directional derivatives which appear), so (6) also gives a natural definition of $dv/d\tau$ when v is not a genuine vector field defined on an open set but rather is only defined along the curve (as is the case when $v = u$), and we extend the notation $D_u v$ to mean the right side of (6) in this situation.

3.3. The electromagnetic field tensor.

The Lorentz force law states that for fixed x, $\hat{F}_x(\cdot)$ is a linear transformation on the tangent space M_x. Then the metric tensor associates to $\hat{F}_x(\cdot)$ the bilinear function $F_x(\cdot,\cdot)$ on M_x defined by

(1) $$F_x(w, y) := <\hat{F}_x(w), y> \quad \text{for all } w, y \in M_x .$$

We shall usually abbreviate F_x as F and shall view F as a tensor field on M of covariant rank 2. The tensor F is called the *electromagnetic field tensor* and is the central mathematical object in electrodynamics, despite the fact that we chose to introduce $\hat{F}$ first.

The Lorentz equation 3.2(3) virtually forces $F_x(\cdot,\cdot)$ to be antisymmetric. To see this, note that for any four-velocity $u(\cdot)$ satisfying the Lorentz equation 3.2(3), we have

(2) $$F(u(\tau), u(\tau)) = <\hat{F}(u(\tau)), u(\tau)>$$
$$= \frac{m}{q}<\frac{du}{d\tau}, u(\tau)> = \frac{m}{q}\frac{1}{2}\frac{d<u(\tau),u(\tau)>}{d\tau} = 0 .$$

Assuming that any timelike unit vector can be the four-velocity of an infinitesimal charged particle in the external field F, (2) implies that

(3) $$F_x(v, v) = 0 \quad \text{for any timelike vector } v \in M_x .$$

This says that the map $v \longmapsto F_x(v, v)$ vanishes on the open set consisting of all timelike vectors. Since this map is a polynomial in the components of v, it must vanish identically, so (3) holds for all vectors v, timelike or not, and (3) for all vectors implies that F is antisymmetric because

$$F(v, w) + F(w, v) = \frac{1}{2}[F(v+w, v+w) - F(v-w, v-w)] = 0 .$$

We have just shown that the Lorentz equation and a very weak physical assumption imply that at each $x \in M$, the field tensor F_x must be antisymmetric, and we assume this from now on. Thus F is now a differential 2-form, and we are in a position to bring the powerful theory of differential forms to bear.

3.4. The electric and magnetic fields.

To see the relation between the electric and magnetic fields encountered in elementary physics courses and the field tensor F defined in the previous section, choose an orthonormal basis e_0, e_1, e_2, e_3 for Minkowski space $\mathbf{M}$, and write the matrix $(F_{ij}) := (F(e_i, e_j))$ of F with respect to this basis as

$$(1) \qquad (F_{ij}) = \begin{bmatrix} 0 & E^1 & E^2 & E^3 \\ -E^1 & 0 & -B^3 & B^2 \\ -E^2 & B^3 & 0 & -B^1 \\ -E^3 & -B^2 & B^1 & 0 \end{bmatrix},$$

where this notation was chosen because $\vec{E} := (E^1, E^2, E^3)$ and $\vec{B} := (B^1, B^2, B^3)$ will turn out to be the classical electric and magnetic field vectors, respectively.

The matrix $(F^i{}_j)$ of $\hat{F}$ with respect to this basis is:

$$(2) \qquad (F^i{}_j) = \begin{bmatrix} 0 & E^1 & E^2 & E^3 \\ E^1 & 0 & B^3 & -B^2 \\ E^2 & -B^3 & 0 & B^1 \\ E^3 & B^2 & -B^1 & 0 \end{bmatrix}.$$

Although $\hat{F}$ is an antisymmetric linear transformation with respect to the Lorentz inner product, its matrix is not antisymmetric; this is one of the peculiarities of indefinite metrics. It is worth remembering that the matrix of $\hat{F}$ (with respect to an orthonormal basis) is gotten from that of F by reversing the sign of all the elements in the "space" rows.

Consider a charged particle with worldline $\tau \longmapsto (t(\tau), \vec{x}(\tau))$ parametrized by proper time τ, and write its four-velocity u as $u = (u^0, \vec{u})$. Recall from 1.7(5) and 1.7(6) that $u^0 = dt/d\tau = \gamma(v)$, where $\vec{v} := d\vec{x}/dt$ is the coordinate velocity, and that $u = u^0(1, \vec{v})$. Then the Lorentz equation components are

$$(3) \qquad \begin{bmatrix} \dfrac{du^0}{d\tau} \\ \dfrac{d\vec{u}}{d\tau} \end{bmatrix} = u^0 \dfrac{q}{m} \begin{bmatrix} \vec{v} \cdot \vec{E} \\ \vec{E} + \vec{v} \times \vec{B} \end{bmatrix}.$$

In particular, equality of the space components says that

$$(4) \qquad m \frac{d\vec{u}}{d\tau} = u^0 q (\vec{E} + \vec{v} \times \vec{B}),$$

and hence

$$(5) \qquad m \frac{d\vec{u}}{dt} = q(\vec{E} + \vec{v} \times \vec{B}).$$

The right side of (5) is immediately recognized as the classical Lorentz force on a charged particle. To establish a direct connection between (5) and the Lorentz force law as presented in elementary physics courses, we must assume that the rest mass m is constant. Recalling that the four-momentum p is defined as $p := mu$, so its space component is $\vec{p} := m\vec{u}$, we may then write (5) in the form

(6) $$\frac{d\vec{p}}{dt} = q(\vec{E} + \vec{v}\times\vec{B}) ,$$

which is the classical Lorentz force law with the relativistic definition 1.8(7) $\vec{p} := m\gamma(\vec{v})\vec{v}$ of 3-momentum in place of the Newtonian $\vec{p} := m\vec{v}$.

Similarly, the time component of (3) states that the coordinate time rate mdu^0/dt at which the relativistic mass-energy of the particle is changing is the rate $q\vec{E}\cdot\vec{v}$ at which the Lorentz force is doing work on the particle. This is not a new relation but instead follows automatically from (6). To see this, note that $du/d\tau$ is automatically orthogonal to u because $<u,u> = 1$, and hence the top component of the left side of (3) is always the inner product of $\vec{v}$ with the bottom component. In other words, if we have *any* force law of the form $d\vec{p}/dt = \vec{F}$, then the relation

(7) $$m\frac{du^0}{dt} = m\frac{d\vec{u}}{dt}\cdot\vec{v} = \frac{d\vec{p}}{dt}\cdot\vec{v} = \vec{F}\cdot\vec{v}$$

is automatic. Classically, $\vec{F}\cdot\vec{v}$ is the rate of change of kinetic energy of the particle, while the left side of (7) is the rate of change of relativistic mass $dM/dt = mdu^0/dt$, so (7) is essentially just the relation 1.8(13) between relativistic mass and classical kinetic energy.

3.5. The first Maxwell equation.

When written in the notation of three-dimensional vector calculus, there are four Maxwell equations, but written in the language of differential forms on a four-dimensional spacetime, there are only two. The first of these simply states that the differential of the electromagnetic field tensor F vanishes everywhere:

(1) (First Maxwell equation) $dF = 0$.

When F is expressed in terms of the classical electric and magnetic vectors as in 3.4(1), this translates into two three-dimensional vector equations:

(2) (i) $\quad \vec{\nabla}\cdot\vec{B} = 0 ,$

 (ii) $\quad \dfrac{\partial \vec{B}}{\partial t} = -\vec{\nabla}\times\vec{E} .$

Equations (i) and (ii) with respect to a particular Lorentz frame in Minkowski space can be shown to be equivalent to requiring that (i) hold in *all* Lorentz frames [Frankel], [Misner, Thorne, and Wheeler, p. 80]. There seems to be no compelling motivation for (i) in the present context, though in Chapter 5 we shall look at a suggestive formulation of continuum electrodynamics (not equivalent to the standard theory) in which (1) is automatic.

3.6 The second Maxwell equation.

Consider a charged dust with four-velocity vector field $u = \gamma(\vec{v})(1, \vec{v})$ and proper charge density ρ. In Section 2.8 we observed that the mathematical expression of the law of conservation of charge is:

(1) $\qquad \delta(\rho u) = 0$.

By the result of Section 2.9, there exists a 2-form K such that

(2) $\qquad \delta K = 4\pi \rho u^*$.

(The 4π is inserted for aesthetic reasons which will not become visible until much later.) If we denote the matrix $(K_{ij}) = (K(e_i, e_j))$ of K as:

(3) $\qquad (K_{ij}) = \begin{bmatrix} 0 & D^1 & D^2 & D^3 \\ -D^1 & 0 & -H^3 & H^2 \\ -D^2 & H^3 & 0 & -H^1 \\ -D^3 & -H^2 & H^1 & 0 \end{bmatrix}$,

then, equation (2) is equivalent to the two three-dimensional vector equations:

(4) (i) $\qquad \vec{\nabla} \cdot \vec{D} = 4\pi\gamma(\vec{v})\rho$

(ii) $\qquad \dfrac{\partial \vec{D}}{\partial t} = \vec{\nabla} \times \vec{H} - 4\pi\gamma(\vec{v})\rho\vec{v}$.

The function $\gamma(\vec{v})\rho$ represents the charge density as measured in the given Lorentz frame (cf. Section 2.8 and Figure 6 of Chapter 2). The quantity $\gamma(\vec{v})\rho\vec{v}$ is called the (three-dimensional) *charge current vector* and is usually denoted $\vec{J}$ in physics texts. When one is working entirely in a fixed Lorentz frame (the "laboratory frame"), it is usually more convenient to define "charge density" as $\rho_{lab} := \gamma(\vec{v})\rho$ and then (4) becomes

(4)' (i) $\qquad \vec{\nabla} \cdot \vec{D} = 4\pi\rho_{lab}$

(ii) $\qquad \dfrac{\partial \vec{D}}{\partial t} = \vec{\nabla} \times \vec{H} - 4\pi\vec{J}$,

which is how it is written in most physics texts, except that the subscript "lab" is generally omitted.

The equation $dF = 0$ and (2) above constitute *Maxwell's equations,* which are more traditionally written in three-dimensional notation as the four equations 3.5(2)(i), 3.5(2)(ii), (4)'(i), and (4)'(ii).

Of course, from the present point of view, the second Maxwell equation (2) plays the role of a partial definition of K rather than a physical law. The usual way to give physical content to (2) is to postulate some additional relation, called a *constitutive* relation, between the 2-form K in (2) and the electromagnetic field tensor F. Different relations lead to various theories describing dielectrics, magnetically permeable materials, etc. For example, many important physical effects can be explained by assuming that $\vec{D}$ and $\vec{H}$ are proportional to $\vec{E}$ and $\vec{B}$, respectively, in a Lorentz frame in which $e_0 = u$. Although the necessity to asume various *ad hoc* couplings between K and F gives a certain flavor of tautology to the theory, it is nevertheless impressive that such a wide range of physical phemonena can be accomodated by this device.

The simplest possible coupling between K and F is to assume that they are equal, and it is this coupling that is used to establish the foundations of electrodynamics. By definition, this coupling describes a charged dust. Since we are concerned here with the foundations rather than the applications of electrodynamics, we assume from now on that $K = F$. In this case, we have

(5) (Maxwell's equations for a charged dust)

$$\text{(i)} \quad dF = 0$$

$$\text{(ii)} \quad \delta F = 4\pi\rho u^*$$

An important special case is the *free-space* Maxwell equations in which ρ is identically zero. From here on, when we refer to Maxwell's equations, we always mean equations (5).

Maxwell's equations relate the electromagnetic field near a spacetime point with the "current" ρu at that point. Usually they are viewed as equations for F in terms of a given current ρu and so viewed describe how the charges in the system affect the field. The result of Section 2.9 shows that a necessary and sufficient condition for local solutions is the condition for conservation of charge: $\delta(\rho u) = 0$. Obviously, such a solution is unique only up to addition of a solution of the free-space Maxwell equations.

If we adjoin to the Maxwell system (5) the Lorentz equation 3.2(3),

$$\text{(6)} \quad \frac{du}{d\tau} = \frac{q}{m}\hat{F}(u),$$

we obtain a simultaneous system of equations for ρ, u, and F which, by definition, are the equations of motion for a charged dust in which only one sign of charge is present. We shall call the system (5) and (6) the *Maxwell-Lorentz equations*, though some authors use this term to mean equations (5) alone. *A priori*, they appear to be equations for three unknown quantities ρ, u, and F, but given u and F, (5)(ii) may be regarded as *defining* ρ. In other words, (5)(ii) may be regarded as stating that δF is a multiple of $4\pi u$, ρ being defined as that multiple, and under this definition the system (5) and (6) has just the two unknowns u and F. It is worth noting that when only one sign of charge is present, the fact that u is a unit vector permits an explicit formula for ρ in terms of u and F:

$$\text{(7)} \quad \rho := \frac{\text{sign}(q)}{4\pi}<\delta F, \delta F>^{1/2}.$$

An appropriate set of equations for a charged dust which contains charges of both signs is obtained by replacing (6) by

$$\text{(8)} \quad |\rho|\frac{du}{d\tau} = k\rho\hat{F}(u),$$

where k is a positive constant which plays the role of $|q|/m$. When seeking C^∞ solutions, this is usually analytically more natural than letting q vary discontinuously in (6).

3.7 Potentials.

The Maxwell-Lorentz system of differential equations is in general quite difficult to solve explicitly. Fortunately, solutions of the full system are rarely needed in practice since few practical applications involve true charged fluids. More typically, one needs to compute the field F assuming that the current $J := \rho u$ is given *a priori*. For example, electrons flowing through a wire are constrained in space by forces within the wire, and in such a situation, it is often appropriate to regard the current J as known. Thus it is important in practice to be able to explicitly solve the equation

(1) $\qquad \delta F = 4\pi J^*$

for the 2-form F given a 1-form J^* satisfying $\delta J^* = 0$.

This is most conveniently done via the introduction of a so-called *potential* 1-form A satisfying $dA = F$. Such a 1-form always exists locally because of the Maxwell equation $dF = 0$, and when F is globally defined on Minkowski space, A exists globally as well (cf. Section 2.9, Lemma 1).

Of course, again by Section 2.9, the equation $dA = F$ only determines a local A up to an additive term $d\phi$, where ϕ is an arbitrary function. The replacement $A \longmapsto A + d\phi$ of the original local solution A by the new local solution $A + d\phi$ is called a *gauge transformation,* or change of gauge. The result of Section 2.9 also shows that the equation $dA = F$ can be locally satisfied subject to the arbitrary specification of δA. The particular specification $\delta A = 0$ is often convenient and is known as the *Lorentz* gauge. For the rest of this section we work entirely in the Lorentz gauge. Note, however, that even the two relations

(2) $\qquad dA = F \quad \text{and} \quad \delta A = 0$

do not completely determine A, because if ϕ is any function such that $\delta d\phi = 0$, then $A + d\phi$ will also satisfy (2).

There are plenty of such functions ϕ, since computation shows that

(3) $\qquad \delta d\phi = \Box \phi$,

where $\qquad \Box := \dfrac{\partial^2}{\partial x_0^2} - \sum_{i=1}^{3} \dfrac{\partial^2}{\partial x_i^2}$

denotes the usual D'Alembertian operator. More generally, one defines the D'Alembertian operator $\Box$ as a linear mapping $\Box : \Lambda_p(M) \longrightarrow \Lambda_p(M)$ on the space of p-form fields α on a semi-Riemannian manifold M by

(4) $\qquad \Box \alpha := \delta d\alpha + d\delta \alpha$,

(where $\delta \alpha$ is defined as zero when α is a 0-form). † Routine computation shows that if $\alpha_{i_1 \cdots i_p}$ are the components of α with respect to some orthonormal basis for Minkowski space $\mathbf{M}$, then

(5) $\qquad (\Box \alpha)_{i_1 \cdots i_p} = \left(\dfrac{\partial^2}{\partial x_0^2} - \sum_{i=1}^{3} \dfrac{\partial^2}{\partial x_i^2} \right) \alpha_{i_1 \cdots i_p}$.

---

† When M is Riemannian, the differential operator $\Box := \delta d + d\delta$, apart from a sign, is called the *Laplacian* and usually denoted Δ.

3.7 Potentials

That is, the D'Alembertian (4) on p-forms in Minkowski space is just the ordinary D'Alembertian for functions operating componentwise on α. (In a general spacetime, terms involving the connection coefficients would appear on the right side of (5).)

Now in the Lorentz gauge, the Maxwell equation $\delta F = 4\pi J^*$ is the same as

(6) $\qquad \Box A = (\delta d + d\delta)A = \delta dA = \delta F = 4\pi J^*$.

Moreover, the other Maxwell equation $dF = 0$ follows automatically from $F = dA$. Thus solving locally the Maxwell equations

(7) $\qquad$ (i) $\quad dF = 0 \quad$ and

$\qquad\qquad$ (ii) $\quad \delta F = 4\pi J^*$

is algebraically equivalent to solving the two equations

(8) $\qquad$ (i) $\quad \Box A = 4\pi J^* \quad$ and

$\qquad\qquad$ (ii) $\quad \delta A = 0$.

This simplifies the problem considerably in Minkowski space, since (8)(i) decouples into the four separate scalar equations $\Box A_i = J_i$, $i = 0, 1, 2, 3$.

For the rest of this section we work exclusively in a given Lorentz coordinatization of Minkowski space $\mathbf{M}$. In this context there are two particular solutions of (8), called the *retarded* and *advanced* solutions, which play a special role in electrodynamics. They are given explicitly by the following formulae, in which A^{ret} denotes the retarded solution and A^{adv} the advanced:

(9) $\qquad A_i^{ret}(t, \vec{x}) := \int_{R^3} \dfrac{J_i(t - |\vec{y}|, \vec{x} + \vec{y})}{|\vec{y}|} d^3\vec{y}$,

(10) $\qquad A_i^{adv}(t, \vec{x}) := \int_{R^3} \dfrac{J_i(t + |\vec{y}|, \vec{x} + \vec{y})}{|\vec{y}|} d^3\vec{y}$.

Since we are only going to use these for motivation, we shall work formally and not worry about convergence of the integrals or interchange of differentiation and integration. Note, however, that the singularity of the denominator at $\vec{y} = 0$ is not a problem because in polar coordinates it looks like $1/r$ and is damped out by a factor $4\pi r^2$ in $d^3 \vec{y}$.

That A^{ret} and A^{adv} satisfy (8)(ii) when $\delta J^* = 0$ is immediate. However, that they satisfy (8)(i) is not trivial. (Consider the surprising fact that the analogs of (9) and (10) don't work in one and two space dimensions.) A good proof can be found in [John]; since we shall use only the result and not the ideas of the proof, we shall not give one here (but cf. Exercise 9, where the proof of a closely related result is outlined).

The geometrical meaning of (9) can be understood as follows. As the variable $\vec{y}$ ranges over R^3, the spacetime point $(t - |\vec{y}|, \vec{x} + \vec{y})$ ranges over the past light cone with vertex at $(t, \vec{x})$. Thus (9) may be viewed as an integral of J over this cone with respect to the measure $d^3\vec{y}/|\vec{y}|$ transferred to the cone via the map $\vec{y} \longleftrightarrow (t - |\vec{y}|, \vec{x} + \vec{y})$. This measure is invariant under Lorentz transformations of the cone (Exercise 7), and so A^{ret} and A^{adv} are independent of the coordinatization used in (9) and (10).

The corresponding fields

$$F^{ret} := dA^{ret} \quad \text{and} \quad F^{adv} := dA^{adv}$$

are called, respectively, the *retarded* and *advanced* fields. The importance of the retarded field is that it is a solution of Maxwell's equations whose value at a spacetime point x depends only on the values of the given current J in the causal past of x. If we view the field F as "produced" by the current J, then use of the retarded field is in accord with the general feeling that "influence" should not be able to propagate faster than the speed of light. Though it does not seem clear that use of the retarded fields is forced by elementary notions of causality, this approach is certainly plausible and natural. ‡

The name "potential" for the 1-form A can be partially explained as follows. Recall that the current four-vector J is defined by $J := \rho u$, and $J_0 = \rho u_0$ is the charge density as measured in the coordinate frame. If ΔV is an infinitesimal spatial volume containing $\vec{z}$, $J_0(s, \vec{z})\Delta V$ is the charge Δe in ΔV at time s, and the contribution of this charge to the component A_0^{ret} of the retarded potential a distance r from $\vec{z}$ at time $s+r$ will be $\Delta e/r$, which is a constant multiple of the classical expression for the electrostatic potential energy. Although the other components of A^{ret} are not generally interpreted as classical potential energies, it is worth noting that the portion of any component A_i^{ret} due to Δe varies according to the familiar $1/r$ law of the classical potential energy. The "space" part of the four-vector A^i is called the "vector potential" in physics texts, while A^0 is the "scalar potential". Exercise 5 reminds the reader of the relation between these quantities and the electric and magnetic fields.

3.8. The energy-momentum tensor.

The concept of energy-momentum which has evolved for electrodynamics is more subtle and complicated and than the simple formulation for uncharged particles developed in Section 1.8. It is widely conceded that the part of electrodynamics which deals with point particles is beset with difficulties, and most of these can be traced back to the definition of energy-momentum, which is borrowed from continuum electrodynamics. The latter, however, is unintuitive, and the extent to which it is forced by the logical structure of the theory depends on one's point of view. In the following pages we shall explain the standard definition of energy-momentum of the electromagnetic field, but the reader may well feel that it has a certain *ad hoc* flavor. Although not as arbitrary as it may seem at first, it is also not inconceivable that a different definition could lead to a more satisfactory theory. Some proposals along these lines are discussed in Chapter 5.

The particular treatment of energy and momentum which has proved relevant in electrodynamics rests on the concept of *energy-momentum tensor,* which can be viewed in several different ways. For the purposes of introduction, it will be convenient to view the energy-momentum tensor as a vector-valued 3-form $S(\cdot, \cdot, \cdot)$. Thus for any triple of vectors v_1, v_2, v_3 in the tangent space M_x at a spacetime point x, $S_x(v_1, v_2, v_3)$ is a vector in M_x, and the map $v_1, v_2, v_3 \longmapsto S_x(v_1, v_2, v_3)$ is trilinear and antisymmetric. The component array of S is $S^i{}_{jkl}$ and of course is antisymmetric in jkl. For the physicist reader who already knows about energy-momentum tensors, we remark that $S^i{}_{jkl}$ is the (vector-valued) Hodge dual of of a tensor $T^i{}_j$, which, apart from index position, is what is usually called the energy-momentum tensor in the physics literature. We shall say more about $T^i{}_j$ later.

In order to physically define S, it is necessary to fix both an orientation and a time orientation on Minkowski space **M**, and we suppose this has been done. The physical meaning of $S(v_1, v_2, v_3)$ varies according to the type of subspace of M_x spanned by v_1, v_2, v_3. We may as well assume that this subspace is three-dimensional, since otherwise one of these

---

‡ A much-quoted formulation of electrodynamics in which the the advanced and retarded fields appear symmetrically is [Wheeler and Feynman]. Expository accounts of this theory can be found in several textbooks, for instance [Panofsky and Phillips].

vectors is a linear combination of the others, and $S(v_1, v_2, v_3)$ vanishes by antisymmetry. These are essentially three different kinds of three-dimensional subspaces, up to Lorentz transformations on M_x. (See Exercise 1.8.)

The first kind is a spacelike subspace (i.e. all vectors in the subspace are spacelike). The prototype is the subspace of **M** spanned by

$$(0, 1, 0, 0), \ (0, 0, 1, 0), \ (0, 0, 0, 1) \ .$$

The second is a subspace which has an orthonormal basis consisting of one timelike and two spacelike vectors; the prototype is the subspace of **M** spanned by

$$(1, 0, 0, 0), \ (0, 1, 0, 0), \ (0, 0, 1, 0) \ .$$

This type of subspace is isomorphic to three-dimensional (one time, two space) Minkowski space.

The third kind is a subspace restricted to which the metric is degenerate (i.e there is a nonzero vector in the subspace which is orthogonal to all vectors in the subspace). It can be shown that such a subspace can be spanned by three orthogonal vectors, two spacelike (which can be normalized to unit vectors) and one null vector. The prototype is the subspace of **M** spanned by

$$(1, 1, 0, 0), \ (0, 0, 1, 0), \ (0, 0, 0, 1) \ .$$

Note that this subspace has no orthonormal basis.

We now state the physical interpretation of $S(v_1, v_2, v_3)$ in each of these cases. Since a 3-form on a three-dimensional space is determined up to a constant multiple by its values on a basis, we may as well assume that v_1, v_2, v_3 is orthonormal when the subspace has an orthonormal basis. For ease of exposition, we also assume that the spacetime M is Minkowski space **M**, and we identify all the tangent spaces $M_x = \mathbf{M}_x$ with **M**. The interpretations to be described are physically reasonable in any spacetime, but the language necessary to describe them correctly is more complicated.

Case 1. Suppose v_1, v_2, v_3 is an orthonormal set of spacelike vectors.

Let v_0 be the unique forward-pointing timelike unit vector such that v_0, v_1, v_2, v_3 is orthonormal, and orient the subspace spanned by v_1, v_2, v_3 by declaring the latter set to be positively oriented in the subspace if and only if v_0, v_1, v_2, v_3 is a positively oriented basis for **M**. Let $s = +1$ *or* -1 according as v_1, v_2, v_3 is positively or negatively oriented. Then, at any point $x \in M$, the physical interpretation of $S_x(v_1, v_2, v_3)$ is:

(1) $\quad S_x(v_1, v_2, v_3) = s \lim_{h_1 \to 0} \lim_{h_2 \to 0} \lim_{h_3 \to 0} \frac{1}{h_1 h_2 h_3}$ (total energy-momentum

in the rectangular solid at x spanned by $h_1 v_1, h_2 v_2, h_3 v_3$).

By "rectangular solid at x ... " we mean the following. Write $x = (t_0, \vec{x}_0)$ relative to the coordinatization defined by v_0, v_1, v_2, v_3. Then the solid lies in the hyperplane of constant coordinate time $t = t_0$, and within this hyperplane it is the solid with corner at $\vec{x}_0$ and spanned by the three vectors $h_1 v_1, h_2 v_2, h_3 v_3$ (see Figure 1). The reason for using a rectangular solid instead of a cube is to make it apparent from the definition that $S(v_1, v_2, v_3)$ is a linearization and hence is automatically linear in v_1, v_2, v_3 separately.

Notice that the physical quantity described on the right side of (1) is independent of v_1, v_2, v_3 so long as this set is orthonormal and positively oriented (meaning that $s = 1$). A

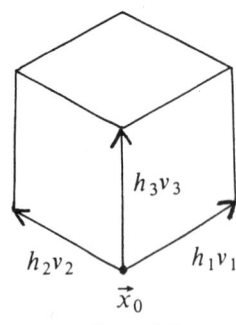

Figure 3-1

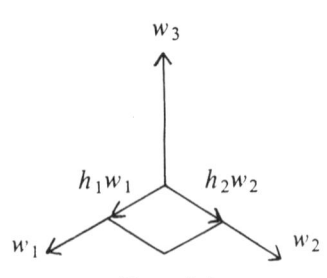

Figure 3-2

3-form on a 3-dimensional space has the same property, which provides a reassuring check on the mathematical appropriateness of the physical definition (1).

Case 2. Suppose v_1, v_2, v_3 is an orthonormal set consisting of two spacelike and one forward timelike vector.

To avoid a possible notational confusion later, we relabel this set as w_0, w_1, w_2, where w_0 is the forward timelike vector. By Exercise 1.2, this may be extended to a positively oriented orthonormal basis w_3, w_0, w_1, w_2 for **M**. Then at any $x \in \mathbf{M}$, the physical interpretation of $S_x(w_0, w_1, w_2)$ is:

(2) $\quad S_x(w_0, w_1, w_2) = \lim_{h_0, h_1, h_2 \to 0} \dfrac{1}{h_0 h_1 h_2}$ (amount of energy-momentum

which crosses the spatial parallelogram P spanned by $h_1 w_1, h_2 w_2$ in the w_3-direction between times t and $t + h_0$, where crossings in the $-w_3$-direction are counted negatively.

(See Figure 2.) Informally, $S_x(w_0, w_1, w_2)$ may be described as the rate at which energy-momentum is crossing a unit area spanned by w_1, w_2 in the w_3-direction.

To check that (2) is physically appropriate is left as an exercise. (For instance, what if another orthonormal basis spanning the same subspace as w_0, w_1, w_2 had been used to calculate the physical quantity on the right?) Many will be willing to accept this on faith, but those who do work through it will find the physical insights rewarding. It is actually rather beautiful and amazing, at least on first sight, that a (vector-valued) 3-form is precisely the right mathematical object to describe the physical quantity on the right side of (2).

One peculiarity of the definition (2) should be mentioned. For simplicity, suppose that the only energy-momentum to be counted is that of two particles of rest mass m, one with spatial velocity $v w_3$ and the other with the opposite velocity $-v w_3$, whose worldlines pass through x (taken as the origin in Figure 3). (We assume that the particles can pass through the same point without interacting.) Then $S(w_0, w_1, w_2) = 2mv\gamma(v)w_3$. The energy crossings from the two particles cancel, but the spatial components of momentum crossing do not cancel, even though the particles have opposite velocities, because they cross in opposite directions. That is, taking $v \geq 0$ for simplicity of exposition,

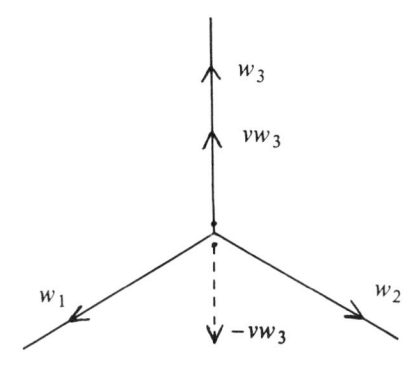

Figure 3-3.

The w_3-components of spatial momentum crossing the w_1-w_2 plane from the two particles do not cancel.

---

total energy-momentum crossing

$$= m\gamma w_0 + mv\gamma w_3 \qquad \text{(from particle 1 crossing positively)}$$

$$-(m\gamma w_0 - mv\gamma w_3) \qquad \text{(from particle 2 crossing negatively)}$$

$$= 2mv\gamma w_3.$$

This usually seems strange the first time one sees it, but some thought will show that all the signs are as they should be. Also notice that interchanging w_1 and w_2 will change w_3 into $-w_3$ because the direction of w_3 was chosen to positively orient a basis containing w_1 and w_2. This changes the signs of all the crossings, and so

$$S(w_0, w_2, w_1) = -2mv\gamma w_3 = -S(w_0, w_1, w_2),$$

as it should.

Case 3. Suppose the restriction of the Minkowski metric to the subspace spanned by v_1, v_2, v_3 is degenerate.

In this case (which is not important in practice), there seems to be no really satisfying physical interpretation of $S(v_1, v_2, v_3)$, although interpretations can be contrived. To see what is happening, take

$$v_1 = (1, 1, 0, 0), \quad v_2 = (0, 0, 1, 0), \quad v_3 = (0, 0, 0, 1), \quad \text{in Minkowski space,}$$

and set

$$e_0 := (1, 0, 0, 0), \quad e_1 := (0, 1, 0, 0).$$

Of course,

$$S(v_1, v_2, v_3) = S(e_0, v_2, v_3) + S(e_1, v_2, v_3)$$

in some sense algebraically reduces the interpretation to Cases 1 and 2.

A more direct and interesting interpretation is obtained by writing

$$w_\epsilon := (1+\epsilon, 1, 0, 0),$$

and noting that

$$\lim_{\epsilon \to 0} S(w_\epsilon, v_2, v_3) = S(v_1, v_2, v_3).$$

For ϵ positive, $S(w_\epsilon, v_2, v_3)$ can be interpreted under Case 2, and for ϵ negative under Case 1, so we have the curious fact that the interpretation of $S(v_1, v_2, v_3)$ is a limiting case of *both* Cases 1 and 2.

From here on we specialize from an arbitrary spacetime to Minkowski space **M**. This is not just for convenience but because what we shall do does not work (at least not in any obvious and straigtforward way) in an arbitrary spacetime. In Minkowski space all the tangent spaces $\mathbf{M}_x$ are identified with **M** (cf. Section 2.4.1), and so it makes sense to integrate the vector-valued 3-form $S(\cdot, \cdot, \cdot)$ over a three-dimensional submanifold N. (This does not make sense in an arbitrary spacetime because the values of $S_x(v_1, v_2, v_3)$ lie in the tangent space M_x, rather than in a fixed vector space, and there is no reasonable way to compare or add vectors in different tangent spaces.) Explicitly, we define

(3) $$\left(\int_N S\right)^i := \int_N S^i,$$

where the components are with respect to some orthonormal basis for **M**, and the S^i on the right denotes the scalar-valued 3-form $v_1, v_2, v_3 \longmapsto S(v_1, v_2, v_3)^i$. It is easy to check that this definition does not depend on the choice of basis.

To see the physical meaning of $\int_N S$, first take N to be a hyperplane of constant time $t = t_1$. Then the integral may be intuitively regarded as a limit of a sum of terms like $S(v_1, v_2, v_3)$, where v_1, v_2, v_3 span one of a number of disjoint parallelipipeds into which N is divided, so $\int_N S$ will represent the total energy-momentum in all space at time $t = t_1$. A similar interpretation holds for any three-dimensional spacelike submanifold N. (A *spacelike* submanifold is one such that the restriction of the metric to each its tangent spaces is negative definite; in otherwords, all tangent vectors are spacelike.)

For another example, let t_1, t_2, and z_0 be constants, and let

$$N = \{(t, x, y, z_0) \mid t_1 \leq t \leq t_2, \ -\infty < x, y < \infty\}$$

be the three-dimensional submanifold generated by letting the plane $z = z_0$ evolve from time t_1 to t_2. Then $\int_N S$ represents the total energy-momentum which passes through the plane $z = z_0$ in the direction of positive z between times t_1 and t_2.

To formulate the integral law of conservation of energy-momentum, consider a four-dimensional "cube"

$$K := \{(t, x, y, z) \mid t_1 \leq t \leq t_2, \ x_1 \leq x \leq x_2, \ y_1 \leq y \leq y_2, \ z_1 \leq z \leq z_2\}.$$

This four-dimensional "cube" has eight three-dimensional "faces", which we denote as K_{ia}, where $i = 0, 1, 2, 3$, $a = 1, 2$, and K_{ia} is the face obtained by fixing the i'th coordinate at the lower (for $a = 1$) or upper (for $a = 2$) limit. For instance,

$$K_{01} = \{ (t, x, y, z) \in K \mid t = t_1 \},$$

and K_{02} is the same with t_1 replaced by t_2. The faces K_{ia} of K will be considered as oriented three-dimensional manifolds (with boundary) with orientations induced from K, which in turn is considered in the natural way as an oriented four-dimensional manifold with boundary. Let C denote the three-dimensional "cube"

$$C := \{ (x, y, z) \mid x_1 \leq x \leq x_2, \, y_1 \leq y \leq y_2, \, z_1 \leq z \leq z_2 \} .$$

Then

$$(-1)^a \int_{K_{0a}} S , \qquad a = 1, 2,$$

represents the total energy-momentum in C at times t_1 and t_2, respectively, where the sign $(-1)^a$ arises from taking into account the relative orientations of K_{01} and K_{02} induced from K. Moreover,

$$\int_{K_{ia}} S , \qquad i = 1, 2, 3, \, a = 1, 2,$$

is the energy-momentum which passes out each of the six two-dimensional faces of C between times t_1 and t_2. Thus

(4) $\qquad \int_{\partial K} S = $ (energy-momentum in C at time t_2) - (e-m in C at time t_1) + (e-m which leaves C between times t_1 and t_2),

and so conservation of energy-momentum demands that (4) vanish.

By Stokes' Theorem, (4) is also equal to

(5) $\qquad \int_K dS ,$

where the differential dS of the vector-valued 3-form S is defined componentwise by $(dS)^i := d(S^i)$, $i = 0, 1, 2, 3$. Hence conservation of energy-momentum is equivalent to demanding that (5) vanish for all four-dimensional "cubes" K, which in turn is equivalent to

(6) $\qquad dS = 0 ,$

and this is one elegant local formulation of a law of conservation of energy-momentum in Minkowski space.

Actually, this law is more commonly expressed in terms of the covariant divergence of the contravariant form of the Hodge dual of S. If we forget temporarily that S is a vector-valued (rather than a scalar-valued) 3-form and apply results

2.8(12) $\qquad \delta\alpha = -\bot d \bot \alpha ,$

and

2.2(13) $\qquad \bot(\bot\alpha) = (-1)^{p+1}\alpha$ for all p-forms α on Minkowski space,

then we obtain

(7) $$\delta(\perp S) = -(\perp d \perp)\perp S = -\perp dS = 0 .$$

If we now raise an index on $\perp S$, whose indicial expression we take as $(\perp S)_i{}^j$,† and call the result T as is traditional (i.e. $T^{ij} := (\perp S)^{ij} := g^{i\alpha}(\perp S)_\alpha{}^j$, then (7) may be rewritten as

(8) $$(\delta T)^j := T^{\alpha j}{}_{|\alpha} = 0 .$$

Equation (8) is the more common local formulation of conservation of energy-momentum. It is not as directly intuitive as (6), but is more convenient in many computations. It is taken over into general relativity as an expression of conservation of energy-momentum even though there is no obvious corresponding integral conservation law in general spacetimes. When physics texts refer to an "energy-momentum tensor", they usually mean T^{ij}.

To verify that $dS = 0$ and $\delta T = 0$ are equivalent for vector-valued S, we need only make an appropriate definition of the Hodge dual for vector-valued forms and check that 2.2(13) and 2.8(12) carry over. The definition is the obvious componentwise one, which is invariant under linear changes of coordinates:

(9) for each i, j, $\quad (\perp S)_i{}^j := (\perp(S^j))_i$,

where the S^j on the right denotes the scalar-valued form $v_1, v_2, v_3 \mapsto S(v_1, v_2, v_3)^j$. These definitions are set up so that taking vector components commutes with both $\perp$ and d, and the analog of 2.2(13) is immediate. The analog of 2.8(12) for vector-valued forms follows from 2.8(33), which implies that in Minkowski space, taking components also commutes with δ. Explicitly, in the notation of Section 2.8, for each fixed j,

$$(\delta T)^j = \delta(T^{\bullet j}) = -\perp d\perp(T_\bullet{}^j) = -\perp d\perp(\perp(S^j)) = -\perp d(S^j) ,$$

so $dS = 0$ is equivalent to $\delta T = 0$.

We shall denote either mixed form $T_j{}^i$ or $T^i{}_j$ of T as $\hat{T}$. All T^{ij} which will occur in this section are symmetric, and in this case, $T_j{}^i = g_{j\alpha}T^{\alpha i} = g_{j\alpha}T^{i\alpha} = T^i{}_j$ for all i and j, so one need not worry which mixed form is meant. In coordinate-free notation, $\hat{T} = \perp S$, which is a linear transformation on **M**. There is a pedantic notational question as to whether one should denote the components of $\hat{T}$ as $\hat{T}^i{}_j$ or $T^i{}_j$, and we shall not commit ourselves but shall use whichever seems more natural in the context. So that there shall be no question, we explicitly state that $\hat{T}^i{}_j = T^i{}_j$ for all i and j.

Keeping track of the various sign changes when taking Hodge duals and raising and lowering indices is annoying, but sometimes necessary, when basing physical interpretations on $\hat{T}$. Unfortunately, the "natural" guess as to sign is not always the correct one, and knowing the following rule of thumb, whose verification is left as an exercise, allows one to concentrate on more important things.

Rule of thumb for signs in interpretation of $\hat{T}(v)$:

To compute $\hat{T}(v)$ in terms of S, where v is a unit vector, embed v as the first member of a positively oriented, orthonormal basis v, e_1, e_2, e_3.

---

† The peculiar index placement $(\perp S)_i{}^j$ instead of the more usual $(\perp S)^j{}_i$ was chosen so that (8) comes out in the most commonly used form in which the covariant divergence is taken with respect to the first component. If one wants to write $\perp S$ as $(\perp S)^i{}_j$, then one should define T by $T^{ij} := (\perp S)^j{}_i$, but this makes the relation between T and S more difficult to say in words. Since T usually turns out to be symmetric in practice (and will always be symmetric in this section), the distinction is usually unimportant. (Actually, the most natural convention seems to be to define covariant divergence with respect to the last, rather than first, contravariant component, but this was only realized after nearly all the type was set.)

(i) If v is timelike, then $\hat{T}(v) = S(e_1, e_2, e_3)$.

(ii) If v is spacelike then $\hat{T}(v) = -S(e_1, e_2, e_3)$.

For example, if e_0, e_1, e_2, e_3 is a positively oriented orthonormal basis with e_0 forward timelike, then $\hat{T}(e_0)^0 = S(e_1, e_2, e_3)^0$ is interpreted as the mass-energy density (always positive in all known classical systems). Also, $\hat{T}(e_3)^0 = -(-S(e_0, e_1, e_2)^0)$, the two minuses appearing because e_3 is spacelike and e_3, e_0, e_1, e_2 is negatively oriented. However, going back to the physical interpretation of S in Case 2, we recall that $+S(e_0, e_1, e_2)^0$ is interpreted as the *negative* of the mass-energy per unit area passing through the e_1-e_2 plane in the direction of e_3 (again because e_3, e_0, e_1, e_2 is negatively oriented). For instance, if the energy-momentum accounted for by $\hat{T}$ is due to a stream of particles moving *upward* (in the e_3-direction) as in Figure 4, then $\hat{T}(e_3)^0$ will be *negative,* contrary to what may seem the natural guess.

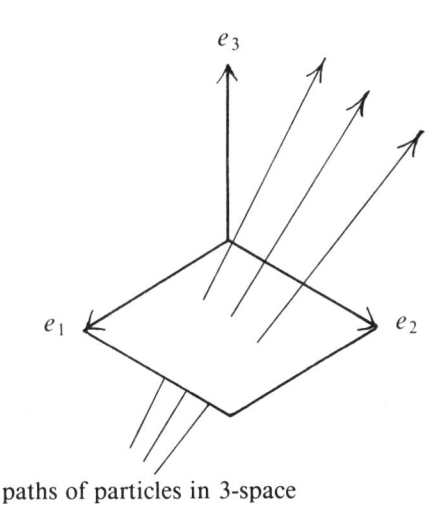

paths of particles in 3-space

Figure 3-4. For a stream of particles crossing the e_1-e_2 plane in the upward direction, the energy component $\hat{T}(e_3)^0$ will be negative, contrary to the natural guess.

A less complicated rule of thumb when working with $\hat{T}$ without reference to S is the following, which avoids the double negative in the above example.

Alternate rule of thumb:

For any unit vector v, the physical interpretation of $\hat{T}(v)$ is the following.

(i) If v is forward timelike, then $\hat{T}_x(v)$ is the energy-momentum per unit volume at $x \in \mathbf{M}$ relative to a Lorentz frame in which $v = (1, 0, 0, 0)$.

(ii) If v is spacelike, then $-\hat{T}_x(v)$ is the energy-momentum per unit area per unit time at x flowing in the v-direction relative to a frame in which v is purely spacelike; i.e. a frame in which $v = (0, \vec{v})$.

Note that (ii) says that the "natural" guess as to sign is always wrong for spacelike vectors.

Example 1.

Consider an uncharged dust with four-velocity field u and proper mass density ρ in Minkowski space $\mathbf{M}$ (i.e. $\rho(x)$ is the mass density as measured in the rest frame of the particle at $x \in \mathbf{M}$.) The appropriate energy-momentum tensor T for this system, viewed as a vector-valued 1-form, is

(10) $\qquad \hat{T} = \rho u \otimes u^*$,

whose components are

(11) $\qquad T^i{}_j = \rho u^i u_j$.

To see that this is appropriate, at any given spacetime point x choose a positively oriented orthonormal basis for $\mathbf{M}_x$ of the form $e_0 := u$, e_1, e_2, e_3, and let

$$S := \bot \hat{T} .$$

Then

$$S(e_1, e_2, e_3) = \hat{T}(u) = \rho u^*(u) u = \rho u = \rho e_0 .$$

The left side of this equation is, by definition, interpreted as the energy-momentum density as measured in the rest frame of the fluid particle at x. The right side represents a pure mass-energy density in the rest frame. Thus taking (10) as the energy-momentum tensor says that in the rest frame of the particle at x, one observes a mass-energy density ρ and zero spatial momentum density, which does indeed correspond to the idea of a continuous distribution of matter moving coherently in the sense that nearby particles have nearly the same velocity.

It is instructive to write the components $\hat{T}^i{}_j$ of T relative to a new basis

$$e_0' := \gamma(v) e_0 + v \gamma(v) e_1 ,$$
$$e_1' := v \gamma(v) e_0 + \gamma(v) e_1 ,$$
$$e_2' := e_2 ,$$
$$e_3' := e_3 ,$$

Related to the old by a boost with velocity v. With respect to the new basis $\{e_i'\}$, the fluid particle at x (which was stationary in the frame defined by the old basis) is seen as moving with velocity $-v$ in the e_1'-direction.

The matrix of $\hat{T}$ with respect to the new basis is easily computed to be:

$$(T^i{}_j) = \begin{bmatrix} \dfrac{\rho}{(1-v^2)} & \dfrac{\rho v}{(1-v^2)} & 0 & 0 \\ \dfrac{-\rho v}{(1-v^2)} & \dfrac{-\rho v^2}{(1-v^2)} & 0 & 0 \\ 0 & 0 & 0 & 0 \\ 0 & 0 & 0 & 0 \end{bmatrix} .$$

The 0-0 component, $\rho/(1-v^2)$, is the mass-energy density as measured in the new frame. The factor $1/(1-v^2)$ may be understood physically by viewing it as a product of one factor of $(1-v^2)^{-1/2}$ due to the increase in relativistic mass with velocity with another due to the adjustment in the measurement of mass density necessitated by the Lorentz length contraction.

According to the definition of S and the rules of sign given above, the *negative* of the 0,1 component,

$$\frac{-\rho v}{(1-v^2)},$$

is interpreted as the mass-energy passing through a unit area in the e'_2-e'_3 plane in the e'_1-direction per unit time, and the negative of the 1,1 component,

$$\frac{\rho v^2}{1-v^2},$$

has a similar meaning with "mass-energy" replaced by "momentum in the e'_1-direction". It is helpful to think carefully through the physical interpretations in a transparent physical case such as this and to refer back to it when confused in more complicated situations.

Notice also that although $\hat{T}$ is a symmetric linear transformation in the sense that $<\hat{T}(v), w> = <v, \hat{T}(w)>$ for all vectors v and w, the matrix for $\hat{T}$ with respect to the orthonormal basis $\{e'_i\}$ is not symmetric. Symmetric linear transformations always have symmetric matrices only in vector spaces with a positive definite or negative definite inner product.

Example 2.

Consider a perfect fluid with four-velocity field u, proper mass density ρ, and pressure p in **M**. The physical picture is a bit more complicated than that of the uncharged dust of Example 1 in that u_x is no longer pictured as the four-velocity of a particular fluid particle at the space-time point x, but instead represents a sort of average fluid four-velocity at x. Previously, we pictured the fluid as moving coherently in the sense that nearby particles had nearly the same velocity, but now the picture is closer to that of a gas in that near any point there are many particles with widely varying velocities, some average of which defines u.

At each spacetime point x, let $Q_x(\cdot)$ denote the orthogonal projection from $\mathbf{M}_x$ onto the orthogonal complement of u_x in $\mathbf{M}_x$. Explicitly, for any z in $\mathbf{M}_x$, $Q_x(z) = z - <z,u>u$. The appropriate energy-momentum tensor $\hat{T}$ for this system is considered to be:

(12) $$\hat{T} = \rho u \otimes u^* - pQ,$$

where p is a scalar function called the *pressure*. Let x be a given spacetime point, and choose a positively oriented, orthonormal basis for $\mathbf{M}_x$ of the form $e_0 := u$, e_1, e_2, e_3. Then $\hat{T}(e_3) = -pe_3$, which says (recalling the rule of thumb for signs) that the rate per unit area at which momentum is crossing the e_1-e_2 plane in the e_3-direction is $+pe_3$. Since the average fluid velocity is zero in the local frame defined by $\{e_i\}$ (because $e_0 = u$), the momentum crossing the e_1-e_2 plane is conceived as due to random fluid particles crossing in both directions. (See Figure 5.) Particles crossing from below are crossing in the "positive" sense, and their momenta are counted positively. Particles crossing from above are crossing in the "negative" sense, and their momenta are counted negatively. Assuming that the distribution of momenta of the particles crossing from below is the same as that of the particles crossing from above, the e_1 and e_2 components of momenta of the particles crossing from above will cancel with those of the particles crossing from below, so we expect that $\hat{T}^1{}_3 = \hat{T}^2{}_3 = 0$. However, the e_3 components do not cancel (cf. the discussion of Case 2 in the physical interpretation of S), and $\hat{T}^3{}_3$ will be proportional to the total amount of e_3-

momentum crossing per unit area per unit time without regard to sign or direction. This explains the physical identification of $\hat{T}^3{}_3$ with the pressure across the e_1-e_2 plane. Similarly, $\hat{T}^1{}_1$ and $\hat{T}^2{}_2$ are identified with the pressures across the e_2-e_3 and e_1-e_3 planes, respectively. It is possible to conceive of fluids in which pressures across different planes might be unequal, but in a perfect fluid they are equal by definition. In fact, (12) implies that

$$\hat{T}^1{}_1 = \hat{T}^2{}_2 = \hat{T}^3{}_3 = -p \ .$$

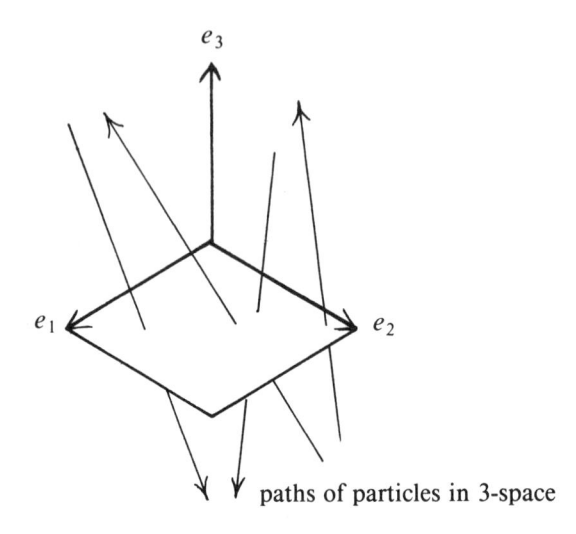

paths of particles in 3-space

Figure 3-5

For real fluids, p is presumably a positive function (as it would have to be in the above model), but of course one can write down (12) for negative p as well. Negative pressure is sometimes referred to as "tension". More generally, if the space part of the energy-momentum tensor $\hat{T}$ is diagonal in some Lorentz frame, one speaks of "tension" in the spatial directions corresponding to the positive diagonal elements and "pressure" in the spatial directions corresponding to the negative ones. The energy-momentum tensor for the electromagnetic field which will be introduced below has tension in some directions.

Now we are ready to consider a *charged* dust, which will lead us to the form of the law of conservation of energy-momentum used in electrodynamics. As usual, let u denote the four-velocity field and ρ the proper charge density of a charged dust in a spacetime M. Let m/q denote the mass/charge ratio of the particles of which the dust is composed, so that $(m/q)\rho$ is the proper mass density. Since our initial remarks do not use the charge density and apply to many "dust-like" physical systems, charged or not, it will be convenient to write

(13) $$\rho_m := \frac{m}{q}\rho$$

for the proper mass density. We have seen in Example 1 above that if the dust were not charged, the appropriate 2-contravariant form of the energy-momentum tensor would be

(14) $$\rho_m u \otimes u \ .$$

3.8 Energy-momentum tensor

It is routine to compute (Exercise 2.21) that

(15) $\quad\quad\quad\quad \delta(\rho_m u \otimes u) = \delta(\rho_m u)u + \rho_m D_u u$.

Since u is a unit vector field, $D_u u$ is orthogonal to u (because $0 = u(<u,u>) = 2<D_u u, u>$), and hence (15) vanishes (i.e. energy-momentum is conserved for the uncharged dust of Example 1) if and only if both

(16) $\quad$ (i) $\quad \delta(\rho_m u) = 0 \quad$ and

$\quad\quad\quad$ (ii) $\quad \rho_m D_u u = 0$.

We observed in Section 2.8 that equation (i) expresses conservation of mass. The reader who is familiar with differential geometry will immediately recognize (ii) as the geodesic equation, which says that the dust particles move along geodesics (which in Minkowski space are straight lines). Another way to understand (16) is to recall from Section 3.2 that in Minkowski space, $D_u u = \dfrac{du}{d\tau}$, so (ii) says that the proper acceleration of each dust particle vanishes. In Newtonian language, (ii) says that no forces act on the dust particles.

To describe a physical system in which forces do act on the particles, one can add an appropriate tensor $T_{force} = T_{force}{}^{ij}$ to (14), obtaining as total energy-momentum tensor T :

$$T = \rho_m u \otimes u + T_{force} .$$

The equation

$$\delta(T) = 0$$

of conservation of energy-momentum then becomes

(17) $\quad\quad\quad \delta(\rho_m u)u + \rho_m D_u u = -\delta(T_{force})$.

Separating (17) into components in the direction of u and orthogonal to u yields:

(17) $\quad$ (i) $\quad \delta(\rho_m u) = -<\delta(T_{force}), u>$

$\quad\quad\quad$ (ii) $\quad \rho_m D_u u = -\delta(T_{force}) + <\delta(T_{force}), u>u$.

Usually, one wants a law of conservation of mass and therefore chooses T_{force} so that the right side of (i) vanishes. The second equation is the equation of motion, analogous to Newton's law

$$\text{mass} \cdot \text{acceleration} = \text{force} ,$$

and one chooses T_{force} so that its divergence gives the desired force law.

In continuum electrodynamics, the desired force law is the Lorentz equation 3.2(3), and so we seek a tensor whose divergence is $-\rho \hat{F}(u)$. The following 2-contravariant tensor T_{elm} (elm for "electromagnetic"), called the *energy-momentum tensor for the electromagnetic field*, can be shown to have that property:

(18) $\quad\quad\quad T^{ij}_{elm} := \dfrac{1}{4\pi}[F^i{}_\alpha F^{\alpha j} + \dfrac{1}{4}g^{ij}F^{\alpha\beta}F_{\alpha\beta}]$.

We shall prove this in a moment, but the proof is algebraically a little easier if done in terms of the mixed tensor (in which the identifying subscript "elm" is dropped because the space is needed for an index):

(19) $$4\pi T^i{}_j := F^i{}_\alpha F^\alpha{}_j - \frac{1}{4}\delta^i{}_j F^\alpha{}_\beta F^\beta{}_\alpha,$$

where $\delta^i{}_j$ denotes the Kronecker delta, which is 1 if $i=j$ and 0 otherwise.

The minus sign in front of the second term of (19) appears because indices were interchanged on one of the F's. This was done because that makes it apparent that (19) can be written in a neat, coordinate-free way as

(20) $$4\pi \hat{T} = \hat{F}^2 - \frac{1}{4}\mathrm{tr}(\hat{F}^2)I .$$

Here $\hat{T}$ and $\hat{F}$ denote the linear transformations with matrices $(T^i{}_j)$ and $(F^i{}_j)$, respectively, tr denotes the trace, and I the identity transformation. Unfortunately, (20) appears to have no obvious physical or heuristic interpretation and in some situations seem actually harder to use than (19) itself. (For example, the reader might compare the computation below of the covariant divergence of (19) using 2.8(6) with a corresponding computation using (20) and 2.8(1).)

Now we shall show how the Maxwell-Lorentz equations imply that the covariant divergence of (18) is $-\rho \hat{F}(u)$. It is enough to prove this for (19) since lowering an index commutes with the covariant derivative and hence with the covariant divergence. For simplicity we shall work with tensor components with respect to a Lorentz coordinatization of Minkowski space, but the same computation will serve without alteration in general spacetimes if one uses a coordinate system with respect to which all the connection coefficients Γ^k_{ij} vanish at the point at which the covariant divergence is taken. The result which we shall establish is:

(21) $$4\pi \delta(T)_j := 4\pi \frac{\partial T^\alpha{}_j}{\partial x_\alpha} = -\delta(F)^\alpha F_{j\alpha} + \frac{1}{2}F^{\alpha\beta}(dF)_{\alpha\beta j},$$

and then Maxwell's equations imply that

(22) $$4\pi \delta(\hat{T})_j = -\delta(F)^\alpha F_{j\alpha} = -4\pi \rho u^\alpha F_{j\alpha} = -4\pi \rho \hat{F}(u)_j .$$

Starting from (19), we have

(23) $$4\pi \delta(T)_j = \frac{\partial F^\beta{}_\alpha}{\partial x_\beta}F^\alpha{}_j + F^\beta{}_\alpha \frac{\partial F^\alpha{}_j}{\partial x_\beta} - \frac{1}{4}\frac{\partial (F^\alpha{}_\beta F^\beta{}_\alpha)}{\partial x_j}$$

$$= (\delta F)_\alpha F^\alpha{}_j ++ F^\beta{}_\alpha \frac{\partial F^\alpha{}_j}{\partial x_\beta} + \frac{1}{4}\frac{\partial (F^{\alpha\beta}F_{\alpha\beta})}{\partial x_j}$$

$$= -(\delta F)^\alpha F_{j\alpha} + F^\beta{}_\alpha \frac{\partial F^\alpha{}_j}{\partial x_\beta} + \frac{1}{2}F^{\alpha\beta}\frac{\partial F_{\alpha\beta}}{\partial x_j} .$$

To handle the middle term (which is the only tricky part of the calculation), first verify by routine calculation the following formula, which is valid for any 2-form and quite useful in its own right:

(24) $$(dF)_{ijk} = \frac{\partial F_{ij}}{\partial x_k} + \frac{\partial F_{jk}}{\partial x_i} + \frac{\partial F_{ki}}{\partial x_j} .$$

From this we obtain

$$(25) \quad F^{\beta\alpha}\frac{\partial F_{\alpha j}}{\partial x_\beta} = F^{\beta\alpha}(dF)_{\alpha j\beta} - F^{\beta\alpha}\frac{\partial F_{j\beta}}{\partial x_\alpha} - F^{\beta\alpha}\frac{\partial F_{\beta\alpha}}{\partial x_j}.$$

From the antisymmetry of F it follows that the middle term on the right side of (25) is the negative of the left side, and hence

$$(26) \quad F^\beta{}_\alpha\frac{\partial F^\alpha{}_j}{\partial x_\beta} = F^{\beta\alpha}\frac{\partial F_{\alpha j}}{\partial x_\beta} = \frac{1}{2}[F^{\beta\alpha}(dF)_{\alpha j\beta} - F^{\beta\alpha}\frac{\partial F_{\beta\alpha}}{\partial x_j}].$$

Substitution of (26) in (23) yields (21).

Thus if we define the 2-contravariant energy-momentum tensor T_{cdust} for a charged dust as:

$$(27) \quad T_{cdust} := \frac{m}{q}\rho u \otimes u + T_{elm},$$

then the Maxwell-Lorentz equations imply that $\delta(T_{cdust}) = 0$. In fact, assuming Maxwell's equations, $\delta(T_{cdust}) = 0$ is *equivalent* to the Lorentz equation. It is natural to think of the two terms on the right of (27) as describing two forms of energy-momentum, that of the matter and that of the field, neither of which is conserved separately but whose sum is conserved. Since the expression (18) for T_{elm} depends only on the field tensor F, it is generally taken to be the "energy-momentum tensor of the electromagnetic field" even in situations in which (27) is not an appropriate energy-momentum tensor for the system as a whole. An example of such a situation would be the extension of Example 2 to a *charged* fluid with pressure, for which the energy-momentum tensor T would be taken as the sum of the 2-contravariant form of the right side of (12) and T_{elm}.

The classical description of the energy-momentum tensor of the electromagnetic field in terms of the electric and magnetic vectors $\vec{E}$ and $\vec{B}$ (cf. Section 3.4) is of considerable interest. The most enlightening description is provided by writing the mixed tensor $\hat{T} = (T^i{}_j)$ as an operator matrix

$$\hat{T} = \begin{bmatrix} M & P \\ N & Q \end{bmatrix},$$

where $M: R^1 \longrightarrow R^1$, $Q: R^3 \longrightarrow R^3$, $N: R^1 \longrightarrow R^3$, and $P: R^3 \longrightarrow R^1$:

$$(28) \quad \hat{T} = \frac{1}{4\pi}\begin{bmatrix} \dfrac{E^2+B^2}{2} & -(\vec{E}\times\vec{B})^* \\ \vec{E}\times\vec{B} & \vec{E}\otimes\vec{E}^* + \vec{B}\otimes\vec{B}^* - \dfrac{E^2+B^2}{2}I \end{bmatrix}.$$

The notation in (28) is entirely classical, and all operations (in particular the operation * which converts vectors into 1-forms) are with respect to the usual positive-definite inner product on R^3. Of course, E^2 means the square of the Euclidean norm of $\vec{E}$, not the second component of $\vec{E}$, etc., and I is the identity transformation on R^3.

The $T^0{}_0$ component, $(E^2+B^2)/8\pi$, is physically interpreted as the mass-energy per unit volume of the electromagnetic field. The vector $\vec{E}\times\vec{B}/4\pi$ which appears in the first column is called the *Poynting vector* and is interpreted as the 3-momentum per unit volume of the field. From its appearance in the first row (and recalling the rule of thumb regarding signs of Section 3.8), it may also be interpreted as the energy-flow vector of the field in the sense that for any vector $\vec{v}$, $(\vec{E} \times \vec{B}) \cdot \vec{v}$ is the rate at which field energy is crossing a unit area to which $\vec{v}$ is normal in the direction of $\vec{v}$. In particular, the surface integral (in the sense of ordinary vector calculus)

$$\text{(29)} \qquad \int_S (\vec{E} \times \vec{B}) \cdot \vec{dS}$$

of the outward component of the Poynting vector over a closed two-dimensional oriented surface S (such as a sphere) will give the rate at which electromagnetic energy is flowing out of the three-dimensional region V bounded by S.

It is instructive to see the classical statement of conservation of energy of field plus charges emerge from (29) via ordinary vector calculus. By Gauss's Theorem,

$$\frac{1}{4\pi} \int_S (\vec{E} \times \vec{B}) \cdot \vec{dS} = \frac{1}{4\pi} \int_V \vec{\nabla} \cdot (\vec{E} \times \vec{B}) \, dV .$$

Recalling the vector identity $\vec{\nabla} \cdot (\vec{E} \times \vec{B}) = \vec{B} \cdot (\vec{\nabla} \times \vec{E}) - \vec{E} \circ (\vec{\nabla} \times \vec{B})$, and using the three-dimensional version 3.5(2), 3.6(4) of Maxwell's equations, this transforms into

$$\text{(30)} \qquad \frac{1}{4\pi} \int_V \vec{\nabla} \cdot (\vec{E} \times \vec{B}) \, dV$$

$$= \frac{1}{4\pi} \int_V \left(-\vec{B} \cdot \frac{\partial \vec{B}}{\partial t} - \vec{E} \cdot \frac{\partial \vec{E}}{\partial t} - 4\pi \vec{E} \cdot \vec{J} \right) dV$$

$$= -\frac{\partial}{\partial t} \int_V \frac{E^2 + B^2}{8\pi} \, dV - \int_V \vec{E} \cdot \vec{J} \, dV .$$

(Recall from Section 3.6 that $\vec{J}$ denotes the three-dimensional charge current vector $\vec{J} := \rho\gamma(v)\vec{v}$ and that $\rho\gamma(v)$ is the charge density as measured in the given Lorentz frame with respect to which all the above 3-vectors are defined.) The first term on the right is the negative of the rate of change of the field energy, and the negative of the second,

$$\int_V \vec{E} \cdot \vec{J} \, dV = \int_V \vec{E} \cdot \vec{v} \gamma(v) \rho \, dV ,$$

is the rate at which the Lorentz force $q(\vec{E} + \vec{v} \times \vec{B})$ is doing work (i.e. $\overrightarrow{force} \cdot \overrightarrow{velocity}$ with $q = \gamma(v)\rho dV$) on the fluid charges. Hence (29) says that the rate at which the field energy is flowing outward across the boundary equals the rate at which the field energy in the volume is decreasing minus the rate at which the field is doing work on the charged particles.

References.

Nearly all the material on electrodynamics in this chapter can be found in any advanced text, though notation and organization are likely to differ. Two of many which can be recommended are [Jackson] and [Panofsky and Phillips]. Unusually detailed explanations of various aspects of the energy-momentum tensor appear in [Misner, Thorne, and Wheeler, Chapter 5]. The latter is fun to browse through for the superb graphics and many physical insights, as is [Feynman, Leighton, and Sands].

Exercises 3

3.1 Let

$$F = (F_{ij}) = \begin{bmatrix} 0 & E^1 & E^2 & E^3 \\ -E^1 & 0 & -B^3 & B^2 \\ -E^2 & B^3 & 0 & -B^1 \\ -E^3 & -B^2 & B^1 & 0 \end{bmatrix},$$

be the matrix of a field tensor. Compute that the matrix of $\perp F$ is

$$\perp F = \begin{bmatrix} 0 & -B^1 & -B^2 & -B^3 \\ B^1 & 0 & -E^3 & E^2 \\ B^2 & E^3 & 0 & -E^1 \\ B^3 & -E^2 & E^1 & 0 \end{bmatrix}.$$

In other words, the matrix of $\perp F$ is obtained from that of F by the substitutions $\vec{E} \longmapsto -\vec{B}$, $\vec{B} \longmapsto \vec{E}$.

3.2 Show that for any mixed field tensor $\hat{F} = (F^i{}_j)$ (cf. 3.4(2)), there exists an orthonormal basis for **M** with respect to which the matrix of $\hat{F}$ is

(*) $$\begin{bmatrix} 0 & E^1 & E^2 & 0 \\ E^1 & 0 & 0 & 0 \\ E^2 & 0 & 0 & B^1 \\ 0 & 0 & -B^1 & 0 \end{bmatrix}.$$

Equivalently, there exists a Lorentz transformation matrix $L = (L^i{}_j)$ such that $L^{-1}(3.4(2))L = (*)$. Observe that L can be chosen as a purely spatial rotation.

3.3 (Continuation of Exercise 2) Show that if $\vec{E} \cdot \vec{B} \neq 0$, then there exists an orthonormal basis with respect to which the matrix of $\hat{F}$ is

(†) $$\begin{bmatrix} 0 & E^1 & 0 & 0 \\ E^1 & 0 & 0 & 0 \\ 0 & 0 & 0 & B^1 \\ 0 & 0 & -B^1 & 0 \end{bmatrix}.$$

Note that $\vec{E}$ and $\vec{B}$ are parallel in this frame.

Hint: If the matrix is of the form (*) in Exercise 2, then $\vec{E} \times \vec{B}$ is in the e_3-direction. Recalling from 3.8(28) that $\vec{E} \times \vec{B}$ describes the flow of field energy in the given frame, this suggests that if we switch to a frame moving with an appropriate velocity in the e_3-direction, we might be able to make the energy flow vanish, in which case $\vec{E}$ and $\vec{B}$ would be parallel in the new frame.

3.4 Compute $\hat{F}^2$ and $\hat{F}(\bot\hat{F})$ with respect to the basis of Exercise 2.

(a) Show that the scalar quantities $\vec{E}\cdot\vec{E} - \vec{B}\cdot\vec{B}$ and $\vec{E}\cdot\vec{B}$ are independent of the orthonormal basis with respect to which the matrix 3.4(1) of a field tensor F is written.

(b) Show that $\hat{F}$ is invertible if and only if $\vec{E}\cdot\vec{B} \neq 0$ and find a simple expression for $\hat{F}^{-1}$.

3.5 Write the potential four-vector $A = (A^i)$ as $A = (\phi, \vec{A})$, where ϕ is traditionally called the *scalar potential* and $\vec{A}$ the *vector potential*.

(a) Show that the electric and magnetic fields $\vec{E}$ and $\vec{B}$ are given in any gauge by

(i) $$\vec{E} = -\nabla\phi - \frac{\partial\vec{A}}{\partial t}$$

(ii) $$\vec{B} = \nabla\times\vec{A} \ .$$

(b) Suppose that we are in an electrostatic situation, which means that $\vec{E}$ and $\vec{B}$ are time-independent. Show that there is a gauge in which A is also time-independent. Conclude that in an electrostatic situation, the electric field is the gradient of a scalar potential ϕ.

3.6 Let F be an electromagnetic field tensor on $\mathbf{M}$ and $A = (A_i)$ a potential 1-form satisfying $dA = F$. As in the preceding problem, write its contravariant form $(A^i) = (\phi, \vec{A})$. In the notation of Section 2.9, $\phi = A_T$ is the "time" part of A, and the "space" part A_S is

$$A_S = \sum_{i=1}^{3} A_i dx^i = -\sum_{i=1}^{3} A^i dx^i \ .$$

(a) Writing F in terms of the electric and magnetic fields $\vec{E}$ and $\vec{B}$, translate equations

2.9(25)   (i) $d_S\beta_S = \gamma_S$

(ii) $\dfrac{\partial\beta_S}{\partial t} = d_S(\beta_T) + \gamma_T$

and

2.9(27)   (i) $d(\bot_S\beta_T) = \bot_S\alpha_T$

(ii) $\dfrac{\partial(\bot_S\beta_T)}{\partial t} = (-1)^p d_S(\bot_S\beta_S) + (-1)^{p-1}\bot_S\alpha_S \ ,$

with $\gamma := F$, $\beta := A$, and $\alpha := \delta(A)$, into 3-dimensional vector equations involving $\vec{A}$, $\vec{E}$, and $\vec{B}$. (Take S as the time-zero hyperplane $t = 0$.)

(b) Show that there exists a solution to $dA = F$ with

$$\nabla\cdot\vec{A} = 0 \ .$$

Such a solution is said to be in the *Coulomb* gauge. (Note: The solution involves knowing how to solve Poisson's equation $\Delta\phi = \psi$ for ϕ given ψ.)

(c) Show that if F is a free field (i.e. if $\delta F = 0$ in addition to $dF = 0$), then there is a Coulomb gauge solution with ϕ identically zero. This observation is the starting point for the usual procedure for "quantizing" the electromagnetic field.

3.7. Identify the forward light cone

$$C := \{ (|\vec{x}|, \vec{x}) \mid 0 \neq \vec{x} \in R^3 \},$$

with $R^3 - \{0\}$ via the map J defined by

$$J(\vec{x}) := (|\vec{x}|, \vec{x}), \quad \vec{x} \in R^3.$$

and let μ be the measure on the cone which corresponds under this identification to the measure

$$\frac{d^3\vec{x}}{|\vec{x}|}$$

on R^3, where $d^3\vec{x}$ denotes Lebesgue measure on R^3. Show that μ is invariant under Lorentz transformations which preserve the time orientation; i.e. for any such Lorentz transformation and any measurable set E, $\mu(L(E)) = \mu(E)$. (Of course, the same holds for the backward light cone.)

3.8. Let $\Box := \delta d + d\delta$ be the D'Alembertian operator on p-form fields defined in 3.7(4). Show that for any C^∞ function f (i.e. a 0-form) on a compact semi-Riemannian manifold M,

$$\int_M \Box f = 0.$$

Hint: For a quick, non-computational proof, use Exercise 2.23.

3.9. (This and the next problem are for those with some knowledge of the theory of distributions, or at least a working understanding of intuitive "Dirac delta function" techniques as commonly practiced by physicists.) A standard way to construct solutions for inhomogeneous, constant-coefficient partial differential equations such as

(1) $$\Box f = h$$

(where h is a given scalar function on M) is to first find a "Green function" G such that

(2) $$(\Box G)(x) = \delta(x), \quad x \in M,$$

where δ denotes the "Dirac delta function". Of course, on the level of rigorous mathematics, (2) should be read as an equation for a distribution G. Then the integral

(3) $$f(x) := \int_M G(x-y) h(y) \, d^4y$$

is, at least on the formal algebraic level, a solution of (1). Comparing (3) with 3.7(9) suggests that a Green function for $\Box$ should be

(4) $$G(t, \vec{x}) = \frac{\delta(t - |\vec{x}|)}{4\pi |\vec{x}|} .$$

Show that (4) is, in fact, a distribution solution of (2). That is, show that for any C^∞ function g on **M** with compact support,

(5) $$\int_\mathbf{M} (\Box G)(t, \vec{x}) \, g(t, \vec{x}) \, dt d^3\vec{x} := \int_\mathbf{M} \frac{\delta(t - |\vec{x}|)}{4\pi |\vec{x}|} \Box g(t, \vec{x}) \, dt d^3\vec{x}$$

$$:= \int_{R^3} \frac{1}{4\pi |\vec{x}|} (\Box g)(|\vec{x}|, \vec{x}) \, d^3\vec{x} = g(0, \vec{0}) .$$

Note that only the last equality requires proof; the others are definitions. (Despite its appearance, the last equality is not completely routine.)

3.10. The distribution

$$\frac{\delta(t - |\vec{x}|)}{2|\vec{x}|} + \frac{\delta(t + |\vec{x}|)}{2|\vec{x}|}$$

is often considered to be "$\delta(<x, x>)$" (where $\delta(\cdot)$ denotes the one-dimensional Dirac delta function and $x = (t, \vec{x})$). It is not so obvious how to give a direct mathematical meaning to $\delta(<x, x>)$, but whatever it is, it should be a "generalized function" (distribution) with support on the light cone and invariant under Lorentz transformations. We have seen in Exercise 9 that the distributions

$$\frac{\delta(t - |\vec{x}|)}{|\vec{x}|} \quad \text{and} \quad \frac{\delta(t + |\vec{x}|)}{|\vec{x}|} ,$$

posess all these properties except that they are not invariant under Lorentz transformations which reverse the time orientation. However, such transformations merely interchange them, and any symmetric linear combination will be invariant under all Lorentz transformations. "Show" by formal delta-function manipulations in the style common in physics texts that

(*) $$\delta(<x, x>) = \frac{\delta(t - |\vec{x}|)}{2|\vec{x}|} + \frac{\delta(t + |\vec{x}|)}{2|\vec{x}|} .$$

Since it is reasonably obvious what the right side of (*) means in terms of rigorous distribution theory, such manipulations motivate (*) as a *definition* of $\delta(<x, x>)$.

3.11. Symmetry (i.e. $T^{ij} = T^{ji}$) is considered an important property of energy-momentum tensors. Analogies with mechanical systems [cf. Misner, Thorne, and Wheeler, Section 5.7] suggest that symmetry is a reasonable property to hope for, but there seems to be no really fundamental reason that more general energy-momentum tensors should be symmetric. In fact, the question of symmetry of the electromagnetic energy-momentum tensor appropriate for the "macroscopic" form of Maxwell's equations 3.6(4) (with $\vec{D} \neq \vec{E}$ and $\vec{H} \neq \vec{B}$ in general) is still generally considered as unresolved [Jackson, Section 6.9]. See also [Dodson and Poston, p. 503] for other interesting insights on the question of symmetry.

One reason that symmetry is considered desirable is that it implies a law of conservation of "angular momentum" which we shall now describe. Given a 2-contravariant energy-momentum tensor field $x \longmapsto T_x = (T_x^{ij})$ on Minkowski space $\mathbf{M}$, define an associated 3-contravariant *angular momentum tensor* field $L_x = (L_x^{ijk})$, $x \in \mathbf{M}$ by

$$L_x^{ijk} = T_x^{ik} x^j - T_x^{ij} x^k .$$

(Some authors reverse the signs on the right.) The relation with classical Newtonian angular momentum is seen most easily by looking at the components

$$L_x^{0jk}$$

with $j, k = 1, 2, 3$ spatial indices relative to an orthonormal basis e_1, e_2, e_3 for space. Assume that T is symmetric, and for simplicity of exposition, take $j = 1, k = 2$. From the interpretation of $T_x^{01} = T_x^{10}$ as e_1-momentum per unit volume $\vec{p}$ at x and setting $\vec{r} := (x^1, x^2, x^3)$, we arrive at the interpretation of

$$L_x^{012} = T_x^{02} x^1 - T_x^{01} x^2 = \vec{r} \times \vec{p} ,$$

as the e_3-component of the classical angular momentum $\vec{r} \times \vec{p}$ per unit volume at x, and similarly L^{0jk} are the other components of $\vec{r} \times \vec{p}$. Note that the primary physical interpretation of T_x^{01} is as the e_1-component of the energy flux vector rather than as e_1-momentum per unit volume, so symmetry of T is essential to the physical interpretation just given. For more detailed physical discussions, see [Misner, Thorne, and Wheeler, Section 5.11] or [Dixon, Section 3.1].

Show that the angular momentum tensor L has zero divergence, i.e.

$$\sum_{\alpha = 0}^{3} \frac{\partial L^{\alpha jk}}{\partial x_\alpha} = 0 ,$$

if and only if T is symmetric. Thus symmetry of T assures a law of conservation of generalized "angular momentum" formulated analogously to the law of conservation of energy-momentum discussed in Section 3.8.

3.12 Show that if the electric field $\vec{E}$ vanishes in a given Lorentz frame, then any charged particle obeying the Lorentz equation has constant speed (not velocity!) in that frame.

3.13. Let a particle of charge q and mass m move as described by the Lorentz equation in a uniform (constant in space and time) magnetic field and no electric field in some Lorentz frame. By the preceding problem, the speed $|\vec{v}|$ in this frame is constant.

(a) Suppose that the motion is confined to a plane orthogonal to the magnetic field $\vec{B}$. Show that the motion is in a circle of radius r related to the speed $|\vec{v}|$ by

$$r = \frac{M |\vec{v}|}{q |\vec{B}|} ,$$

where $M := m \gamma(\vec{v})$ is the relativistic mass.

(b) Dropping the assumption that the motion is confined to a plane, show that it is a helix whose projection on any plane orthogonal to $\vec{B}$ is the uniform circular motion of (a).

(c) The reader may remember from previous courses in electromagnetism that any *accelerated* charged particle loses energy by radiation. A particle moving in a circle in a uniform pure magnetic field is, of course, accelerated toward the center, but according to Exercise 12, its relativistic energy remains constant. Is this a contradiction? What conclusions can be drawn?

3.14 This exercise explores the effect of changing time units in electrodynamics. For simplicity, we work in Minkowski space **M**, but the same general analysis will apply to arbitrary spacetimes. The reader who is not familiar with issues involved in analyzing units may want to read the Appendix on units before working this exercise.

Within the framework of a system of units in which the velocity of light is unity, changing the time units by a factor of $k > 0$ may be regarded as equivalent to multiplying the metric tensor by k^2, so define a new metric tensor $\bar{g}$ by

$$\bar{g} := k^2 g \ .$$

In this formulation, the coordinatization of the set **E** of events is unchanged. (Of course, one could equally well change the coordinatization while keeping the same metric on **M**.) Worldlines $s \longmapsto z(s) \in \mathbf{M}$ are unchanged as point-sets, but proper time changes. The new proper time $\bar{\tau}$ is of course defined as the arc length with respect to the new metric:

$$\bar{\tau} := \int_0^t \bar{g}(\frac{dz}{dt}, \frac{dz}{dt})^{1/2} \, dt \ = \ k \int_0^t g(\frac{dz}{dt}, \frac{dz}{dt})^{1/2} \, dt \ = \ k\tau \ .$$

(a) Let $\bar{u}$ denote the new four-velocity, and show that

$$\bar{u} = k^{-1} u \quad \text{and} \quad \bar{u}^{\bar{*}} = k u^* \ ,$$

where $\bar{*}$ denotes the index-lowering operation with respect to the new metric:

$$(v^{\bar{*}})_i := \bar{g}_{i\lambda} v^\lambda \ .$$

(b) Let $\bar{m}$, $\bar{q}$, and $\bar{F} = (\bar{F}_{ij})$ denote respectively the mass, charge, and field tensor in the new system. Assume that

$$\bar{F} \ = \ \phi(k, \frac{\bar{q}}{q}, \frac{\bar{m}}{m}) F$$

for some scalar function ϕ, and take the new Lorentz equation to be

(1) $$\bar{m} \frac{d\bar{u}}{d\bar{\tau}} \ = \ \bar{q} \hat{\bar{F}}(\bar{u}) \ .$$

The index-raising "hat" operation on the right is with respect to the *new* metric $\bar{g}$. In coordinates, (1) reads

(2) $$\bar{m} \frac{d\bar{u}^i}{d\bar{\tau}} \ = \ \bar{q} \, \bar{g}^{i\lambda} \bar{F}_{\lambda\nu} \bar{u}^\nu \ .$$

Recall that the matrix (g^{ij}) is the inverse of the matrix (g_{ij}), so $\bar{g}^{ij} = k^{-2} g^{ij}$.

Using (2), the new inhomogenous Maxwell equation

(3) $$\bar{\delta}\bar{F} = 4\pi\bar{\rho}\bar{u}^{\bar{*}}$$

(where $\bar{\delta}$ denotes the new covariant divergence operation on forms defined with respect to the new metric $\bar{g}$), and the *a priori* relation (see appendix, equation (11))

$$\bar{\rho} := k^{\beta-3}\rho \; ,$$

show that the basic relation between units of time, mass, and charge in the new system is

(4) $$\frac{\bar{m}}{m} = k^{-1}\left(\frac{\bar{q}}{q}\right)^2 \; .$$

That is, the new units of time, mass, and charge may be chosen arbitrarily subject to the necessary relation (4).

(c) Show that for any particular particle, it is possible to choose new time, mass, and charge units such that for this particle,

$$\bar{m} = \bar{q} = 1 \; .$$

What is the resulting unit of time in seconds for the electron, whose mass is 9.1×10^{-28} grams and charge is 4.8×10^{-10} statcoulombs?

3.15 The preceding exercise analyzed the transformation properties of the Maxwell-Lorentz equations under the metric replacement $g \longmapsto \bar{g} := k^2 g$ with k a constant. It is also interesting to consider a *conformal* transformation

$$g_x \longmapsto \bar{g}_x := k(x)^2 g_x \; , \quad x \in M \; ,$$

in which the metric $g_x(\cdot,\cdot)$ at a point x in an arbitrary spacetime M is replaced by a positive multiple of itself, the multiple varying from point to point.

Show that Maxwell's equations are invariant under such conformal transformations in the sense that if $F = (F_{ij})$, ρ, u satisfy

$$\delta F = \rho u^* \; ,$$

and if we define

$$\bar{F}_x := F_x \; , \quad \bar{\rho}_x := k(x)^{-3}\rho_x \; , \quad \bar{u}_x := k(x)^{-1} u_x \; ,$$

then

$$\bar{\delta}\bar{F} = \bar{\rho}\bar{u}^{\bar{*}} \; ,$$

where $\bar{\delta}$ denotes the covariant divergence operation with respect to the new metric $\bar{g}$, and $\bar{*}$ is the index-lowering operation with respect to this metric. This rather remarkable mathematical fact seems to have no obvious physical interpretation.

Chapter 4

The Electrodynamics of Point Charges

4.1. Introduction.

It is widely recognized that the current theory of the electrodynamics of point charges with nonzero mass (*point electrodynamics* for short) contains many difficulties. In this chapter we shall describe the most widely accepted version of this theory and discuss some of the problems with it.

We begin by recalling the Maxwell-Lorentz equations from Section 3.6:

(1) $$\frac{du}{d\tau} = \frac{q}{m}\hat{F}(u) \ ,$$

(2) $$dF = 0 \ , \text{ and}$$

(3) $$\delta F = 4\pi\rho u^* \ .$$

Suppose we want to adapt these equations to apply to a collection of a finite number of charged particles with four-velocities $u_1, u_2, \cdots, u_n$, positive masses $m_1, m_2, \cdots, m_n$, and charges $q_1, q_2, \cdots, q_n$, respectively. Equation (1) was initially postulated only for particles of infinitesimal mass, but this was because we had *a posteriori* knowledge that it would lead to difficulties for massive particles. In the absence of this knowledge, we would surely postulate

(4) $$m_i \frac{du_i}{d\tau} = q_i \hat{F}(u_i),$$

and we shall temporarily work within the framework of this equation even though we know that we shall ultimately have to abandon it. The first Maxwell equation (2) requires no modification, but the other Maxwell equation (3) is troublesome because ρ is zero except on the worldlines of the particles, where it is infinite. (We can't simply ignore (3) on the worldlines, for this would lead to an underdetermined system. Given *any* solution F to the free-space Maxwell equations, we can solve (1) for the particle worldlines in terms of initial positions and velocities. Thus ignoring (3) on the worldlines corresponds to an "external field" in which the field determines the particle motions, but the particles do not affect the field.)

One possiblility for adapting (3) to the case of point particles is to treat the current J as some kind of four-vector--valued distribution, often poetically described by notation similar to

(5) $$J(x) = \sum_{i=1}^{n} q_i \int_{-\infty}^{\infty} \delta_4(x - z_i(\tau_i))\, u_i(\tau_i) d\tau_i \ ,$$

where $z_i(\cdot)$ is the worldline of the i'th particle parametrized by proper time τ_i, $J(x)$ the current at x, and δ_4 the four-dimensional "Dirac delta function". That is, J is supposed to be the vector-valued distribution † whose value on a scalar test function f, symbolically

---

† The reader who does not know what a "distribution" is can skip to the next paragraph without loss of continuity.

denoted

$$\int_{R^4} J(x)f(x)d^4x \; ,$$

is

$$\int_{R^4} J(x)f(x)d^4x := \sum_{i=1}^{n} e_i \int_{-\infty}^{\infty} f(z_i(\tau_i))\, u_i(\tau_i)d\tau_i \; .$$

This general approach is common in the physics literature, though not usually carried out in a rigorous and systematic way.

A related idea turns out to be easier to handle. We already have the formula 3.7(9) for the retarded potential A^{ret} produced by a given current $J = \rho u$. By imagining ρ as spatially concentrated to approximate a point particle, we can induce a formula for the retarded potential which should be produced by a point particle. In a frame in which the particle is instantaneously at rest, this turns out to be the classical Coulomb potential q/r expanding spherically outward from the particle at the speed of light. We *define* this to be the potential of a point particle and in terms of this potential A^{ret} can compute the corresponding retarded field $F^{ret} := dA^{ret}$. For a particle which is not accelerated, F^{ret} is the classical Coulomb field $-q/r^2$. In general, it is considerably more complicated, but in all cases it is infinite on the particle's worldline, and so the Lorentz equation (4) cannot be taken literally. The most obvious way to remedy this would seem to be to use in the Lorentz equation for the i'th particle, not the total retarded field F produced by all the particles, but rather the retarded field produced by all but the i'th particle. Unfortunately, this approach leads to an apparent violation of conservation of energy-momentum if the energy-momentum tensor of the field is taken to be the usual one given by equation 3.8(18).

The central problem in the logical foundations of point electrodynamics is how to save the law of conservation of energy-momentum. Various approaches suggest themselves. For instance, one might try a different energy-momentum tensor (or possibly even a completely different approach to the conservation law) or use fields other than the retarded fields or modify the Maxwell-Lorentz equations.

The most widely accepted resolution is suggested by the very computation of energy-momentum balance which exposes the problem. In this computation many coalescences and cancellations occur, and only two kinds of terms survive to spoil the conservation law. One of the two is abolished, partly by *fiat,* in a process known as *mass renormalization.* The other can be disposed of by adding an extra term to the Lorentz equation, obtaining what is known as the *Lorentz-Dirac* equation. However, the cost of this flamboyant rescue of the conservation law is high because it is hard to reconcile the Lorentz-Dirac equation with traditional concepts of causality. (Some argue that it actually violates causality, though on a time scale which is classically unobservable.) Apart from this, the resulting theory suffers from the serious inconvenience that there seem to be no satisfactory general existence and uniqueness theorems for the equations of motion of a system of more than one particle. Worse, there are physically reasonable initial conditions for which all solutions seem physically unreasonable. This chapter is devoted to an exposition of the chain of ideas which leads to the Lorentz-Dirac equation. In the next chapter we shall discuss further difficulties with this equation and examine some other interesting approaches.

4.2 The retarded potentials and fields of a point particle.

For the rest of this chapter we work entirely in Minkowski space **M**, and most of what we do is not easily adapted (if at all) to general spacetimes. The metric tensor components with respect to an orthonormal basis e_i are denoted g_{ij}, so that $g_{ij} := <e_i, e_j>$. We shall use notation closer to that traditionally used in physics texts than that used up to now because it is more suited to the kind of calculation which we shall be doing. Our adoption of the physicists' helpful practice of naming an object via subscripts or superscripts, as in A^{ret}, causes a rather pedantic notational problem. If A is to be a covariant object, then the name almost has to be a superscript in order to leave room for the component index downstairs as in A_i^{ret}. But then how does one refer to the corresponding contravariant object? The conventions we have been using dictate $(A^{ret})^i$, but A_{ret}^i is surely more natural, and we shall use the latter.

Formula 3.7(9), rewritten here in contravariant form with a simple change of variable, gives the contravariant retarded potential A_{ret} produced by the current $J := \rho u$ of a charged dust as

$$A_{ret}(t, \vec{x}) := \int_{R^3} \frac{u(t - |\vec{x} - \vec{y}|, \vec{y})\rho(t - |\vec{x} - \vec{y}|, \vec{y})}{|\vec{x} - \vec{y}|} d^3\vec{y} .$$

The contribution of a small spatial volume ΔV centered at $\vec{y}$ to this integral is approximately

$$\frac{u(t - |\vec{x} - \vec{y}|, \vec{y})}{|\vec{x} - \vec{y}|} \rho(t - |\vec{x} - \vec{y}|, \vec{y}) \Delta V .$$

If ρ were the charge density as measured in the given coordinatization, rather than the proper charge density, then the factor $\rho \Delta V$ would approximate the charge Δq in the spatial volume ΔV. For a *fixed* $\vec{y}$, it is always possible to choose a frame in which the particle at $(t - |\vec{x} - \vec{y}|, \vec{y})$ is at rest, and in that frame, the above contribution is

(1) $$\frac{u(t - |\vec{x} - \vec{y}|, \vec{y})}{|\vec{x} - \vec{y}|} \Delta q = \frac{(1, \vec{0})}{|\vec{x} - \vec{y}|} \Delta q .$$

The right side is recognized as the classical Coulomb potential, and the entire formula can be interpreted as saying that a charge Δq at a particular time, say time 0 in its rest frame, emits a "potential" which travels outward at the speed of light and which has the value $\frac{(1, \vec{0})}{r}\Delta q$ on a sphere of radius r at time r, where the coordinates are with respect to the rest frame of the charge. (See Exercise 9 of Chapter 3 for a mathematically fancier motivation of (1).)

We shall presently *define* the retarded potential due to a point charge Δq by (1), but first we write it in coordinate-independent form. Consider a particle with charge q and worldline $\tau \longmapsto z(\tau)$ parametrized by proper time τ, and let x be an arbitrary point of **M**, which we call the *field point*, since our considerations are directed toward evaluating the field tensor at that point. The backward light cone from x will intersect the worldline in exactly one point, $z_{ret} = z(\tau_{ret})$, called the *retarded point*. (See Figure 1 and Exercise 1.) The corresponding proper time τ_{ret} is called the *retarded time*. Thus, given the worldline $z(\cdot)$, τ_{ret} and z_{ret} are functions of the field point x, and we write $\tau_{ret}(x)$ and $z_{ret}(x) := z(\tau_{ret}(x))$ for their values at x. We define another function, the *retarded distance* $x \longmapsto r_{ret}(x)$, to be the spatial distance between x and $z_{ret}(x)$ as measured in a Lorentz frame in which the particle at $z_{ret}(x)$ is instantaneously at rest; i.e. a frame in which $z'(\tau_{ret}(x)) = (1, \vec{0})$.

4.2 Retarded fields

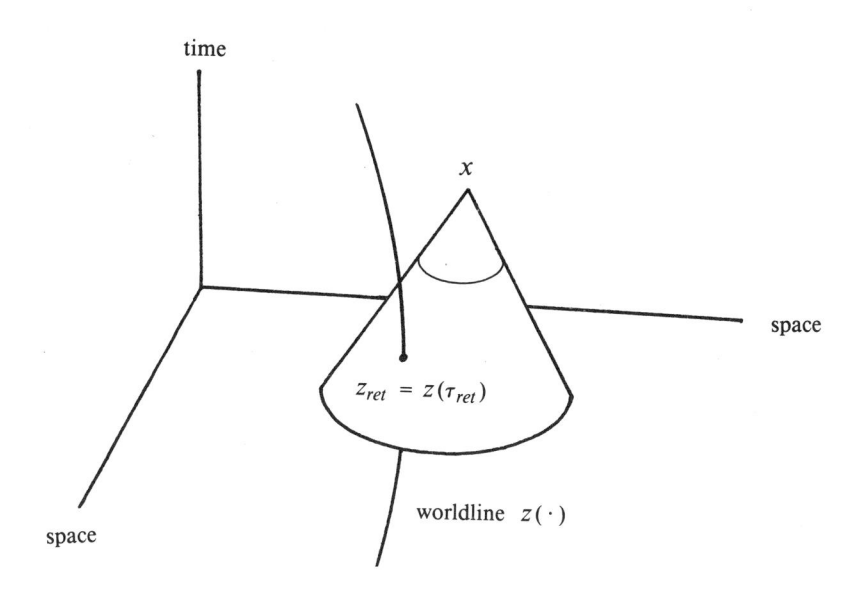

Figure 4-1. The backward light cone from the field point x intersects the worldline at the retarded point $z_{ret} = z(\tau_{ret})$.

Let $u(\tau) := z'(\tau)$ and define a vector function $x \longmapsto u_{ret}(x)$ on **M** by

$$u_{ret}(x) := u(\tau_{ret}(x)).$$

In other words, $u_{ret}(x)$ is found by shooting a light cone backwards from x and taking $u_{ret}(x)$ as the particle's four-velocity where the cone intersects the worldline. Note that u_{ret} may also be viewed as a *vector field* on **M**, and later it will be useful to think of it in this way.

When x is given, we sometimes write z_{ret} for $z_{ret}(x)$ for brevity, and similarly for u_{ret} and r_{ret}. The vector $x - z_{ret}$ can be uniquely written as

$$(2) \qquad x - z_{ret} = \langle x - z_{ret}, u_{ret}\rangle u_{ret} + r_{ret} w$$

with w a spacelike unit vector orthogonal to u_{ret} and $r_{ret} > 0$; this equation both defines w and r_{ret} and shows that r_{ret} is the spatial distance from z_{ret} to x in the given rest frame. From (2) follows

$$0 = \langle x - z_{ret}, x - z_{ret}\rangle = \langle x - z_{ret}, u_{ret}\rangle^2 - r_{ret}^2,$$

and since in the rest frame, $\langle x - z_{ret}, u_{ret}\rangle = (x^0 - z^0) > 0$, we obtain the following simple formula for r_{ret}:

$$\boxed{(3) \qquad r_{ret} \; = \; <x-z_{ret}, u_{ret}> \; .}$$

It is worth noting that

$$\boxed{(4) \qquad x \; - \; z_{ret} \; = \; r_{ret}(u_{ret} + w) \; .}$$

We now define the contravariant retarded potential A_{ret} due to a particle of charge q as

$$\boxed{(5) \qquad A_{ret}(x) \; := \; q \, \frac{u_{ret}(x)}{r_{ret}(x)} \; .}$$

This is just (1) written in coordinate-free notation. The "advanced" potential A_{adv} is defined in the same way using the forward light cone from x in place of the backward cone, and (2), (3), and (4) are appropriately modified. The potential (5) for a particle in uniform motion (i.e. $u(\tau)$ is constant) is known as the *Liénard-Wiechert* retarded potential.

The retarded field F^{ret} is, of course, defined as

$$(6) \qquad F^{ret} \; := \; dA^{ret} \; ,$$

where A^{ret} is the covariant form of (5). Notice that the definition of $A^{ret}(x)$ uses only the tangent $u(\tau_{ret})$ to the worldline at the retarded point, but to compute $F^{ret}(x)$, derivatives of $u(\tau_{ret})$ are also needed; in particular, the acceleration $D_u u$ is needed, so the retarded field produced by a particle will depend on its acceleration.

Before calculating F^{ret}, we review several simple facts from differential geometry. Let $\{e_i\}$ be a basis for **M**, and $\{x_i\}$ the associated linear coordinatization of **M**, so that $\{dx^i\}$ is the dual basis to $\{e_i\}$. Recall from 2.5(14) that the differential $d\alpha$ of any p-form α may be computed as

$$(7) \qquad d\alpha \; = \; \sum_{i=0}^{3} dx^i \wedge D_{e_i} \alpha \; .$$

Also recall from 3.2(6) the definition of the covariant derivative $D_u v$ of a vector field v defined along a curve $z(\tau)$ with tangent $u(\tau) := z'(\tau)$. In the present context, this is just componentwise differentiation: if $v = v^i e_i$, then

$$D_u v \; = \; \frac{dv^i}{d\tau} e_i \; .$$

There is also a natural abstract definition of the derivative $D_u \alpha$ of a 1-form α defined along $z(\cdot)$, which we need not give because it reduces to componentwise differentiation in the present setting, and this is the form in which we shall use it:

$$(8) \qquad D_u(\sum f_i dx^i) \; := \; \sum \frac{df_i}{d\tau} dx^i \; .$$

Now consider the following situation. Given such a 1-form α defined along $z(\cdot)$, and a real-valued function ϕ on **M**, define a 1-form $\tilde{\alpha}$ on all of **M** by

(9) $\quad \tilde{\alpha}_x := \alpha_{z(\phi(x))} \quad$ for all $x \in M$.

That is, if we write $\alpha_{z(\tau)} = \sum \alpha_i(\tau) dx^i$, then the components of $\tilde{\alpha}$ are the components of α composed with ϕ:

$$\tilde{\alpha}_x = \sum (\alpha_i \circ \phi)(x) dx^i = \sum \alpha_i(\phi(x)) dx^i .$$

It follows immediately from (7), (8), and the chain rule that

$$d\tilde{\alpha}_x = \sum_j dx^j \wedge D_{e_j} \tilde{\alpha}_x = \sum_j dx^j \wedge \sum_i \frac{\partial(\alpha_i \circ \phi)}{\partial x_j}(x) dx^i$$

$$= \sum_{i,j} dx^j \wedge \frac{d\alpha_i}{d\tau} \frac{\partial \phi}{\partial x_j} dx_i$$

$$= (\sum_j \frac{\partial \phi}{\partial x_j} dx^j) \wedge \sum_i u(\alpha_i) dx_i ,$$

which may be summarized as

(10) $\quad (d\tilde{\alpha})_x = (d\phi)_x \wedge D_{u_{z(\phi(x))}} \alpha .$

With these tools in hand, we begin the calculation of F^{ret}:

(11) $\quad F^{ret} := dA^{ret} = d(q \frac{u_{ret}*}{r_{ret}})$

$$= q[d(\frac{1}{r_{ret}}) \wedge u_{ret}* + \frac{1}{r_{ret}} d(u_{ret}*)]$$

$$= q[-\frac{1}{r_{ret}^2} dr_{ret} \wedge u_{ret}* + \frac{1}{r_{ret}} d(u_{ret}*)] .$$

To continue, we shall need to compute $d(r_{ret})$ and $d(u_{ret}*)$.

From (10), with $u*$ as α and $\tau_{ret}(x)$ as $\phi(x)$,

(12) $\quad d(u_{ret}*) = d\tau_{ret} \wedge D_{u_{ret}}(u*) = d\tau_{ret} \wedge (D_{u_{ret}} u)*$

$$= d\tau_{ret} \wedge a_{ret}* ,$$

where the second equality was established in 2.7(2), and

$$a_{ret} := D_{u_{ret}} u$$

denotes the four-acceleration of the particle at the retarded time τ_{ret}. Now we need $d\tau_{ret}$, which may be obtained by taking the differential of

$$<x - z_{ret}, x - z_{ret}> = 0 ,$$

using the easily established relations

(i) $\quad d(<x, x>) = \frac{\partial(x^j g_{jk} x^k)}{\partial x_i} dx^i = 2x_i dx^i = 2x* ,$

(ii) $$d(\langle x, z_{ret}\rangle) = \frac{\partial(x^j g_{jk} z_{ret}^k)}{\partial x_i} dx^i$$

$$= z_{ret}{}^* + x^j g_{jk} dz_{ret}^k$$

$$= z_{ret}{}^* + x^j g_{jk} \frac{dz^k}{d\tau}(\tau_{ret}) d\tau_{ret}$$

$$= z_{ret}{}^* + \langle x, u_{ret}\rangle d\tau_{ret} \quad,$$

and similarly,

(iii) $$d(\langle z_{ret}, z_{ret}\rangle) = 2\langle z_{ret}, u_{ret}\rangle d\tau_{ret} \;.$$

This yields

$$d\tau_{ret} = \frac{x^* - z_{ret}{}^*}{\langle x - z_{ret}, u_{ret}\rangle} \quad,$$

and combining this with (4) and (3) gives the result we shall need:

(13) $$d\tau_{ret} = u_{ret}{}^* + w^* \;.$$

The above index calculations are routine to check, but it is also worth saying what is happening in coordinate-free language. If $\tau \longmapsto y(\tau) \in \mathbf{M}$ is any vector function of proper time, we may define a corresponding "retarded" vector field $x \longmapsto y_x$ on $\mathbf{M}$ by sending a light cone backwards from x and observing the value of $y(\tau)$ at the retarded time $\tau = \tau_{ret}(x)$:

$$y_x := y(\tau_{ret}(x)) \;, \quad \text{for all } x \in \mathbf{M} .$$

The prototype, of course, is the way we obtained the vector fields u_{ret} and a_{ret} from the corresponding vector functions on the worldline. Then, from the chain rule, for any vector $v \in \mathbf{M}_x$,

$$D_v y = \frac{dy}{d\tau}(\tau_{ret}(x)) \cdot v(\tau_{ret}) = \frac{dy}{d\tau}(\tau_{ret}(x)) \cdot d\tau_{ret}(v) \;.$$

If $y, \bar{y}$ are any vector fields, then for any vector $v \in \mathbf{M}_x$,

(iv) $$(d\langle y, \bar{y}\rangle)(v) := v(\langle y, \bar{y}\rangle) = \langle D_v y, \bar{y}\rangle + \langle y, D_v \bar{y}\rangle \;.$$

If the vector fields y and $\bar{y}$ are derived from vector functions on the worldline in the manner just described, then

$$(d\langle y, \bar{y}\rangle)(v) = [\langle \frac{dy}{d\tau}, \bar{y}\rangle + \langle y, \frac{d\bar{y}}{d\tau}\rangle] d\tau_{ret}(v) \quad,$$

which may be more compactly written as

(v) $$d\langle y, \bar{y}\rangle = [\langle \frac{dy}{d\tau}, \bar{y}\rangle + \langle y, \frac{d\bar{y}}{d\tau}\rangle] d\tau_{ret} \;.$$

4.2 Retarded fields

Formulas (i), (ii), and (iii) are simple variants of (iv) and (v).

Similarly, dr_{ret} is obtained by taking the differential of (3):

$$d(r_{ret}) = d<x-z_{ret}, u_{ret}>$$
$$= u_{ret}{}^* + <x, a_{ret}>d\tau_{ret} - <u_{ret}, u_{ret}>d\tau_{ret} - <z_{ret}, a_{ret}>d\tau_{ret}$$
$$= u_{ret}{}^* + (<x-z_{ret}, a_{ret}> - 1)d\tau_{ret} ,$$

and from (2), (13), and the relation

$$<u, a> = <u, \frac{du}{d\tau}> = \frac{1}{2}\frac{d<u, u>}{d\tau} = 0 ,$$

we obtain

(14)
$$dr_{ret} = u_{ret}{}^* + (r_{ret}<w, a_{ret}> - 1)d\tau_{ret}$$
$$= r_{ret}<w, a_{ret}>u_{ret}{}^* + (r_{ret}<w, a_{ret}> - 1)w^* .$$

Finally, combining (11), (12), and (14) yields the final expression for F^{ret}, in which the subscripts "*ret*" on the right side have been dropped for brevity, it being understood that *all quantities on the right are evaluated at the retarded point*:

(15)
$$F^{ret} = q\left[\frac{1}{r^2} w^* \wedge u^* + \frac{1}{r}\left(u^* \wedge a^* + w^* \wedge a^* - <w,a> w^* \wedge u^*\right)\right].$$

A variant of this formula which will be convenient for later applications is obtained by writing

$$a = -<a,w>w + a_\perp$$

with $a_\perp$ orthogonal to w (and u):

(16)
$$F^{ret} = q\left[\frac{1}{r^2} w^* \wedge u^* + \frac{1}{r}\left(u^* \wedge a_\perp{}^* + w^* \wedge a_\perp{}^*\right)\right] .$$

The terms in parentheses can of course be combined to $(u+w)^* \wedge a_\perp{}^*$, but for certain physical interpretations one wants to keep them separate.

For reference, we also include the corresponding formula for the advanced field F^{adv}, in which *all quantities on the right are evaluated at the advanced point* $z(\tau_{adv})$ at which the worldline intersects the *future* light cone from x:

$$(17) \qquad F^{adv} = q\left[\frac{1}{r^2} w^* \wedge u^* + \frac{1}{r}\left(u^* \wedge a_\perp^* - w^* \wedge a_\perp^*\right)\right],$$

where the relation between x, z_{adv}, $r = r_{adv}$, $u = u_{adv}$, and $w = w_{adv}$ is now defined as

$$x - z_{adv} = r_{adv}(w_{adv} - u_{adv}) \qquad \text{with} \qquad r_{adv} > 0.$$

(The physical meaning of r_{adv} and w_{adv} are the same as before: w_{adv} is the spatial unit vector pointing from the advanced point z_{adv} to the field point x in the rest frame at the advanced point.)

Since the quantity $F^{ret} - F^{adv}$ turns out to be of interest, we emphasize that in subtracting (16) and (17) one must keep in mind that the evaluation points for u and a are different in (16) and (17), and that terms of the difference which might appear to cancel in the above abbreviated notation do not necessarily cancel.

The physical interpretation of (16) is fairly simple in a Lorentz frame in which the particle is at rest at the origin at the retarded time. The first term, $(q/r^2) w^* \wedge u^*$, represents a spherically symmetric, radially outward electric field

$$(18) \qquad \vec{E}_{coul} := \frac{q\vec{w}}{r^2},$$

where $\vec{w}$ is is the unit vector in the direction of the field point (i.e. $w = (0, \vec{w})$ in the rest frame); this is just the familiar Coulomb field. The other term, which falls off like $1/r$ as $r \to \infty$, is called the *radiation field*. The first term of the radiation field, $(q/r) u^* \wedge a_\perp$ represents another electric field

$$(19) \qquad \vec{E}_{rad} := -\frac{q}{r}\vec{a}_\perp,$$

where $\vec{a}_\perp$ is the space component of $a_\perp$. The remaining term, $q/r\, w^* \wedge a_\perp^*$ corresponds to a magnetic field

$$(20) \qquad \vec{B}_{rad} := \frac{q}{r} \vec{a}_\perp \times \vec{w}.$$

Note that the electric and magnetic vectors of the radiation field in the given rest frame are always orthogonal and of equal magnitude $|q/r\, \vec{a}_\perp|$.

The total electric field in the rest frame is

$$\vec{E} = \vec{E}_{coul} + \vec{E}_{rad},$$

and the total magnetic field $\vec{B}$ is $\vec{B}_{rad}$. Recall from Section 3.8 that the Poynting vector

$$\vec{P} := \frac{\vec{E} \times \vec{B}}{4\pi}$$

describes the flux of field energy in the sense that for any unit vector $\vec{n}$, $\vec{P} \cdot \vec{n}$ is the rate at which field energy is flowing across a unit area normal to $\vec{n}$. The Poynting vector decomposes into a part

$$(21) \qquad \frac{\vec{E}_{coul} \times \vec{B}_{rad}}{4\pi}$$

which falls off like r^{-3} as $r \to \infty$ and a term

(22) $$\vec{P}_{rad} := \frac{\vec{E}_{rad} \times \vec{B}_{rad}}{4\pi} = \frac{q^2}{4\pi r^2} (\vec{a}_\perp \cdot \vec{a}_\perp) \vec{w} \ .$$

which is only $O(r^{-2})$. The $O(r^{-3})$ term (21) is orthogonal to $\vec{w}$, which implies that it makes no contribution to the radially outward flow of field energy, relative to the rest frame. The $O(r^{-2})$ term P_{rad} is in the direction of $\vec{w}$ and accounts for the total outflow of energy through a sphere S of radius r (relative to the rest frame at the retarded time) with the particle at the center at the retarded time:

(23) rate of energy outflow through a "retarded sphere" S of radius r

$$= \int_S \vec{P} \cdot d\vec{S}$$

$$= \int_S \vec{P}_{rad} \cdot d\vec{S} = q^2 \int_S \vec{a}_\perp \cdot \vec{a}_\perp \frac{dS}{4\pi r^2} = \frac{2}{3} q^2 \vec{a} \cdot \vec{a} \ .$$

This equation is usually interpreted as stating that the particle is radiating energy at a proper-time rate of

(24) rest-frame rate of energy radiation $= \frac{2}{3} q^2 \vec{a} \cdot \vec{a} = -\frac{2}{3} q^2 <a, a>$.

Equation (24) is known as the *Larmor radiation law,* after its discoverer.

It should be noted that (23) is relativistically exact only relative to the rest frame at the retarded time. If one has a particle moving relative to a fixed "laboratory" frame and wants to calculate the energy radiation rate through a fixed surface in the laboratory frame, the problem is considerably more complicated. A common approximation used for "slowly moving" particles in practical work is to simply ignore the distinction between the laboratory frame and the retarded frames. This works well most of the time but can be dangerous if the worker does not think carefully through the validity of each approximation. It is often quite easy to write down plausible-looking approximations which lead to erroneous results.

4.3. Radiation reaction and the Lorentz-Dirac equation.

Now we are ready to examine the question of energy-momentum balance in point electrodynamics. Recall that if T denotes the energy-momentum tensor 3.8(17) for the electromagnetic field (which we previously called T_{elm}), then, assuming Maxwell's equations, the law of conservation of energy-momentum

$$\delta(T + \frac{m}{q}\rho u \otimes u^*) = 0$$

for a charged dust is equivalent to the Lorentz equation

$$m\frac{du}{d\tau} = q\hat{F}(u) \ .$$

We shall take over this energy-momentum tensor from the charged dust model and use it with the retarded fields to formulate a law of conservation of energy-momentum for point particles. Though it is natural to expect to be able to recover the Lorentz equation from this law, it turns out that a very different equation of motion, known as the *Lorentz-Dirac* equation, emerges. The procedure which leads to this strange result is conceptually simple, though the actual calculations are lengthy and nontrivial. We shall explain the ideas in this section and carry out the difficult part of the calculations in the next.

Consider a single charged particle with worldline $\tau \longmapsto z(\tau)$ parametrized by proper time τ. Let F_{ret} denote the retarded field 4.2(16) associated with this worldline, and let F_{ext} ("ext" for "external") denote the "rest" of the field. If we like, we may think of F_{ext} as the sum of the retarded fields produced by all the other charged particles in the universe, though it will not matter where F_{ext} comes from. The total field F is then

(1) $\qquad F := F_{ext} + F_{ret} \ .$

We shall study the motion of the given particle as related to the total field F. To this end we surround the particle with a closed two-dimensional spatial surface whose geometry is specified, such as the surface of a sphere of fixed radius r_0. Of course, to make this meaningful we must also state the Lorentz frame with respect to which these geometrical specifications are defined. Later we shall do this precisely, but the details are not important at the moment. Just think of any closed surface with the particle inside.

When we let this two-dimensional spatial surface "move" through time with the particle, it generates a three-dimensional hypersurface which surrounds the worldline. This hypersurface is depicted in Figure 2, in which one space dimension is of course suppressed, as a tube surrounding the worldline, and we shall also call the three-dimensional version a "tube". If we "cap" the tube by cutting it with two spacelike hypersurfaces, then the part of the tube between the caps together with the caps themselves is expected to be the three-dimensional boundary of a four-dimensional manifold. (Strictly speaking, we should round off the "edges" where the hyperplanes cut the tube, but we ignore such details.) Now let us cap the tube by choosing a proper time τ together with a slightly later proper time $\tau+\Delta\tau$ and using for the caps the intersections of the region inside the tube with the two spacelike hyperplanes orthogonal to the worldline at the points $z(\tau)$ and $z(\tau+\Delta\tau)$. We assume that r_0 and $\Delta\tau$ are small enough so that no other particle's worldline crosses the tube or caps. The 3-manifold consisting of the original tube together with its caps will also be called a "tube", for lack of a more descriptive term, and the old tube is the "side" of the new tube.

The integral of $\perp T$ over the cap corresponding to proper time $\tau+\Delta\tau$ gives the total energy-momentum due to the electromagnetic field F inside the original sphere at proper time $\tau+\Delta\tau$ (relative to the rest frame of the particle at that proper time), and similarly for the integral over the cap at τ, except that the sign is reversed owing to the relative

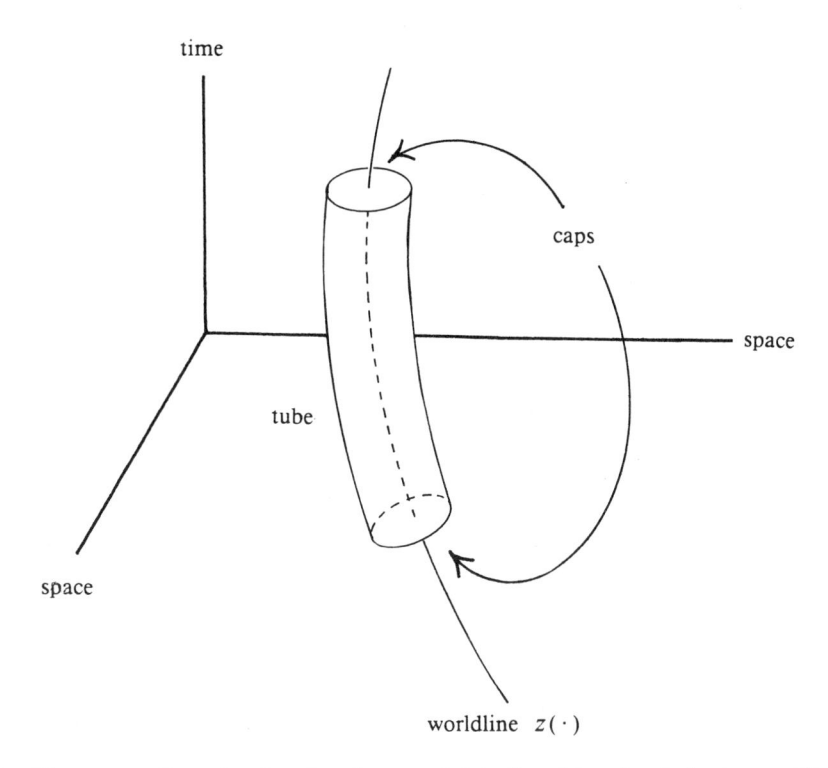

Figure 4-2. When a two-dimensional surface (represented in the picture by circles in spacelike planes) surrounding a particle evolves through time it generates a three-dimensional "tube" surrounding the worldline.

orientations of the caps. The integral of $\perp T$ over the "side" of the tube may be thought of as the total energy-momentum flowing out of the sphere between times τ and $\tau+\Delta\tau$. The most naive formulation of the principle of conservation of energy-momentum requires that the integral of $\perp T$ over the entire cut tube (side and caps) plus the change in energy-momentum of the particle between proper times τ and $\tau+\Delta\tau$ must vanish. Given the worldline and F_{ext}, we can compute F_{ret}, T, and the integral of $\perp T$ over the tube, and the above requirement of conservation of energy-momentum will give a relation between the change in energy-momentum of the particle and F_{ext}, which in the limit $r_0 \to 0$, $\Delta\tau \to 0$, is expected to yield an equation of motion for the particle in terms of F_{ext}.

In actually carrying out this program, several difficulties are encountered. The first is that the integral of $\perp T$ over the caps diverges. That is, the electromagnetic energy-momentum inside the sphere at any given time is infinite. This is clearly illustrated by the case of a stationary particle with $F_{ext} = 0$. In this case, 4.2(16) gives F_{ret} as the familiar time-independent Coulomb field, whose expression in three-dimensional notation is

$$\vec{E} = q \frac{\vec{w}}{r^2}$$

$$\vec{B} = 0,$$

where we are working in a Lorentz frame in which the particle is stationary at the origin, r is the distance of the field point from the origin, and $\vec{w}$ is the spatial unit vector in the direction of the field point. The component $T^0{}_0$ of T is then

$$\frac{|\vec{E}|^2 + |\vec{B}|^2}{8\pi} = \frac{q^2}{8\pi}\frac{1}{r^4},$$

so the integral of $\perp T$ over either of the caps has energy component

(2) $$\lim_{\epsilon \to 0}\int_\epsilon^{r_0}\frac{q^2}{8\pi}\frac{1}{s^4}4\pi s^2 ds = \lim_{\epsilon \to 0}\frac{q^2}{2}\left[\frac{1}{\epsilon} - \frac{1}{r_0}\right],$$

which diverges like $1/\epsilon$. (The other components of the integral vanish because the Poynting vector $\vec{E} \times \vec{B}$ vanishes.) However, in this special case one can argue that even though the integral over each cap is infinite, it is the "same" infinity for each, so it is reasonable to regard the contributions over the two caps as cancelling.

In general, it is not clear what to do about the infinite integrals over the caps, but there is one important case in which there is a more or less plausible way to proceed. Consider a so-called "scattered particle", which comes in from spatial infinity on a straight line as a "free particle" (i.e. asymptotically constant velocity), interacts for a finite time with a spatially localized electromagnetic field (perhaps produced by other nearby particles), and finally sails off asymptotic to a usually different straight line. Cap the tube with hyperplanes as before, but this time in the remote past and remote future. Let $u(\tau)$ denote the particle's four-velocity at proper time τ, and let

$$u_{in} := \lim_{\tau \to -\infty} u(\tau) \quad \text{and} \quad u_{out} := \lim_{\tau \to \infty} u(\tau).$$

Temporarily work in a Lorentz frame in which the particle is at rest in the infinite past, so that in this frame u_{in} has components $(1, 0, 0, 0)$, and formally take the past cap at $\tau = -\infty$ to avoid having to write $\lim_{\tau \to -\infty}$ repeatedly. Then $\int_{pastcap} \perp T$ has 0-component formally given by the infinite expression (2) and other components zero. When written in coordinate-free language, this says that formally, in the limit as $\tau \to -\infty$,

(3) $$\int_{pastcap} \perp T = [\lim_{\epsilon \to 0}\frac{1}{2}\int_\epsilon^{r_0} s^{-2}ds\,]u_{in}.$$

Similarly, in the formal limit as $\tau \to +\infty$,

(4) $$\int_{futurecap} \perp T = [\lim_{\epsilon \to 0}\frac{1}{2}\int_\epsilon^{r_0} s^{-2}ds\,]u_{out}.$$

The difference of the left sides of (4) and (3), ignoring for the moment the embarrassment that both are infinite, is the difference of the energy-momentum associated with the electromagnetic field in the future cap and that in the past cap. The difference of the right sides looks formally like a difference

$$m_f u_{out} - m_f u_{in}$$

of particle momenta, where m_f is a mass (the "field mass").

The equation of conservation of energy-momentum should read

(5) $$\int_{side} \perp T + \int_{caps} \perp T + mu_{out} - mu_{in} = 0,$$

where m is the rest mass of the scattered particle. If we formally write

$$\int_{caps} \perp T \;=\; m_f(u_{out} - u_{in}) \;,$$

and continue to ignore the fact that the "field mass"

$$m_f \;:=\; \lim_{\epsilon \to 0} \frac{1}{2} \int_{\epsilon}^{r_0} s^{-2} ds$$

is infinite, then (5) may be rewritten as

(6) $$m_{ren} u_{out} - m_{ren} u_{in} \;=\; - \int_{side} \perp T \;,$$

with the so-called "renormalized mass" m_{ren} symbolically "defined" as

$$m_{ren} \;:=\; m_f + m \;.$$

Having gotten this far, many authors simply replace the "old" mass m with m_{ren}, considering the latter as the "experimentally observed" mass. (The question of what, if any, operational meaning is then to be attributed to the old mass is generally ignored.)

The process of getting rid of the infinite contribution of the caps by replacing m by m_{ren} is an example of a class of similar procedures commonly known as "renormalization". Before we finish the "derivation" of the Lorentz-Dirac equation, we shall need to perform yet another mass renormalization.

Obviously, mass renormalization is open to criticism, but let us accept it for the moment and see where it leads. In this case, (6) becomes the condition for conservation of energy-momentum, and so we seek an equation of motion which will guarantee (6). Let $S_{r_0}(\tau_1, \tau_2)$ denote the portion of the tube (of radius r_0) between two hyperplanes orthogonal to the worldline at $z(\tau_1)$ and $z(\tau_2)$. One simple way to obtain (6) is to assume the differential version

(7) $$m_{ren} \frac{du}{d\tau} \;=\; - \lim_{\Delta\tau \to 0} \frac{1}{\Delta\tau} \lim_{r_0 \to 0} \int_{S_{r_0}(\tau, \tau+\Delta\tau)} \perp T \;,$$

whose integral with respect to τ gives the limit as $r_0 \to 0$ of (6). (Taking the limit as the radius r_0 of the sphere goes to zero is necessary in (7) because the left side of (7) can't depend on r_0.)

Unfortunately, the right side of (7) turns out to be infinite, but the approach is saved by the fact that

$$\lim_{\Delta\tau \to 0} \frac{1}{\Delta\tau} \int_{S_{r_0}(\tau, \tau+\Delta\tau)} \perp T$$

can be written in the form

$$h(r_0) + f(r_0) \frac{du}{d\tau} \;,$$

where h and f are functions such that $\lim_{r_0 \to 0} h(r_0)$ is finite and $\lim_{r_0 \to 0} f(r_0) = \infty$. The term proportional to $\frac{du}{d\tau}$ can be disposed of by a second mass renormalization, and, assuming this, a lengthy and nontrivial computation which will be done in the next section shows (7) to be equivalent to:

$$\text{(8)} \qquad \overline{m}\frac{du}{d\tau} = q\hat{F}_{ext}(u) + \frac{2}{3}q^2 <a,a>u ,$$

where $\overline{m}$ is the doubly renormalized mass and $a := du/d\tau$ the proper acceleration. † That is, the right side of (8) is the right side of (7) with infinite terms "proportional" to $du/d\tau$ discarded. (However, this is not yet the Lorentz-Dirac equation.)

The trouble with equation (8) is that it is *impossible* when $a \neq 0$ because u is always orthogonal to both $du/d\tau$ (because $<u,u> = 1$) and $\hat{F}_{ext}(u)$ (because $\hat{F}_{ext}$ is antisymmetric). In more physical terms, equation (8) says that the change in energy-momentum of the particle per unit proper time differs from that expected from the Lorentz equation by the term $\frac{2}{3}q^2<a,a>u$. In a Lorentz frame in which the particle is instantaneously at rest, this term represents a pure mass-energy loss per unit time proportional to the square of the proper acceleration. In other words, the particle is radiating energy, which is flowing outward through the sphere, and nothing in sight can account for this energy loss.

Keep in mind, however, that the desired equation of conservation of energy-momentum is not (7), but rather its integral (6). In particular, (7) implies (6) but not conversely, so there is still hope to salvage something from (8) and a natural thing to try is to integrate (8) over all time, obtaining

$$\text{(9)} \qquad \overline{m}\, u_{out} - \overline{m}\, u_{in} = q\int_{-\infty}^{\infty} \hat{F}_{ext}(u(\tau))\, d\tau$$

$$+ \frac{2}{3}q^2 \int_{-\infty}^{\infty} <a(\tau),a(\tau)>u(\tau)\, d\tau .$$

Now

$$<a(\tau),a(\tau)> = <\frac{du}{d\tau},\frac{du}{d\tau}> = \frac{d}{d\tau}<\frac{du}{d\tau},u> - <\frac{d^2u}{d\tau^2},u> ,$$

and since $<du/d\tau,u> = 0$ because u is a unit vector, we have

$$\text{(10)} \qquad <a,a>u = -<\frac{da}{d\tau},u>u .$$

The right side is recognized as the negative of the projection of $da/d\tau$ on u, which motivates writing the silly-looking relation

$$\text{(11)} \qquad \frac{da}{d\tau} = <\frac{da}{d\tau},u>u + [\frac{da}{d\tau} - <\frac{da}{d\tau},u>u] ,$$

in which the term in parentheses on the right is orthogonal to u. Integrating (10) and (11) over all time and rearranging gives

$$\text{(12)} \qquad \int_{-\infty}^{\infty} <a(\tau),a(\tau)>u(\tau)\, d\tau = -\int_{-\infty}^{\infty} <\frac{da}{d\tau},u>u\, d\tau$$

$$= \int_{-\infty}^{\infty} \frac{da}{d\tau}\, d\tau + \int_{-\infty}^{\infty} [\frac{da}{d\tau} - <\frac{da}{d\tau},u>u]\, d\tau$$

---

† Actually, the equation (8) obtained depends on the shape of the tube chosen, but as this turns out not to affect the essentials of the argument, we defer discussion of this point to the next section.

$$= -a(\infty) + a(-\infty) + \int_{-\infty}^{\infty} [\frac{da}{d\tau} - <\frac{da}{d\tau}, u>u]\, d\tau \ .$$

For a scattered particle it is natural to assume that the acceleration vanishes in the infinite past and future, and then we have

(13) $$\int_{-\infty}^{\infty} <a,a>u\, d\tau = \int_{-\infty}^{\infty} [\frac{da}{d\tau} - <\frac{da}{d\tau}, u>u]\, d\tau \ .$$

The significance of (13) is that (9), and hence (6) (in the limit $r_0 \to 0$ and with the doubly renormalized mass $\bar{m}$ in place of m_{ren}), follows not only from (8) but, assuming that

$$a(-\infty) = a(\infty) \ ,$$

also from

(14) $$\bar{m}\frac{du}{d\tau} = q\hat{F}_{ext}(u) + \frac{2}{3}q^2[\frac{da}{d\tau} - <\frac{da}{d\tau}, u>u] \ ,$$

and (14) is just (8) with the impossible term $\frac{2}{3}q^2<a,a>u$ replaced by a term orthogonal to u. Equation (14) is essentially the Lorentz-Dirac equation, though it is usually written via (10) in the form:

(15) (Lorentz-Dirac equation)

$$\bar{m}\frac{du}{d\tau} = q\hat{F}_{ext}(u) + \frac{2}{3}q^2[\frac{d^2u}{d\tau^2} + <\frac{du}{d\tau},\frac{du}{d\tau}>u] \ ,$$

or, more briefly,

$$\bar{m}a = q\hat{F}_{ext}(u) + \frac{2}{3}q^2[\frac{da}{d\tau} + <a,a>u] \ .$$

In the rest of the text we shall usually write m in place of $\bar{m}$ in the Lorentz-Dirac equation because there is really no practical distinction between them; to the extent that either has physical meaning, both are measured in the same way.

In the above motivation of the Lorentz-Dirac equation, the only omitted step is the passage from (7) to (8). We shall save this calculation for the next section because it is quite complicated and the details are not necessary for an understanding of the fundamental issues. All we need to know at this point is that the calculation uses only the expression 4.2(16) for the retarded field and the definition 3.8(18) of the energy-momentum tensor. It does not use the Lorentz equation and therefore applies also to a charged particle acted on by nonelectromagnetic forces. For example, we can imagine attaching a tiny rocket engine to an electron and computing the energy radiated as the engine pushes it around the universe. In this case, the Lorentz equation will not hold, of course, and for simplicity we may as well take $F_{ext} = 0$. Then, asuming that the engine is only turned on for a finite time, (6) through (9) are interpreted as stating that the total energy-momentum flowing outward through an infinitesimal sphere surrounding the particle over all time is

(16) $$-\frac{2}{3}q^2 \int_{-\infty}^{\infty} <a,a>u\, d\tau \ .$$

That is, (16) is the total energy-momentum radiated by the electron, and this energy-momentum must be furnished by the rocket engine *in addition to* the energy-momentum change

$$\overline{m}u_{out} - \overline{m}u_{in}$$

of the particle itself.

A crude physical picture is that if you push an electron, it radiates, and you therefore have to push it harder to cause a given change in its motion than would be necessary for a neutral particle. The problem with this picture is that it is impossible for it to hold instantaneously for the same reason that (8) is impossible. This can even be seen on the classical level in a frame in which the particle is at rest. The classical power (energy per unit time) transmitted to a particle of velocity $\vec{v}$ by a force $\vec{F}$ is $\vec{F} \cdot \vec{v}$, and must vanish if the particle is at rest. However, if we interpret (16) as stating that energy is radiated at a rate proportional to the square of the acceleration, then the radiated power is nonzero when the particle is accelerated even if $\vec{v} = 0$.

Recall that the Larmor radiation law, which states that energy is radiated in the particle's rest frame at a rate proportional to the square of the acceleration, was obtained at the end of Section 4.2 in a plausible way independently of the above considerations. This makes the interpretation of (16) just described even more reasonable. Note that (16) is weaker than the Larmor law in that it is integrated but is stronger in that it refers to energy-momentum rather than just rest-frame energy.

Although (8) cannot hold (at least within the framework in which we have been working in which it is assumed that the rest mass of the electron is constant), (16) presumably gives the total energy-momentum which must be supplied by the rocket engine for the radiation. Assuming that the engine is only turned on for a finite time, the term

$$(17) \qquad \frac{2}{3}q^2 \frac{da}{d\tau} + <a, a>u$$

in the Lorentz-Dirac equation, called the *radiation reaction* term, has the same integral as $<a, a>u$, and since (17) is orthogonal to u as a genuine "force" should be, it is usually assumed to be the additional instantaneous force which must be supplied by the rocket for the radiation. No definitive experimental test of this assumption has yet been carried out.

The Lorentz-Dirac equation is unusual in that it is a *third-order* differential equation for the worldline $\tau \longmapsto z(\tau)$. (It cannot be viewed as merely a second-order equation for $u(\tau)$ because $z(\tau)$ enters implicitly through F_{ext}.) In particular, to specify a solution one needs not only an initial spacetime point $z(\tau_0)$ and an initial four-velocity $u(\tau_0)$, but also an initial acceleration $a(\tau_0)$. Since the motivation for the equation assumed that $a(-\infty) = 0 = a(\infty)$, it is reasonable to take as partial initial conditions $a(\infty) = 0$ and $u(\infty) = u_{out}$, where u_{out} is given. *A priori*, it might seem more natural to use $a(-\infty) = 0$ and $u(-\infty) = u_{in}$ as initial conditions, but the former is usually chosen for reasons which will appear presently.

Rewriting (15) as

$$(18) \qquad \frac{2}{3}q^2 \frac{da}{d\tau} - \overline{m}a = -q\hat{F}_{ext}(u) - \frac{2}{3}q^2 <a, a>u,$$

setting $\epsilon := \dfrac{2q^2}{3\overline{m}}$, and multiplying by the integrating factor $\dfrac{e^{-\tau/\epsilon}}{\epsilon \overline{m}}$ yields

$$(19) \qquad \frac{d(ae^{-\tau/\epsilon})}{d\tau} = -\left[\frac{3}{2q}\hat{F}_{ext}(u) + <a,a>u\right]e^{-\tau/\epsilon}.$$

An integral of (19) formally satisfying $a(\infty) = 0$ is

$$\text{(20)} \qquad a(\tau) = e^{\tau/\epsilon} \int_{\tau}^{\infty} [\frac{3}{2q}\hat{F}_{ext}(u(s)) + \langle a(s), a(s)\rangle u(s)] e^{-s/\epsilon} ds .$$

More precisely, (20) is equivalent to (18) under the subsidiary condition that

$$\text{(21)} \qquad \lim_{\tau\to\infty} a(\tau)e^{-\tau/\epsilon} = 0 .$$

The corresponding integral of (18) for the initial condition $a(-\infty) = 0$ is

$$\text{(22)} \qquad a(\tau) = -e^{\tau/\epsilon} \int_{-\infty}^{\tau} [\frac{3}{2q}\hat{F}(u(s)) + \langle a(s), a(s)\rangle u(s)] e^{-s/\epsilon} ds ,$$

but the fact that $e^{-s/\epsilon}$ blows up as $s \to -\infty$ calls into question the convergence of the integral. The subsidiary condition under which (22) is equivalent to (18) is

$$\text{(23)} \qquad \lim_{\tau\to-\infty} a(\tau)e^{-\tau/\epsilon} = 0 ,$$

which is much more stringent than (21), since (21) only requires that $a(\tau)$ not increase exponentially too fast as $\tau\to\infty$, whereas (23) requires that $a(\tau)$ actually *decrease* exponentially as $\tau\to-\infty$. Neither (21) nor (23) guarantees, *a priori*, the *other* natural initial condition (respectively, $a(-\infty) = 0$ and $a(\infty) = 0$). In fact, we shall see in Section 5.5 that there are physically realistic situations in which these conditions cannot be simultaneously realized. Since the condition

$$a(-\infty) = a(\infty)$$

was a hypothesis for the replacement of (8) by the Lorentz-Dirac equation (15), as well as for the justification of the mass renormalization associated with the caps (which requires that both vanish), this is not a desirable state of affairs.

Most authors assume that (20) is the physically correct integral of (18) without mentioning (22), presumably for reasons similar to those cited above. However, (20) brings new problems. Many authors interpret (20) as violating "causality" because in order to compute the acceleration $a(\tau)$ at a particular proper time τ via (20), we need to know the external field F_{ext} at proper times s with $s > \tau$; i.e. at *later* proper times. (There is no such difficulty, however, with (22).) This is sometimes picturesquely described by saying that applying an external field $F_{ext}(s)$ at a time $s > \tau$ sends a signal backwards in time to the particle at the earlier time τ. The particle can start accelerating before the external field is applied! This "phenomenon" (which nobody seems to believe can actually occur in any physically observable way) is known as *preacceleration*.

Many authors try to resolve the problem in the following *ad hoc* manner. Due to the strongly damping factor $e^{-s/\epsilon}$ in the integral, violations of causality should be observed only over time intervals roughly on the order of magnitude of ϵ. However, ϵ turns out to be exceedingly small for real particles (on the order of 10^{-23} sec. for the electron), and in fact is so small that times of this size cannot be observed within the accepted domain of validity of classical physics. In this way the whole problem is pushed into the domain of quantum electrodynamics (whose logical foundations, incidentally, are far more murky than those of the classical theory).

The problem has not been solved there in any lucid and generally accepted way either. In fact, it is not even clear that this particular problem exists in that framework except insofar as classical electrodynamics is supposed to be recoverable as a limit of the quantum theory. Since quantum electrodynamics is troubled by many divergences and difficulties of its own, there is a school of thought (cf. [Dirac, 1951]) that one should try to resolve the difficulties on the classical level with the hope that a quantization of a better classical theory might also clean up the quantum theory. Others feel that the heart of the matter lies on the quantum side and that the idealizations of classical point electrodynamics may simply be unattainable as a limit of quantum theory, just as the Heisenberg uncertainty principle tells us that the idealization of a particle having simultaneously precisely defined position and velocity cannot be realized.

In the discussion leading up to the Lorentz-Dirac equation, we worked within the framework of a single particle interacting with a given external field. However, a complete physical theory should also be able to handle the problem of predicting the motion of a finite collection of charged particles in terms of appropriate initial data. It is not clear that the theory described so far can do this, and this is a serious drawback.

For simplicity, let us consider the problem of predicting the motion of two electrons which come in from spatial infinity with four-velocities u_1, u_2 asymptotic in the infinite past to given four-velocities u_1^{in}, u_2^{in}, respectively. To simplify even further, suppose that the motion is in a straight line with

$$u_1^{in} = \gamma(\vec{v})(1, \vec{v}) \quad \text{and} \quad u_2^{in} = \gamma(\vec{v})(1, -\vec{v}) ,$$

so that the incoming asymptotic velocities are equal and opposite. Further suppose that the only electromagnetic fields present are the retarded fields due to the particles:

$$F = F_1^{ret} + F_2^{ret} ,$$

where F_i^{ret} is the retarded field due to particle i. Physical intuition suggests that u_1 and u_2 should become asymptotic to $\gamma(\vec{v}_{out})(1, \vec{v}_{out})$ and $\gamma(\vec{v}_{out})(1, -\vec{v}_{out})$, respectively, in the infinite future for some $\vec{v}_{out}$. A complete physical theory should enable us to determine, at least to an arbitrary approximation, the worldline of each particle. Short of this, one would like to be able to compute $\vec{v}_{out}$ in terms of $\vec{v}_{in}$. (Presumably, $\vec{v}_{out}$ should be less than $\vec{v}_{in}$ due to loss of energy by radiation. However, we cannot conclude this rigorously from the above derivation of the Lorentz-Dirac equation due to the infinite mass renormalization and the possibility that the asymptotic conditions $a(-\infty) = 0 = a(\infty)$ may be unrealizable.) Many interesting things are known about this special case (which is surprisingly difficult) [Eliezer][Baylis and Huschilt][Hsing and Driver], but natural questions still remain, and only fragmentary knowledge exists concerning existence and behavior of solutions for more general motion.

The way in which the theory surrounding the Lorentz-Dirac equation applies to this problem can be described as follows. Given two putative worldlines $z_1(\cdot)$ and $z_2(\cdot)$ for the two electrons, we can compute the associated retarded fields F_1^{ret} and F_2^{ret}. Taking F_2^{ret} as the external field in the Lorentz-Dirac equation for electron 1 and F_1^{ret} for electron 2, we can check whether the two worldlines satisfy the respective Lorentz-Dirac equations. It is easy to see that two completely arbitrary curves $z_1(\cdot)$, $z_2(\cdot)$ will not necessarily satisfy these equations, so the theory does provide a constraint on the worldlines. However, it is not clear that these equations uniquely specify the two worldlines given the incoming four-velocities nor that there exists a pair of worldlines which satisfy the equations for arbitrary asymptotic initial conditions.

An approximation scheme which is sometimes suggested is to first compute the motion using the classical Coulomb repulsion only for the force. Using this first approximation to the motion, compute the radiation fields from 4.2(16), and then compute a second approximation to the worldlines by adding these radiation fields to the force. Iterating the process produces a succession of worldlines which might plausibly converge to a solution of the problem just described, but there is no proof.

If we pose the same symmetric one-dimensional problem for opposite charges, some of the solutions are startling. For instance, if $\vec{v}_1^{in} = 0 = \vec{v}_2^{in}$, then the particles never collide, and they end up traveling in opposite directions with velocities asymptotic to that of light (i.e. $|\vec{v}_1^{out}| = 1 = |\vec{v}_2^{out}|$), and with proper acceleration bounded away from 0. (This is true for all initial accelerations.) As mentioned above, this does not contradict the derivation of the Lorentz-Dirac equation, even in spirit, because the asymptotic condition $a(\infty) = 0$ is not realizable in this situation. Proofs of the these facts, the main ideas of which were discovered by [Eliezer, 1943], are presented in Section 5.5.

This example also shows that the Lorentz-Dirac equation does not necessarily guarantee conservation of energy in any practical sense. (Since conservation of energy-momentum was the very motivation for the equation, this is certainly disturbing!) In fact, the equation predicts that we can, in principle, extract infinite kinetic energy from a pair of oppositely charged particles of equal mass by releasing them at rest in some inertial frame and waiting until they shoot off in opposite directions with sufficiently high velocities. (Presumably, no one believes that this can actually be done; some argue that the equation is basically wrong and others that quantum effects intervene when the particles get sufficiently close.)

4.4 Calculation of the energy-momentum radiated by a point particle.

In this section we show the equivalence (within the geometrical framework to be described and modulo mass renormalization) of equations 4.3(7) and 4.3(8) of the previous section, thereby completing the chain of reasoning leading to the Lorentz-Dirac equation. We shall do this by computing the limit

(1) $$\lim_{r \to 0} \int_{S_r(\tau_1, \tau_2)} \bot T$$

of the integral of the Hodge dual of the energy-momentum tensor T corresponding to the retarded field F_{ret} of a point charge plus an external field F_{ext} over a "Bhabha tube" $S_r(\tau_1, \tau_2)$ of "radius" r surrounding the world line of a charged particle between proper times τ_1 and τ_2. In three-dimensional language, the integral in (1) represents the total energy-momentum radiated through a closed two-dimensional surface surrounding the particle, which may be roughly thought of as a sphere of radius r. The precise definition of the tube will be given later, and then it will be apparent in exactly what sense it is generated by a sphere of radius r.

Before embarking on this calculation, we review a few simple ideas from the theory of integration of forms. Let

$$\psi : M \longrightarrow N$$

be a C^∞ mapping from a manifold M into a manifold N. Then at any point p of M, ψ induces a natural linear transformation

$$\psi_p'(\cdot) : M_p \longrightarrow N_{\psi(p)} ,$$

from the tangent space M_p to the tangent space $N_{\psi(p)}$ called the *derivative* of ψ at p, and defined by:

(2) $\qquad \psi_p'(v)(f) := v(f \circ \psi) \qquad$ for all $v \in M_p$ and all $f \in C^\infty(N)$.

The definition (2) is algebraically simple, but it is more enlightening to geometrically picture the action of $\psi_p'(\cdot)$ as follows. Given $v \in M_p$, choose a curve $s \longmapsto c(s)$ through p (say $c(0) = p$), such that $c'(0) = v$ (cf. Section 2.4). Then ψ maps the curve $s \longmapsto c(s)$ into the curve $s \longmapsto (\psi \circ c)(s) = \psi(c(s))$ in N, and $\psi_p'(v)$ is just the tangent vector $(\psi \circ c)'(0)$ to the image curve $\psi \circ c$. Another useful observation is that in the special case in which M and N are vector spaces and in which $v = \partial/\partial x_i$ is partial differentiation with respect to a coordinate, we have

$$\psi_p'\left(\frac{\partial}{\partial x_i}\right) = \frac{\partial \psi}{\partial x_i}(p)$$

$$:= \lim_{h \to 0} \frac{\psi(p^1, \cdots, p^i + h, \cdots, p^n) - \psi(p^1, \cdots, p^i, \cdots, p^n)}{h}.$$

Sometimes for brevity we denote ψ_p' as simply ψ'.

Recall that any linear map $T: V \longrightarrow W$ between vector spaces V and W induces a natural adjoint linear transformation

$$T^\dagger: W^* \longrightarrow V^*$$

from the dual space W^* of W to the dual space V^* of V defined by

$$(T^\dagger(\alpha))(v) := \alpha(Tv) \qquad \text{for all } \alpha \in W^* \text{ and } v \in V.$$

In particular, $\psi_p'^\dagger$ is a linear map from the space $N_{\psi(p)}^*$ of linear forms on $N_{\psi(p)}$ to the space M_p^* of linear forms on M_p. For brevity we shall drop the prime and denote $\psi_p'^\dagger$ as $\psi_p^\dagger$, or simply $\psi^\dagger$ when we don't need to specify p. ‡ For any linear form $\alpha \in N_{\psi(p)}^*$, its image $(\psi_p^\dagger)(\alpha)$ under the adjoint of ψ_p' is called the *pullback* of α. Similarly, any q-covariant tensor $\alpha \in (N_{\psi(p)})_q^0$, $q \geq 1$, can be "pulled back" to the q-covariant tensor $\psi_p^\dagger(\alpha) \in (M_p)_q^0$ defined by

$$\psi_p^\dagger(\alpha)(v_1, \cdots, v_q) := \alpha(\psi_p'(v_1), \cdots, \psi_p'(v_q)) \qquad \text{for all } v_1, \cdots, v_q \in M_p.$$

Note that $\psi_p^\dagger$ commutes with the tensor and wedge products, e.g.

$$\psi_p^\dagger(\alpha \wedge \beta) = \psi_p^\dagger(\alpha) \wedge \psi_p^\dagger(\beta).$$

If $p \longmapsto \alpha_p$ is a form *field* on N, then the map $p \longmapsto \psi_p^\dagger(\alpha)$ is a form field on M, which we naturally abbreviate as $\psi^\dagger(\alpha)$. Thus,

$$\psi^\dagger(\alpha)_p := \psi_p^\dagger(\alpha).$$

It is straightforward to check that if $p \in M$ and $\{y^i\}_{i=1}^n$ is a local coordinatization for N at $\psi(p)$ (so that the $y^i(\cdot)$ are real-valued functions on N), then $\psi^\dagger$ sends the local form fields dy^i to $d(y^i \circ \psi)$:

$$\psi^\dagger(dy^i) = d(y^i \circ \psi).$$

Moreover, for any $f \in C^\infty(N)$,

$$\psi^\dagger(f dy^i) = (f \circ \psi) d(y^i \circ \psi),$$

and so if ψ is a diffeomorphism (i.e. C^∞-invertible) and we define a local coordinate system $\{x^i\}$ for M at p by

---

‡ The more common notation is ψ^*, but when ψ is a vector-valued function, this would clash with our use of "*" for the index-lowering operation.

4.4 Radiation calculation

$$x^i := y^i \circ \psi,$$

then

(3) $$\psi^\dagger(f dy^1 \wedge dy^2 \wedge \cdots \wedge dy^n) = (f \circ \psi) dx^1 \wedge dx^2 \wedge \cdots \wedge dx^n.$$

Now assume that M and N are connected and orientable of dimension n. Then (3) makes plausible (and is almost a proof of) the basic fact that for any n-form α with compact support,

(4) $$\int_N \alpha = \pm \int_M \psi^\dagger(\alpha),$$

where "+" holds if ψ is orientation-preserving and "−" otherwise. (To say that ψ is orientation-preserving means that at all points $p \in M$, ψ_p' sends a positively oriented basis for M_p to a positively oriented basis for N_p. Since M is connected, "all points" may be replaced by "some point".)

A more geometrical way to understand (4) intuitively is illustrated in Figure 3, in which M and N are taken to have dimension 2. Recall that

$$\int_N \alpha$$

is heuristically pictured as an infinite sum of numbers obtained by applying α to vectors spanning infinitesimal "rectangles" into which N is subdivided. The application of α to the rectangle spanned by the images $\psi_p'(v), \psi_p'(w)$ of two vectors $v, w \in M_p$ is, by definition of "pullback", given by

$$\alpha(\psi_p'(v), \psi_p'(w)) = \psi_p^\dagger(\alpha)(v, w).$$

If M is thought of as subdivided into infinitesimal rectangles, then their "images" under ψ_p' will similarly subdivide N, and the geometrical meaning of (4) is evident.

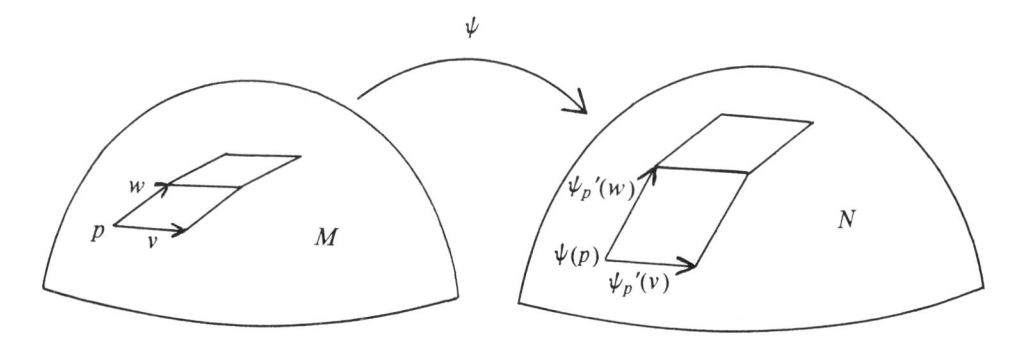

Figure 4-3.

Now we recapitulate the framework in which we shall be working. Consider a point particle with charge q and worldline $\tau \longmapsto z(\tau)$ parametrized by proper time τ. We do not assume the Lorentz equation or any other equation of motion, so $z(\cdot)$ is simply an arbitrary curve whose tangent vector

$$u(\tau) := \frac{dz}{d\tau}$$

satisfies $<u,u> = 1$. This means that our calculation of the radiated energy-momentum will apply to any charged particle regardless of the force which shapes its worldline.

As described in the preceding section, we are going to surround the worldline by a three-dimensional "tube" of "radius" r, which may be crudely pictured as generated by the time evolution of a spatial two-dimensional sphere of radius r surrounding the particle. Our first task is to precisely define the geometry of the tube.

It will be convenient to construct the tube as an immersion †

$$\phi: R^1 \times S^2 \longmapsto \mathbf{M},$$

where R^1 is the real line and S_r^{std} the "standard" two-dimensional unit sphere of radius r in R^3,

$$S_r^{std} := \{(x,y,z) \in R^3 \mid x^2 + y^2 + z^2 = r^2\},$$

and $R^1 \times S_r^{std}$ is considered as a Riemannian manifold with the obvious product structure. This will allow us to compute the integrals over the tube $\phi(R^1 \times S_r^{std})$ of the four 3-forms $\bot T^i$, $i = 0, 1, 2, 3$, as integrals over $R^1 \times S_r^{std}$ of their pullbacks $\phi^\dagger T^i$:

$$\int_{\phi(R^1 \times S_r^{std})} \bot T^i = \int_{R^1 \times S_r^{std}} \phi^\dagger(\bot T^i).$$

This equation was written in terms of the four scalar 3-forms $\bot T^i := \bot(T^i{}_\bullet)$ (cf. Section 2.8), because (4) was stated for scalar n-forms, but there is no other reason to carry these indices in the notation, and from now on we drop them. In other words, we shall be using the obvious extension of (4) to n-forms with values in a fixed vector space.

Before proceeding it may be helpful to present a close analogy in ordinary Euclidean 3-space R^3 where the geometry is easily visualized. Consider a curve $t \longmapsto \vec{x}(t)$ in R^3 which never intersects itself and will be the analog of the worldline. At each point $\vec{x}(t)$ on this curve, construct a circle $C_r(t)$ of fixed radius r with the plane of the circle perpendicular to the curve, as in Figure 4. The union of all the circles $C_r(t)$ forms a curved tube B. Suppose that we want to calculate the surface area of the tube between two circles $C_r(t_1)$ and $C_r(t_2)$.

In general, the tube B can wind around and intersect itself and so need not be an embedded submanifold. In fact, it need not even be an immersed submanifold. (To see what can go wrong, think of revolving a circle of radius r about another circle of radius R to form a torus. If $0 < r < R$, there is no problem, but think about what happens if $r = R$ or $r > R$.) However, if $\vec{x}(\cdot)$ is sufficiently regular and r is small enough, it is reasonable to expect the tube to be an immersed two-dimensional submanifold of R^3. Moreover, it should be obtainable as an embedding of a product of a line and a circle as follows. Let

$$S_r := \{(x,y) \in R^2 \mid x^2 + y^2 = r^2\}$$

be a fixed circle of radius r, considered as a one-dimensional Riemannian manifold in the obvious way. For each fixed t, the circle $C_r(t)$ is (isometrically) isomorphic, as a Riemannian manifold, to S_r via some map

$$\psi_t(\cdot): S_r \longrightarrow C_r(t),$$

which depends on t. If the curve is sufficiently regular, and r is small enough, it is

---

† i.e. a C^∞ map whose derivative is everywhere nonsingular. By the Open Mapping Theorem, such a map is locally a bijection.

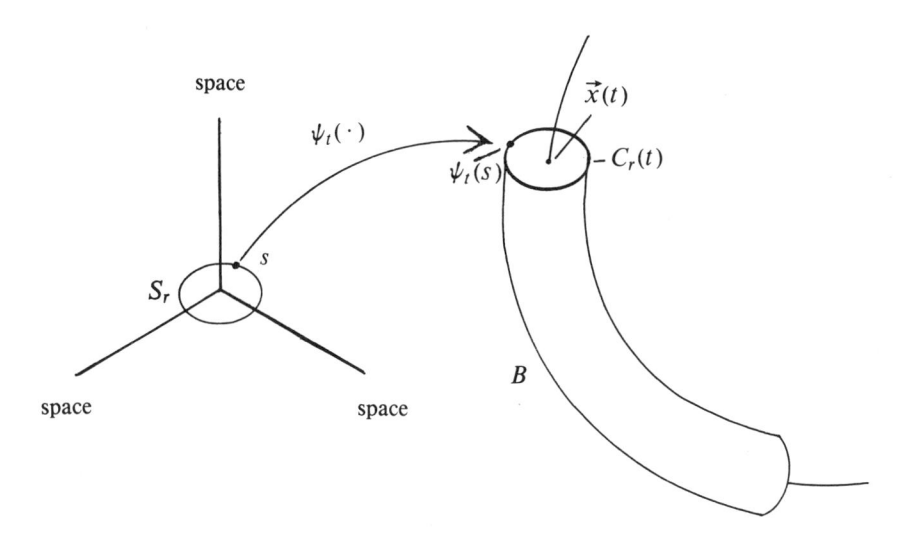

Figure 4-4. A curved tube in R^3 generated by moving a circle of radius r along a curve, keeping the center on the curve and the plane of the circle perpendicular to the curve. The map $s \longmapsto \psi_t(s) = \psi(t, s)$ identifies a fixed circle S_r of radius r with the generating circle $C_r(t)$ at $\vec{x}(t)$

reasonable to expect to be able to choose the maps ψ_t so that for sufficiently small intervals (t_1, t_2), the map

$$t, s \longmapsto \psi_t(s)$$

from $(t_1, t_2) \times S_r$ to B is C^∞-invertible. Let us take this as a hypothesis and write $\psi(t, s)$ in place of $\psi_t(s)$, so we are now assuming that there exists a C^∞-invertible map

$$\psi(\cdot, \cdot): (t_1, t_2) \times S_r \longrightarrow B$$

such that for each fixed t,

(5) $\qquad s \longmapsto \psi(t, s)$ maps S_r isometrically onto $C_r(t)$.

The usual Riemannian metric on R^3 induces in an obvious way a Riemannian structure on the tube B; the tube metric at any point is just the restriction of the Riemannian metric to the tangent space of the tube. This metric in turn, induces an "area 2-form" σ for the tube, defined up to a sign. A natural definition of the "area" of the tube in this context is:

$$\text{area of tube} := \left| \int_B \sigma \right|.$$

By (4),

(6) $\qquad \int_B \sigma = \pm \int_{(t_1, t_2) \times S_r} \psi^\dagger(\sigma).$

This reduces the problem to computing $\psi^\dagger(\sigma)$ and integrating it over the product $(t_1, t_2) \times S_r$. Given an explicit curve $\vec{x}(t)$, it is usually easy to write down an explicit ψ satisfying (5), and given ψ, the calculation of the pullback $\psi^\dagger(\alpha)$ is the same as the problem of calculating, in an obvious notation,

$$\sigma(\psi'(\frac{\partial}{\partial t}), \psi'(\frac{\partial}{\partial s})),$$

which in turn is just the problem of computing the area of the parallelogram in R^3 spanned by the vectors $\partial\psi/\partial t$ and $\partial\psi/\partial s$. Given the pullback, integration over the product is elementary (precisely because the domain of integration *is* a product). In this way the original problem of computing the area of the tube, which would usually be very messy if done as a Riemann surface integral in R^3, is reduced, modulo the problem of constructing ψ, to a series of elementary calculus calculations. Note that it is not necessary for (6) that the maps (5) be isometric, though an analog of this property will be important in a different way in the calculation below.

Returning to the relativistic case, we are ready to define the "tube" surrounding the worldline $z(\cdot)$. Let a "radius" $r > 0$ and a point $z(\tau)$ on the worldline be given. Choose a Lorentz frame in which event $z(\tau)$ has coordinates $(\tau, \vec{0})$, and in which the particle is at rest at time τ. All times in this paragraph will refer to coordinate times in this frame. At time τ, emit light pulses of infinitesimal duration in all directions, so that the light forms an expanding sphere with radius r at time $\tau+r$. The tube of "radius" r, which we name $S_r(\tau_1, \tau_2)$, will be the union of all such spheres of radius r corresponding to proper times τ between τ_1 and τ_2. More formally,

(7) $\qquad S_r(\tau_1, \tau_2) := \bigcup_{\tau_1 < \tau < \tau_2} S_r(\tau) \;,$

where $S_r(\tau)$ is the sphere of radius r defined by

(8) $\qquad S_r(\tau) := \{z(\tau) + r(u(\tau) + w) \mid \;\; <w, w> = -1 \text{ and } <u(\tau), w> = 0 \} \;.$

We shall call the $S_r(\tau)$ *retarded spheres*. They are, of course, the analogs of the circles $C_r(t)$ in the three-dimensional example. The analogy is not as exact as it could be because the Minkowski space vectors connecting the emission point $z(\tau)$ with the corresponding sphere $S_r(\tau)$ are not orthogonal to the worldline. The analogy would be more nearly exact if in place of the $S_r(\tau)$ we used spheres

$$\bar{S}_r(\tau) := \{ z(\tau) + rw \mid \;\; <w, w> = -1 \text{ and } <u, w> = 0\} \;,$$

which can be thought of as spheres moving with the particle at the center. (This was the setup for the original calculation of [Dirac, 1938].) The retarded spheres $S_r(\tau)$ *are* genuine spheres within the hyperplane $t = \tau+r$ in the emission frame which contain the particle, but the particle is not generally at the center at this time, having moved since the time of emission. The spheres $\bar{S}_r(\tau)$ are easier to visualize, than the $S_r(\tau)$, but we build the tube from the latter because it leads to more natural calculations with our method. We shall call the tube $S_r(\tau_1, \tau_2)$ a *Bhabha tube* (after the physicist H. J. Bhabha, who was apparently the first to realize how much it simplified the integrations [Bhabha]), and its analog constructed with the $\bar{S}_r(\tau_1, \tau_2)$ a *Dirac tube*. Both are pictured in Figure 5.

Note that if the particle has constant velocity between times τ and $\tau+r$, then $\bar{S}_r(\tau+r) = S_r(\tau)$. This implies that if we are dealing with a scattered particle which, by definition, has asymptotically constant velocity in the infinite past and future, then a Bhabha tube will asymptotically approximate a Dirac tube, and hence the integrals of $\bot T$ over the two tubes are expected to be the same because $\delta(T)$ vanishes in the region between them.

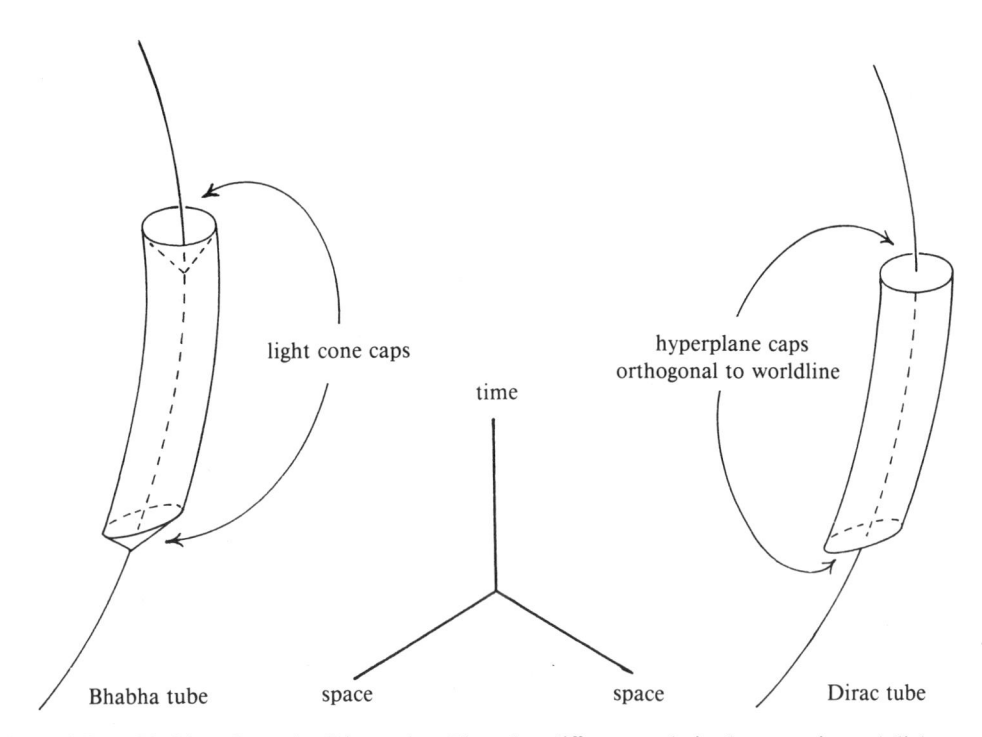

Figure 4-5. A Bhabha tube and a Dirac tube. The tubes differ not only in the caps pictured (light cones are the natural caps for a Bhabha tube and hyperplane caps for a Dirac tube), but also in the shape of their sides. In general, a Bhabha tube cannot be smoothly fitted together with a Dirac tube.

This

means that, apart from there being no fundamental reason to prefer one kind of tube over the other, for our purposes it should make no difference which is used.

As Riemannian manifolds, the $S_r(\tau)$ are all anti-isometric † to the standard sphere S_r^{std} of radius r in R^3. Fix proper times $\tau_1 < \tau_2$, and let $P_r(\tau_1, \tau_2)$ denote the product manifold

$$P_r(\tau_1, \tau_2) := (\tau_1, \tau_2) \times S_r^{std}.$$

Given $r > 0$, choose a C^∞ map

$$\phi : P_r(\tau_1, \tau_2) \longrightarrow M$$

such that for each $\tau \in R^1$, the map

$$\phi(\tau, \cdot) : S_r^{std} \longrightarrow S_r(\tau)$$

is an anti-isometry from S_r^{std} onto $S_r(\tau)$. That this can be done is a lemma which we relegate to Exercise 4.5. ‡ Of course, ϕ depends on r, but we do not indicate this in the

---
† "Anti" because the metrics differ by a sign; a bijection ϕ between semi-Riemannian manifolds is called an *anti-isometry* if its derivative satisfies $\langle \phi'(v_1), \phi'(v_2) \rangle = -\langle v_1, v_2 \rangle$ at each point.
‡ It is immediate from the definition of the $S_r(\tau)$ that for each fixed τ there is an anti-isometry

notation because for most of the discussion r will be fixed.

As in the three-dimensional example, we shall calculate the integral

$$\int_{S_r(\tau_1,\tau_2)} \alpha$$

of a 3-form $\alpha = \perp T^i$ by using

$$\int_{S_r(\tau_1,\tau_2)} \alpha = \int_{P_r(\tau_1,\tau_2)} \phi^\dagger(\alpha) .$$

It will be convenient to regard $P_r(\tau_1, \tau_2)$ as a Riemannian manifold with the product metric ‡‡ (though the Riemannian structure is not needed for the above formula).

For each $\tau, s \in P_r(\tau_1, \tau_2)$, abbreviate the tangent vector $\partial/\partial\tau$ as $v_{\tau,s}$, or simply v. In other words, $v_{\tau,s}$ is the vector which sends a C^∞ function f to

$$\frac{\partial f}{\partial \tau}(\tau, s) .$$

To compute the pullback $\phi^\dagger(\alpha_{\phi(\tau,s)})$, it is enough to compute the images under ϕ' of the triple

$$v, q_1, q_2 ,$$

where q_1, q_2 is a basis for the tangent space of S_r^{std} at s. Since the definition of ϕ guarantees that $\phi'(q_1), \phi'(q_2)$ is a basis for the tangent space of $S_r(\tau)$, and since α is alternating, only the component of $\phi'(v)$ orthogonal to $\phi'(q_1)$ and $\phi_r'(q_2)$ will be needed to compute

$$\phi^\dagger(\alpha)(v, q_1, q_2) := \alpha(\phi'(v), \phi'(q_1), \phi'(q_2)) .$$

These images can be obtained as follows. For each τ, s, we may write

(9) $\qquad \phi(\tau, s) = z(\tau) + r(u(\tau) + w(\tau, s)) ,$

where this defines $w(s, \tau)$ as a spacelike unit vector orthogonal to $u(\tau)$. We shall sometimes abbreviate $w(\tau, s)$ as w. Note that for fixed τ, the map $s \longmapsto rw(\tau, s)$ is an anti-isometry because it is a translate of the anti-isometry $s \longmapsto \phi(\tau, s)$. Recall that

$$\phi'(v) = \phi'(\frac{\partial}{\partial \tau}) = \frac{\partial \phi}{\partial \tau} .$$

We are going to compute the components of $\dfrac{\partial \phi}{\partial \tau}$ along $u(\tau)$ and $w(\tau)$, and the result will appear in terms of the particle's proper acceleration $a(\tau) := du/d\tau$. Differentiating the relation

$$\langle \phi(\tau, s) - z(\tau), w(\tau, s) \rangle = -r$$

$\phi(\tau, \cdot): S_r^{std} \longrightarrow S_r(\tau)$; the technical point which requires proof is that the $\phi(\tau, \cdot)$ can be chosen to be C^∞ in τ. One easy way to do this is to Fermi-Walker transport a basis and use it to identify local space sections with R^3. For the reader who knows what Fermi-Walker transport is, this is all one need say; the reader who does not will find the exercise of independent interest.

‡‡ i.e. the metric with respect to which the natural inclusions of (τ_1, τ_2) and S_r^{std} in $P_r(\tau_1, \tau_2)$ are isometries and such that vectors tangent to S_r^{std} are orthogonal to those tangent to (τ_1, τ_2).

4.4 Radiation calculation

with respect to τ gives

$$\left\langle \frac{\partial \phi}{\partial \tau}, w \right\rangle - \left\langle \frac{dz}{d\tau}, w \right\rangle + r\left\langle u+w, \frac{\partial w}{\partial \tau} \right\rangle = 0 ,$$

and since $u := dz/d\tau$ and w are orthogonal unit vectors and

$$\left\langle u, \frac{\partial w}{\partial \tau} \right\rangle = -\left\langle \frac{du}{d\tau}, w \right\rangle = \langle a, w \rangle ,$$

we obtain

(10) $$\left\langle \frac{\partial \phi}{\partial \tau}, w \right\rangle = r\langle a, w \rangle .$$

Similarly, differentiating

$$\langle \phi(\tau, s) - z(\tau), u(\tau) \rangle = r$$

with respect to τ yields

(11) $$\left\langle \frac{\partial \phi}{\partial \tau}, u \right\rangle = 1 - r\langle w, a \rangle .$$

From (10) and (11) we conclude that

(12) $$\phi'(v_{\tau,s}) = \frac{\partial \phi}{\partial \tau}$$
$$= (1 - r\langle a(\tau), w(\tau, s) \rangle)u(\tau) - r\langle a(\tau), w(\tau, s) \rangle w(\tau, s) +$$

vector orthogonal to $u(\tau)$ and $w(\tau, s)$.

The component orthogonal to u and w, which depends on the particular identifications of the spheres $S_r(\tau)$ with S_r^{std}, will drop out at the end of the calculation.

Also observe that ϕ' sends vectors tangent to S_r^{std} to vectors orthogonal to u and w. To see this, note that if $\eta \longmapsto \tau, s(\eta)$ is any curve in $\{\tau\} \times S_r^{std}$, then its image under ϕ is the curve

(13) $$\eta \longmapsto z(\tau) + r(u(\tau) + w(\tau, s(\eta))) ,$$

and the fact follows immediately upon taking the inner product of (13) with $u(\tau)$ or $w(\tau, s(0))$ and differentiating with respect to η at $\eta = 0$.

Since $s \longmapsto \phi(\tau, s)$ has nonsingular derivative because it is an anti-isometry, this last fact together with (12) implies that ϕ' is invertible at all points τ, s. A standard result [Spivak, Vol. 1, p. 58, Thm. 9] then guarantees that $S_r(\tau_1, \tau_2)$ is at least an immersed submanifold, which is a sufficient analytic hypothesis for our purposes.

Given (12), we can easily obtain some pullbacks which we shall need. First we compute the pullback $\phi^\dagger(\perp u^*)$ at a point $\phi(\tau, s)$. Let q_1, q_2 be a basis for the tangent space to S_r^{std} at s. By definition of $\phi^\dagger$ and 2.2(21),

(14) $$\phi^\dagger(\perp u^*)(v, q_1, q_2) = (\perp u^*)(\phi'(v), \phi'(q_1), \phi'(q_2))$$
$$= \Omega(u, \phi'(v), \phi'(q_1), \phi'(q_2)) ,$$

where Ω is the volume 4-form on **M**. Substituting (12) in (14) yields

$$\phi^\dagger(\perp u^*)(v_{\tau,s}, q_1, q_2) = -r<a,w>\Omega(u, w, q_1, q_2).$$

If q_1, q_2 is chosen as an orthonormal basis for the tangent space of S_r^{std} at s, then $v_{\tau,s}, q_1, q_2$ is an orthonormal basis for the tangent space of $P_r(\tau_1, \tau_2)$ at τ, s, and since a 3-form on a three-dimensional space is determined by its value on any basis, this says that at any point $\tau, s \in P_r(\tau_1, \tau_2)$,

(16) $$\phi^\dagger(\perp u(\tau)^*) = r<a, w(\tau, s)> \Omega_P,$$

where Ω_P denotes the volume 3-form on $P_r(\tau_1, \tau_2)$ relative to the orientation in which $v_{\tau,s}, q_1, q_2$ is a positively oriented basis for $P_r(\tau_1, \tau_2)$ if and only if $w(\tau, s), u(\tau), \phi'(q_1), \phi'(q_2)$ is a positively oriented basis for **M**.

Similar calculations show that

(17) $$\phi^\dagger(\perp w(\tau, s)^*) = (1 - r<a(\tau), w(\tau, s)>)\Omega_P,$$

and that if $y \in \mathbf{M}$ is any vector orthogonal to both $u(\tau)$ and $w(\tau, s)$, then

(18) $$\phi^\dagger(\perp y^*) = 0.$$

For reference, we summarize (16), (17), and (18) as: for any $h \in \mathbf{M}$,

(19) $$\phi^\dagger(\perp h^*) = [<h, u>\cdot r<a, w> - <h, w>(1-r<a, w>)]\Omega_P.$$

Having obtained the crucial formula (19), which enables us to compute the pullback to $P_r(\tau_1, \tau_2)$ of any 3-form on $S_r(\tau_1, \tau_2)$, the rest of the computation is entirely straightforward, though tedious. We write it out in considerable detail so that the browser can see explicitly the sorts of renormalizations and cancellations which occur. Only a few of the many terms contribute to the final result.

Recall that the field tensor F is assumed of the form

$$F = F_{ext} + F_{ret},$$

where F_{ret}, which is given by 4.2(16), contains factors of r^{-1} and r^{-2}. The linear transformation associated with the 2-form F is as usual denoted $\hat{F}$, and similarly for F_{ext} and F_{ret}. The energy-momentum tensor T corresponding to F, considered as the linear transformation (or vector-valued 1-form), defined by 3.8(20) as

$$T = \frac{1}{4\pi}[\hat{F}^2 - \frac{1}{4}\text{tr}(\hat{F}^2)I],$$

is expanded, after considerable labor, in powers of r^{-1} below. The terms of order r^{-1} or lower are not written explicitly because in the integration over $P_r(\tau_1, \tau_2)$ they become multiplied by the area $4\pi r^2$ of the sphere and contribute nothing in the limit $r \to 0$.

4.4 Radiation calculation

(20) $$T = r^{-4}[u \otimes u^* - w \otimes w^* - \frac{1}{2}I]\frac{q^2}{4\pi}$$

$$+ r^{-3}[(u+w) \otimes a_\perp^* + a_\perp \otimes (u+w)^*]\frac{q^2}{4\pi}$$

$$+ r^{-2}[-<a_\perp,a_\perp>(u+w) \otimes (u+w)^*\frac{q^2}{4\pi}$$

$$+ (\hat{F}_{ext}(w) \otimes u^* + u \otimes \hat{F}_{ext}(w)^* - \hat{F}_{ext}(u) \otimes w^* - w \otimes \hat{F}_{ext}(u)^* + <\hat{F}_{ext}(u),w>I)\frac{q}{4\pi}]$$

$$+ \text{ terms } O(r^{-1}).$$

We want to compute

(21) $$\int_{S_r(\tau_1,\tau_2)} \perp T = \pm \int_{P_r(\tau_1,\tau_2)} \phi^\dagger(\perp T).$$

First we need to determine the proper sign in (21), which turns out to be "+". In other words, we assert that ϕ is orientation-preserving when $S_r(\tau_1,\tau_2)$ is given the natural orientation induced from Minkowski space in terms of which our previous physical interpretations of $\int_{tube} \perp T$ were predicated (cf. Sections 3.8 and 2.6). This orientation is one in which a basis y_1, y_2, y_3 for the tangent space of some point $p \in S_r(\tau_1, \tau_2)$ is positively oriented if and only if n, y_1, y_2, y_3 is a positively oriented basis for $\mathbf{M}_p$, where n is a vector which points out of the four-dimensional region bounded by the tube and caps. A normal n at $\phi(\tau, s)$ can be written down immediately from inspection of (12):

$$n := -r<a, w>u(\tau) + (1 - r<a, w>)w(\tau, s).$$

Note that $n \to w(\tau, s)$ as $r \to 0$, so the normal is outward for small r. This implies that if q_1, q_2 is a basis for the tangent space for the 2-sphere S_r^{std} at s, then $\phi'(v_{\tau,s}), \phi'(q_1), \phi_r'(q_2)$ will be a positively oriented basis for $S_r(\tau_1, \tau_2)$ at $\phi(\tau, s)$ if and only if for sufficiently small r,

(22) $$w(\tau, s), \phi'(v_{\tau,s}), \phi'(q_1), \phi'(q_2)$$

is a positively oriented basis for $\mathbf{M}$. But from (12), $\phi'(v_{\tau,s}) \to u(\tau)$ as $r \to 0$, and the orientation of $P_r(\tau_1, \tau_2)$ was chosen in the remark after (16) so that $v_{\tau,s}, q_1, q_2$ is positively oriented if and only if $w(\tau, s), u(\tau), \phi'(q_1), \phi'(q_2)$ is positively oriented. This shows that ϕ is orientation-preserving.

Now we compute

$$\int_{P_r(\tau_1,\tau_2)} \phi^\dagger(\perp \text{ terms in (20)}),$$

the sum of which will be $\int_{S_r(\tau_1,\tau_2)} \perp T$. The contributions from the various terms are reduced to elementary integrals by the formulas (16), (17), and (18) for the pullbacks. Most of the terms integrate to zero. For instance,

$$(23) \quad \int_{S_r(\tau_1,\tau_2)} r^{-4} \perp(u \otimes u^*) = \int_{S_r(\tau_1,\tau_2)} r^{-4} u \cdot \perp u^*$$

$$= \int_{P_r(\tau_1,\tau_2)} r^{-4} u \cdot \phi^\dagger(\perp(u^*)) = -\int_{P_r(\tau_1,\tau_2)} r^{-4} u \cdot r <a, w> \Omega_P$$

$$= \int_{\tau_1}^{\tau_2} d\tau \int_{S_r^{std}} r^{-4} u(\tau) r <a(\tau), w(\tau,s)> d\sigma_r(s) \;,$$

where $d\sigma_r(s)$ denotes the usual positive area measure on S_r^{std} (which assigns total area $4\pi r^2$ to S_r^{std}). Since for fixed τ, the map $s \longmapsto w(\tau,s)$ is an anti-isometry from S_r^{std} onto $S_r(\tau)$, it is odd in s. That is, if $-s$ denotes the antipodal point of $s \in S_r^{std}$, then

$$w(\tau, -s) = -w(\tau, s) \;,$$

and hence (23) vanishes. Many other terms can also easily be seen to vanish by virtue of being odd in s when pulled back to $P_r(\tau_1,\tau_2)$.

A more interesting integral comes from the next term in (20):

$$(24) \quad \frac{q^2}{4\pi} \int_{S_r(\tau_1,\tau_2)} r^{-4} \perp(-w \otimes w^*) = -\frac{q^2}{4\pi} \int_{P_r(\tau_1,\tau_2)} r^{-4} w \cdot (1 - r<a, w>) \Omega_P$$

$$= \frac{q^2}{4\pi} \int_{\tau_1}^{\tau_2} d\tau \int_{S_r^{std}} r^{-3} w(\tau,s) <a(\tau), w(\tau,s)> d\sigma_r(s) \;,$$

where the term which is odd in s was dropped. Recall that the map $s \longmapsto rw(\tau,s)$ anti-isometrically identifies S_r^{std} with a translate of the sphere $S_r(\tau)$. It is easy to show that such a map must preserve orthogonality (of points on the sphere considered as vectors in R^3) and so can be extended to a linear anti-isometry from Euclidean R^3 to the spacelike subspace of vectors in M orthogonal to $u(\tau)$. (Alternatively, the solution to Exercise 5 explicitly provides such an extension for the particular map used there.) Hence the inner integral identifies with the integral written in the notation of classical vector calculus as

$$(25) \quad \int_{S_r^{std}} r^{-3} \frac{\vec{s}}{r} \left(-\vec{a}(\tau) \cdot \frac{\vec{s}}{r}\right) d\sigma_r(\vec{s}) = -\frac{4\pi}{3r} \vec{a}(\tau) \;,$$

where $\vec{a}(\tau)$ is the vector in R^3 which maps to $a(\tau)$ under the above extension. Passing back from (25) to Minkowski space gives the value of (24) as

$$(26) \quad \frac{q^2}{4\pi} \int_{\tau_1}^{\tau_2} d\tau \int_{S_r(\tau_1,\tau_2)} r^{-4} \perp(-w \otimes w^*) = -\frac{q^2}{4\pi} \int_{\tau_1}^{\tau_2} d\tau \, \frac{4\pi}{3r} a(\tau)$$

$$= -\frac{q^2}{3r} (u(\tau_2) - u(\tau_1)) \;,$$

and this is precisely the form that we have agreed to ignore according to the principle of mass renormalization.

The term $-\dfrac{1}{2r^4} I$ gives nothing essentially new, as is seen from writing

4.4 Radiation calculation

(27) $$I = u \otimes u^* - w \otimes w^* - e_2 \otimes e_2^* - e_3 \otimes e_3^*$$

with u, w, e_2, e_3 an orthonormal basis. The terms with e_i^* pull back to zero, and the term with w^* simply cancels half of the mass renormalization of (26). The total mass renormalization due to the r^{-4} terms is

(28) $$-\frac{q^2}{6r}(u(\tau_2) - u(\tau_1)) .$$

Before examining the other terms in (20), an informal observation may clarify what is really going on. In view of the elementary decomposition

(29) $$a_\perp = a + <a, w> w ,$$

the terms in (20) which do not contain $\hat{F}_{ext}$ can be written as a linear combination of terms like

$$u \otimes something , \quad w \otimes something , \quad \text{or} \quad a \otimes something .$$

The terms $u \otimes something$ and $a \otimes something$ obviously integrate over the sphere to multiples of u and a, respectively. Since $w(\tau, s)$ is always orthogonal to u, its integral over the sphere must be orthogonal to u also. That is, its integral is a purely spatial vector in the rest frame at proper time τ. But there is only one distinguished spatial direction appearing in T, and that is the direction of the proper acceleration a. Hence the integral of the $w \otimes something$ terms should also be multiples of a. Any term which integrates over the sphere to a multiple of a will contribute a mass renormalization after the τ integration is performed. Thus all that is left are the terms which integrate over the sphere to multiples of $u(\tau)$ and the terms containing $\hat{F}_{ext}$. The former describe a pure mass-energy loss in the rest frame (i.e. radiation), and the latter turn out to account for the Lorentz force.

Passing to the terms of order r^{-3} in (20), equation (18) shows that $(u+w) \otimes a_\perp^*$ vanishes when pulled back to $P_r(\tau_1, \tau_2)$. The term

$$r^{-3} \frac{q^2}{4\pi} a_\perp \otimes_\perp (u^* + w^*)$$

pulls back to

$$\frac{q^2}{4\pi} \int_{P_r(\tau_1,\tau_2)} r^{-3} a_\perp \, \Omega_P = \frac{q^2}{4\pi} \int_{P_r(\tau_1,\tau_2)} r^{-3} (a + <a,w> w) \, \Omega_P ,$$

which, in view of (24) and (26), contributes to (1) another infinite mass renormalization

$$\frac{q^2}{4\pi} \int_{\tau_1}^{\tau_2} (4\pi r^{-1} a - \frac{4\pi}{3} r^{-1} a) \, d\tau = \frac{2q^2}{3r}(u(\tau_2) - u(\tau_1)) .$$

This is the last mass renormalization that occurs, and so the total of the mass renormalization terms in (1) is the sum of this and (28):

(30) $$\frac{q^2}{2r}(u(\tau_2) - u(\tau_1)) .$$

Finally, we examine the terms of order r^{-2} in (20), which yield both the Lorentz force and the radiation terms. First note that in this order, the terms in the pullbacks containing $r<a,w>$ contribute nothing in the limit $r \to 0$, since $d\sigma_r$ has a factor of r^2 built in. In the following we ignore such terms, which means that integrals over $S_r(\tau_1,\tau_2)$ of *anything* $\otimes \perp u^*$ may be dropped and integrals of *anything* $\otimes \perp w^*$ can be replaced by integrals over $P_r(\tau_1,\tau_2)$ of *anything* Ω_P. Thus of the terms not containing $\hat{F}_{ext}$, only

$$-r^{-2}\frac{q^2}{4\pi}<a_\perp, a_\perp>(u+w)\otimes w^*$$

contributes, and its contribution is

(31)
$$-\frac{q^2}{4\pi}\lim_{r \to 0} r^{-2} \int_{P_r(\tau_1,\tau_2)} <a_\perp, a_\perp>(u+w)\,\Omega_P$$

$$= -q^2 \int_{\tau_1}^{\tau_2} <a_\perp(\tau), a_\perp(\tau)>\, u(\tau)\, d\tau$$

where the part which is odd in s was dropped. To evaluate the remainder, substitute (29) in $<a_\perp(\tau), a_\perp(\tau)>$ to obtain

(32)
$$<a_\perp(\tau), a_\perp(\tau)> = <a(\tau), a(\tau)> + <a(\tau), w(\tau,s)>^2,$$

and insert this in (31). The term containing $<a(\tau), w(\tau,s)>^2$ is evaluated similarly to (24), and the result, in the same notation, is

$$-\frac{q^2}{4\pi}\lim_{r \to 0} r^{-2} \int_{P_r(\tau_1,\tau_2)} <a, w>^2 \cdot u \cdot \Omega_P$$

$$= -\frac{q^2}{4\pi}\lim_{r \to 0} r^{-2} \int_{\tau_1}^{\tau_2} d\tau\, u(\tau) \int_{S_r^{std}} \left(\vec{a}(\tau) \circ \frac{\vec{s}}{r}\right)^2 d\sigma_r(\vec{s})$$

$$= \frac{1}{3}q^2 \int_{\tau_1}^{\tau_2} u(\tau)<a(\tau), a(\tau)>\, d\tau.$$

Hence (31) becomes

(33)
$$\frac{q^2}{4\pi}\lim_{r \to 0} r^{-2} \int_{S_r(\tau_1,\tau_2)} -<a_\perp, a_\perp>(u+w)\perp(u^*+w^*)$$

$$= -\frac{2}{3}q^2 \int_{\tau_1}^{\tau_2} u(\tau)<a(\tau),a(\tau)>\, d\tau.$$

This is the radiation term, which is interpreted as the total energy-momentum radiated by the particle between proper times τ_1 and τ_2.

Finally, we examine the terms in (20) of order r^{-2} containing $\hat{F}_{ext}$. The first, $\hat{F}_{ext} \otimes u^*$, pulls back to zero in the limit $r \to 0$. The observation that $\hat{F}_{ext}(w)$ is always orthogonal to w (because $\hat{F}_{ext}$ is antisymmetric and w is a unit vector) shows that the second term, $u \otimes \hat{F}_{ext}(w)^*$, also pulls back to zero in the limit. The third term,

$$-\frac{q}{4\pi}\hat{F}_{ext}(u)\otimes w^*$$

is constant in the sphere variable when pulled back to $P_r(\tau_1,\tau_2)$, and so contributes

$$\text{(34)} \qquad -q \int_{\tau_1}^{\tau_2} \hat{F}_{ext}(u(\tau))\, d\tau$$

to (1). This is the term responsible for the Lorentz force.

The last two terms contain a remarkable cancellation. The pullback of

$$-w \otimes \hat{F}_{ext}(u)*$$

is seen, by writing

$$\hat{F}_{ext}(u) = -<\hat{F}_{ext}(u), w> w + y$$

with y orthogonal to w, to give the same contribution in the limit as the pullback of

$$\text{(35)} \qquad <\hat{F}_{ext}(u), w> w \otimes w*\ .$$

The corresponding term of (20) has a factor of $\frac{q}{4\pi}$ in front and contributes (cf. (24) and (25))

$$\text{(36)} \qquad \frac{q}{4\pi} \int_{P_r(\tau_1,\tau_2)} <\hat{F}_{ext}(u(\tau)), w(\tau,s)> w(\tau,s)\, \Omega_P = -\frac{q}{3} \int_{\tau_1}^{\tau_2} \hat{F}_{ext}(u(\tau))\, d\tau$$

to (1), which is one third the Lorentz force term. However, this term cancels with the contribution of

$$\text{(37)} \qquad <\hat{F}_{ext}(u), w> I\ .$$

To see this, write I as the sum of one-dimensional projections given in (27), which shows that the entire contribution of (37) comes from

$$-<\hat{F}_{ext}(u), w> w \otimes w*\ ,$$

which is just the negative of (35).

To summarize,

(38) $$\int_{S_r(\tau_1,\tau_2)} \perp T \;=\;$$

(i) (mass renormalization) $\quad\displaystyle \frac{q^2}{2r}(u(\tau_2) - u(\tau_1))$

(ii) (radiation) $\quad\displaystyle -\frac{2}{3}q^2 \int_{\tau_1}^{\tau_2} <a(\tau), a(\tau)> u(\tau)\, d\tau$

(iii) (Lorentz force) $\quad\displaystyle -q \int_{\tau_1}^{\tau_2} \hat{F}_{ext}(u(\tau))\, d\tau$

(iv) $\quad + \; O(r)\;,$

Notes.

(1) It would be interesting to find a way to see, *a priori,* that the Lorentz force term will occur with the correct coefficient in (38). Is the cancellation of (36) and (37) a remarkable accident, or is there a deeper reason?

(2) In our discussion of the difference between integration over a Bhabha tube $S_r(\tau_1, \tau_2)$ and a Dirac tube $\bar{S}_r(\tau_1, \tau_2)$, we skimmed over a subtle but important point. It is true as stated there that within the context of our motivation of the Lorentz-Dirac equation, which assumed asymptotically free particles (i.e. $a(-\infty) = 0 = a(\infty)$), it is reasonable to expect the integrals

$$\int_{S_r(-\infty,\infty)} \perp T \;=\; \int_{\bar{S}_r(-\infty,\infty)} \perp T$$

over *infinite* Bhabha and Dirac tubes to be equal, and thus it should not matter which kind of tube is used. However, the integrals over *finite* tubes are *not* in general the same:

$$\int_{S_r(\tau_1,\tau_2)} \perp T \quad \text{does not necessarily equal} \quad \int_{\bar{S}_r(\tau_1,\tau_2)} \perp T\;,$$

even in the limit $r \to 0$. In fact, [Dirac, 1938] computed that

(39) $$\int_{\bar{S}_r(\tau_1,\tau_2)} \perp T \;=\; -q \int_{\tau_1}^{\tau_2} \hat{F}_{ext}(u(\tau))\, d\tau$$

$$-\frac{2}{3}q^2 \int_{\tau_1}^{\tau_2} \left(\frac{da}{d\tau} + <a(\tau), a(\tau)> u(\tau)\right) d\tau \;+\; \frac{q^2}{2r}(u(\tau_2) - u(\tau_1)) \;+\; O(r)\;,$$

which differs from the corresponding integral (38) for a Bhabha tube in the limit $r \to 0$ by

$$-\frac{2}{3}q^2 \int_{\tau_1}^{\tau_2} \frac{da}{d\tau}\, d\tau \;=\; -\frac{2}{3}q^2 \left(a(\tau_2) - a(\tau_1)\right)\;.$$

If the acceleration vanishes at τ_1 and τ_2 (more generally, if $a(\tau_2) = a(\tau_1)$), then of course (38) and (39) coincide in the limit $r \to 0$. If the acceleration does not vanish at the ends, then the end $S_r(\tau_i)$ for a Bhabha tube is not necessarily the same as the corresponding end $\bar{S}_r(\tau_i)$ for a Dirac tube; the two tubes may not fit together at the ends, and the previous argument that the integrals must be the same because the divergence of T vanishes no longer applies.

If one attempts to obtain an equation of motion by differentiating (39) with respect to τ_2 as in 4.3(7) while ignoring mass renormalization terms, one does obtain the Lorentz-Dirac equation directly from (39). However, this procedure is open to at least two substantial criticisms.

First of all, when the caps are taken at finite times with nonvanishing proper acceleration, it is not clear that the integrals over the caps give only mass renormalization terms. In fact, it seems that these integrals have never been computed exactly, and, since they diverge badly, there may even be problems in defining them. Thus differentiating (39) can hardly be considered a complete derivation in the absence of a rigorous calculation of the integrals over the caps. Secondly, one could presumably get equations other than the Lorentz-Dirac equation or 4.3(8) by appropriately perturbing the tubes. The trouble with the energy-momentum tensor T is that it is unreasonably singular. Though the r^{-4} and r^{-3} terms drop down to r^{-2} and r^{-1} after integration over a sphere of radius r, an $O(r)$ perturbation in the tube can still interact with an $O(r^{-1})$ singularity to produce a nonzero change in the equation. To put it more picturesquely, denting the tube ever so slightly can change the equation obtained. In the absence of a fundamental reason to use one particular kind of tube, it is hard to see how one could have much confidence in an equation "derived" in this way.

4.5 Summary of the logical structure of electrodynamics.

This is an appropriate point to pause and look back on the road we have travelled. We began with the Lorentz force law

(1) $$m\frac{du}{d\tau} = q\hat{F}(u)$$

and observed that nature has made a truly remarkable simplification in declaring that the force $\hat{F}$ shall be a *linear* function of the four-velocity u. The linearity of $\hat{F}$ allows us to define the associated 2-form F, and then the Maxwell equations

(2) $$dF = 0 \quad,$$

and

(3) $$\delta F = 4\pi\rho u^* \quad,$$

while not forced, are natural and in some sense define the simplest possible theory.

The first Maxwell equation allows us to chose a potential 1-form A with $dA = F$. Using the freedom to specify δA to impose the subsidiary condition $\delta A = 0$ leads to the replacement of Maxwell's equations by the single equation

(4) $$\Box A = 4\pi\rho u^* \quad.$$

From the point of view of causality, it is natural to use the retarded solution A_{ret} to (4), and the corresponding solution $F_{ret} := dA_{ret}$ for Maxwell's equations.

If we did not want a law of conservation of energy-momentum, we could stop at this point and be content with an elegant and consistent theory. It is the search for such a law that leads to difficulties. Given the Lorentz force law and Maxwell's equations, it follows algebraically that the covariant divergence of the energy-momentum tensor

$$4\pi T_{elm} := \hat{F}^2 - (1/4)\,tr(\hat{F}^2)I$$

is the negative of the covariant divergence of the energy-momentum tensor $(m/q)\rho\, u \otimes u$ for a pressureless perfect fluid with mass density $(m/q)\rho$. The sum of these tensors has zero divergence and so is a natural candidate for the energy-momentum tensor of the complete system (charged fluid plus electromagnetic field).

Unfortunately, if we use the tensor T_{elm} with the retarded field F_{ret} to compute the radiation from a charged point particle, two serious difficulties arise:

(i) The integrals of T_{elm} over a tube of small radius r surrounding the wordline of the particle contain terms of the general form

$$\frac{K(r)}{r} \int_{-\infty}^{\infty} \frac{du}{d\tau}\, d\tau \; ,$$

with $\lim_{r \to 0} K(r)$ finite and nonzero, and

(ii) even if we use the rationale of mass renormalization to wave away these divergent terms, we find that the equation for the total change in energy-momentum of a scattered particle is not the

(5) $$\int_{-\infty}^{\infty} m\frac{du}{d\tau}\, d\tau = \int_{-\infty}^{\infty} q\hat{F}(u(\tau))\, d\tau$$

which the Lorentz equation implies, but rather has an additional term

(6) $$\frac{2}{3}q^2 \int_{-\infty}^{\infty} <\frac{du}{d\tau}, \frac{du}{d\tau}> u\, d\tau$$

on the right side. † Of course, the appearance of (6) is contrary to the original motivation for adopting T_{elm} as the energy-momentum tensor of the electromagnetic field. (Recall that T_{elm} was adopted precisely because, assuming Maxwell's equations, it makes conservation of energy-momentum for a charged dust equivalent to the Lorentz equation.) Since the theories of charged fluids and of point particles depend on one another only for motivation, this is not a logical contradiction, but it is disconcerting that the two approaches turn out to be essentially different.

Another important difference is that the retarded fields play no special role in the theory of charged fluids, and the extent to which the retardation condition is consistent with the Maxwell-Lorentz equations is unclear. We shall examine this point further in the next chapter.

References.

Several excellent advanced texts treat the Lorentz-Dirac equation only cursorily. Perhaps the most extensive treatment is [Rohrlich, Chapter 6]. This text presents a particular

---

† The remote possibility that the theory could be saved by proving or postulating that (6) vanishes is demolished by the observation that any two forward-pointing timelike unit vectors u, v satisfy the "backward Schwartz" inequality $<u,v> \geq 1$ (Exercise 1.9), so the inner product of (6) with any timelike v is nonzero unless $du/d\tau$ vanishes identically.

point of view which is widely, though not universally, accepted.† Another interesting and useful reference is [Barut]. Dirac's original paper [1938] motivating what is now called the Lorentz-Dirac equation is still one of the clearest sources, though the actual calculations using a Dirac tube are quite tedious. (Exercise 4.8 gives a feel for these.) The great simplification obtained by using a Bhabha tube was recognized in [Bhabha, 1939], but surprisingly, has not been generally adopted in textbooks even up to the present. Variants of this method in the spirit of the treatment of Section 4.4 appear in papers of [Synge, 1970] and [Hogan].

Exercises 4

4.1 Show that the worldline $t \longmapsto (t, \vec{x}(t))$ of a particle whose coordinate velocity $\vec{v}(t) := d\vec{x}/dt$ satisfies $|\vec{v}(t)| < \alpha < 1$ for some α and all times t will intersect exactly once any backward light cone whose vertex is not on the worldline. Give an example showing that this conclusion can fail if it is only assumed that $v < 1$ rather than that v is bounded away from 1. (Hint: For the counterexample, recall Exercise 1.22.)

4.2 In the treatment of Section 3, the Bhabha tube $S_{r_0}(\tau_1, \tau_2)$ was capped with hyperplanes, and the integral of $\perp T$ over the caps could be "evaluated" only when the particle was free (i.e. unaccelerated) at all points on the worldline connected to the caps by forward light rays (so that the field in the caps is a Coulomb field). Suppose that instead of hyperplane caps we use light cone caps

$$C(\tau_i) := \{ r(u(\tau_i) + w) \mid 0 < r < r_0, \ \langle w, u(\tau_i) \rangle = 0 \} \qquad i = 1, 2 \ .$$

Here r_0 is fixed and plays the role of the radius of the tube. One advantage of light cone caps over hyperplane caps for a Bhabha tube is that we know exactly where the worldline cuts the cap (i.e. at the vertex of the cone!). Since we expect the integrals over the caps to be infinite, it is reasonable to consider the integral over

$$C(\tau; \epsilon) := \{ r(u(\tau) + w) \mid \epsilon < r < r_0, \ \langle w, u(\tau) \rangle = 0 \} \ ,$$

and then take the limit $\epsilon \to 0$ after extracting an ϵ-dependent mass renormalization.

(a) Show that

$$\int_{C(\tau, \epsilon)} \perp (u \otimes u^*) = 4\pi [\frac{1}{\epsilon} - \frac{1}{r_0}] u(\tau) \ .$$

(b) Show that if we take $F_{ext} := 0$ in 4.4(20), then

$$\int_{C(\tau, \epsilon)} \perp T = \frac{1}{2} q^2 u(\tau) (\frac{1}{\epsilon} - \frac{1}{r_0}) \ ,$$

so we obtain the infinite mass renormalization associated with the caps directly and *exactly* by this method even when the particle is in arbitrary accelerated motion.

(c) Show that for T the usual energy-momentum tensor defined by 4.4(20) and

---

† I first came to understand the fundamental issues of point electrodynamics by studying this text and the present work owes a considerable debt to it even though the point of view and some of the conclusions are different.

for arbitrary F_{ext},

$$\lim_{\epsilon \to 0} \left[\int_{C(\tau, \epsilon)} \bot T - \frac{1}{2} q^2 u(\tau)(\frac{1}{\epsilon} - \frac{1}{r_0}) \right] = r_0 q F_{ext}(u(\tau)) + O(r_0^2) ,$$

so the terms involving F_{ext} contribute nothing to the limit $r_0 \to 0$. How might this last fact be guessed, *a priori*, before performing any integrations?

4.3 Given $x \in \mathbf{M}$, let $z_{adv} = z_{adv}(x)$ denote the point where the forward light cone from x cuts the worldline, and let u_{adv}, a_{adv} denote the corresponding four-velocity and proper acceleration. Define r_{adv} and w_{adv} by

$$x - z_{adv} = r_{adv}(w_{adv} - u_{adv}) ,$$

$$<w_{adv}, u_{adv}> = 0 \quad \text{,and} \quad <w_{adv}, w_{adv}> = -1 .$$

(See diagram.) Let τ_{adv} be the proper time at the point of intersection of the cone with the worldline: $z(\tau_{adv}) = z_{adv}$. Compute $d\tau_{adv}$ and dr_{adv}.

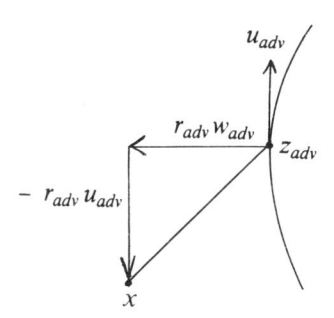

Diagram 4.3. The geometry of the advanced quantities.

4.4 Give an intuitive argument based on time reversal symmetry which deduces the formula 4.2(17) for the advanced field from the corresponding formula 4.2(16) for the retarded field. Alternatively, check 4.2(17) directly using the results of Exercise 3.

4.5 Consider a worldline $\tau \longmapsto z(\tau)$ in a spacetime M with four-velocity $u(\tau) := dz/d\tau$ and proper acceleration $a(\tau) := du/d\tau$. Let $e_1(0), e_2(0), e_3(0) \in M_{z(0)}$ be an orthonormal basis for "space" relative to the rest frame at $z(0)$; i.e. the e_i are orthonormal and orthogonal to $u(0)$. We want to make a mathematical formulation of the intuitive idea of carrying the basis vectors along the worldline without changing their "direction", so that the original basis $\{e_i(0)\}$ for $M_{z(0)}$ is "transported" to an orthonormal basis $\{e_i(\tau)\}$ for $M_{z(\tau)}$. Physically, this might correspond to carrying a gyroscope along the worldline.

Note that ordinary parallel transport (i.e. requiring that $D_u e_i = 0$) will not do, as is easily seen by taking $u(0) = (1, 0, 0, 0)$, $e_1(0) = (0, 1, 0, 0)$, $e_2(0) = (0, 0, 1, 0)$, $e_3(0) = (0, 0, 0, 1)$ in Minkowski space. The parallel transport of these $e_i(0)$ along any worldline is $e_i(\tau) = e_i(0)$, but the $e_i(\tau)$ are not generally orthogonal to $u(\tau)$; i.e. the parallel transport of spatial vectors in a rest frame will not remain spatial in the new rest frames.

To say that the $e_i(\tau)$ are spatial in the rest frame at $z(\tau)$ means precisely that

$$<u(\tau), e_i(\tau)> = 0 .$$

Differentiating, we see that

$$0 = \frac{d}{d\tau}<u(\tau), e_i(\tau)> = u(<u, e_i>) = <D_u u, e_i> + <u, D_u e_i>$$
$$= <a, e_i> + <u, D_u e_i> ,$$

so that the e_i must satisfy

(2) $\qquad <D_u e_i, u> = -<a, e_i> ,$

which says that

$$D_u e_i = -<a, e_i>u + \text{ vector orthogonal to } u .$$

The closest we can come to parallel transport under the given conditions is to require that

(3) $\qquad D_u e_i = -<a, e_i>u ;$

in other words, the e_i change as little as possible while maintaining their spacelike character. The analog of (3) for an arbitrary vector function $\tau \longmapsto v(\tau) \in M_{z(\tau)}$ defined along the worldline is

(4) $\qquad D_u v = -<a, v>u ,$

which makes mathematical sense for an arbitrary vector function $\tau \longmapsto v(\tau)$ but makes physical sense only when $v(\tau)$ is spacelike in the rest frame at $z(\tau)$; i.e. when $<v(\tau), u(\tau)> = 0$ for all τ. Such a *spacelike* vector function $v(\cdot)$ is said to be *Fermi-Walker parallel* along $z(\cdot)$, and its values $v(\tau)$ are said to be the *Fermi-Walker transports* of the initial value $v(0)$.

In view of the fact that Fermi-Walker transport preserves the inner product for spatial vectors, it is natural to require that u itself be Fermi-Walker parallel along $z(\cdot)$, so that $u(\tau)$ together with the $e_i(\tau)$ will form an orthonormal basis. However, u does not necessarily satisfy (4) because $D_u u = a$, which is orthogonal to u. An appropriate generalization of (4) which u also will satisfy is the following. Define an arbitrary vector function $\tau \longmapsto v(\tau) \in M_{z(\tau)}$ to be *Fermi-Walker parallel* if

(5) $\qquad D_u v = (a \otimes u^* - u \otimes a^*)(v) .$

Given any vector $v_0 \in M_{z(0)}$, there exists a Fermi-Walker parallel function $\tau \longmapsto v(\tau)$ with $v(0) = v_0$. This follows from writing (5) in components as a system of ordinary differential equations and applying the standard existence theorem.

(a) Define the linear transformation G on $M_{z(\tau)}$ by the right side of (5):

$$G(v) := (a \otimes u^* - u \otimes a^*)(v) = a \cdot <u, v> - u \cdot <a, v> .$$

Check that G is antisymmetric (i.e. $<G(v), w> = -<v, G(w)>$ for all $v, w \in M_{z(\tau)}$), and hence if $v(\cdot)$ and $w(\cdot)$ are Fermi-Walker parallel along $z(\cdot)$, then for any τ, the map

$$v(0), w(0) \longmapsto v(\tau), w(\tau)$$

is an isometry from $M_{z(0)}$ to $M_{z(\tau)}$.

(b) Show that, in the notation of the text in the paragraphs following equation 4.4(8), there exists a C^∞ map $\tau, s \longmapsto \phi(\tau, s)$ from $(\tau_1, \tau_2) \times S_r^{std}$ to **M** such that for each τ, the map

$$\phi(\tau, \cdot) : S_r^{std} \longmapsto S_r(\tau)$$

is an anti-isometry.

4.6 This exercise is meant to give the reader a feel for the size of the radiation reaction terms in the Lorentz-Dirac equation 4.3(15):

(1) $$a = \frac{q}{m}\hat{F}_{ext}(u) + \frac{2}{3}\frac{q^2}{m}\left[\frac{da}{d\tau} + <a, a>u\right].$$

Suppose that we can directly observe the motion of a particle under the influence of an external electromagnetic field F_{ext}. With observational apparatus available in 1985, the particle will appear to obey the Lorentz equation

(2) $$a = \frac{q}{m}\hat{F}_{ext}(u).$$

We can imagine gradually refining the apparatus until it becomes sufficiently accurate that the radiation reaction term

(3) $$\frac{2}{3}\frac{q^2}{m}\left[\frac{da}{d\tau} + <a, a>u\right]$$

can just be observed. It would be natural to compare its size with that of the dominant Lorentz term

$$\frac{q}{m}\hat{F}_{ext}(u),$$

which in turn is approximately a. From this point of view, it is reasonable to measure the size of (3) by comparing it with the magnitude of a.

The constant $\frac{2}{3}\frac{q^2}{m}$ has dimensions of time in our system of units, and its value is approximately 6×10^{-24} sec (appendix, equation (20)). An acceleration of one gravity, to use a unit with which everyone is familiar, has a numerical value of about 3×10^{-8} sec^{-1} in our units. From this it is immediately apparent that $(2q^2/3m)<a, a>$ is indeed very small relative to the magnitude of a unless the acceleration is enormous relative to the accelerations which macroscopic objects in everyday life can experience. (If we work in the particle's rest frame, we do not need to concern ourselves with the "size" of u; otherwise there is also a possible γ-factor which enters.)

The term

(4) $$\frac{2}{3}\frac{q^2}{m}\frac{da}{d\tau}$$

is not as easy to relate to a. One might try to make this term very large by turning the field on or off very fast. Use back-of-envelope style calculations in the above framework to estimate how fast the field might have to be turned off to make (4) on the same order of magnitude as a.

4.7 This and the next problem are intended to give a feel for the kind of computations involved in integrating over a Dirac tube and to show where the extra term $da/d\tau$ mentioned in note (2) comes from.

Construct the Dirac tube $\bar{S}_\epsilon(\tau_1, \tau_2)$ as the image of the embedding

$$\psi : P_\epsilon(\tau_1, \tau_2) := (\tau_1, \tau_2) \times S_\epsilon^{std} \longmapsto \mathbf{M} \text{ defined by}$$

$$\psi(\tau, s) := z(\tau) + \epsilon y(\tau, s) ,$$

where $y(\tau, s)$ is given by Exercise 5 as a spacelike unit vector orthogonal to $u(\tau)$ and such that for each fixed τ, the map $s \longmapsto y(\tau, s)$ is an anti-isometry. In this problem, $y(\tau, s)$ plays the same role as the $w(\tau, s)$ of the text. We change the name to avoid confusion with the "retarded" w which appears in the formula 4.2(16) for the retarded fields. For a Bhabha tube, the two w's coincide, which is one reason the integration is so simple, but this is not true for a Dirac tube. For the same reason, we use ϵ in place of r for the "radius" of the tube to avoid confusing it with the retarded distance $r = r_{ret}$.

Compute the analog of 4.4(19) for the pullback $\psi^\dagger(\bot h^*)$ of an arbitrary 3-form $\bot h^*$ to $P_\epsilon(\tau_1, \tau_2)$.

4.8 The previous exercise showed how to pull back, and hence integrate, any 3-form $\bot h^*$ as long as we can compute the component of h along y. The complication in using a Dirac tube is that the formula 4.2(16) for the retarded fields is not related to y in any simple way. In fact, even the "u" that appears in 4.2(16) is not orthogonal to $y(\tau, s)$, since at a point $x = z(\tau) + \epsilon y(\tau, s)$ on the tube, that u is not $u(\tau)$, but rather $u(\tau_{ret}(x))$. Also, the "radius" ϵ of the tube is no longer exactly $r = r_{ret}(x)$. Thus to actually perform the integration we must find a way to relate the retarded quantities to the tube parameters. One way to do this is by power series expansions. (Assume that everything in sight is real-analytic.) Let a proper time τ be given, which will be fixed throughout the discussion. Given

$$x = z(\tau) + \epsilon y(\tau, s)$$

on the tube, define

$$\sigma = \sigma(x) := \tau - \tau_{ret}(x) \quad , \text{ so } \quad \tau_{ret}(x) = \tau - \sigma .$$

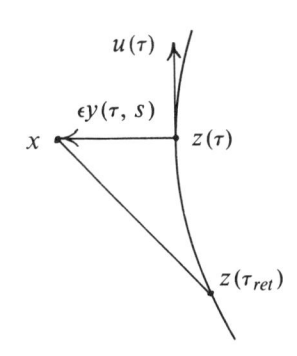

Diagram 4.8

(a) By expanding the retarded quantities as power series in σ, show that

(i) $$\sigma = \epsilon + \frac{1}{2}<a(\tau), y(\tau, s)>\epsilon^2 + O(\epsilon^3) \quad \text{and}$$

(ii) $$r_{ret} = \sigma - \sigma\epsilon<a(\tau), y(\tau, s)> + O(\sigma^3)$$
$$= \epsilon - \frac{1}{2}\epsilon^2<a, y> + O(\epsilon^3) .$$

Here (i) and (ii) are to be interpreted as holding uniformly in $s \in S_\epsilon^{std}$ for all $x = z(\tau) + \epsilon y(\tau, s)$ on the tube with τ fixed. This will be used in part (b) to justify the replacement of the r_{ret}'s which appear in the denominators in the expression 4.4(20) for the energy-momentum tensor by σ's or ϵ's in the limit $\epsilon \to 0$.

(b) Consider the term $$\frac{a_\perp \otimes w^*}{4\pi r^3}$$

which appears in the energy-momentum tensor 4.4(20). Show that there exist constants K_1, K_2, K_3 such that

$$\lim_{\epsilon \to 0} \left[\int_{\bar{S}_\epsilon(\tau_1, \tau_2)} \frac{\perp(a_\perp \otimes w^*)}{4\pi r^3} + \left(\frac{K_1}{\epsilon} + K_2\right)(u(\tau_2) - u(\tau_1))\right]$$

$$= \frac{2}{3}\int_{\tau_1}^{\tau_2} -\frac{da}{d\tau} d\tau + K_3 \int_{\tau_1}^{\tau_2} <a, a>u \, d\tau .$$

(There is no difficulty in computing the constants K_i, but the algebra is very tedious. One can see the form of the terms without actually doing the integrations.)

4.9 In view of equation 4.4(39), one might be tempted to try to interpret the term

(1) $$\frac{2}{3}q^2(\frac{da}{d\tau} + <a, a>u)$$

in the Lorentz-Dirac equation as the instantaneous rest-frame rate of energy-momentum radiation. In the rest frame, this is purely spatial momentum, but in a fixed "laboratory" frame it can have an energy component as well, so there might be hope of describing energy radiation which is actually observed in this way.

(a) As a test case, consider a particle moving at constant speed (i.e. $|\vec{v}|$ is constant) in the laboratory frame, and suppose that the rest-frame rate of energy-momentum radiation is given by (1). Show that the total pure *energy* (not energy-momentum) radiated in any finite proper time interval $\tau_1 \leq \tau \leq \tau_2$ as seen from the lab frame is the same as that predicted by the generalized Larmor law 4.3(16):

$$-\frac{2}{3}q^2 \int_{\tau_1}^{\tau_2} <a, a>u \, d\tau .$$

Thus if we are concerned only with lab-frame *energy* radiation in this special situation (which, as we shall see in Section 5.6, is one of the key experimental tests of the usual radiation laws), energy-momentum radiation rate (1) predicts the same result as the generalized Larmor law.

(b) As a second test case, consider a charged particle moving on a line under the influence of an electric field directed along that line and no magnetic field (all with respect to the laboratory frame). Show that (1) and 4.3(16) do *not* predict the same lab-frame energy radiation for this case.

4.10 This problem uses the notation of Section 4.4. Suppose that at time 0 a charged particle at rest in a laboratory is subjected to a brief, sharp acceleration, as if it were tapped by a hammer. The acceleration will cause it to radiate, the radiation spreading spherically outward at the speed of light. In the idealization in which the acceleration is of infinitesimal duration, the radiation pulse at any given coordinate time is concentrated on a sphere which is both a retarded sphere $S_r(0)$ and a laboratory sphere, since the particle was at rest at the time of acceleration. According to the basic physical interpretation of the energy-momentum tensor presented in Section 3.8, the rate at which energy-momentum is crossing such a sphere is

$$\int_{S_r(0)} \perp T(u, \cdot, \cdot) \ ,$$

where $u = u(0)$ is the four-velocity at the emission time, and the integrand represents the 2-form obtained by contracting u with $\perp T$; i.e. the 2-form $v_1, v_2 \mapsto \perp T(u, v_1, v_2)$.

(a) Show that when T is the energy-momentum tensor for the retarded field (4.4(20) with $F_{ext} := 0$), we have

(1) $$\int_{S_r(0)} \perp T(u, \cdot, \cdot) = \frac{2}{3}q^2 \left[-<a, a>u + \frac{1}{r}a \right] \ .$$

The term $(2q^2/3r)a$ is purely spatial and represents a flow of field 3-momentum. It is the mass renormalization due to the term $a_\perp \otimes w^*/4\pi r^3$ in 4.4(20). (The other mass renormalization produced by the $w \otimes w^*$ terms in the calculation of Section 4.4 do not contribute in this context. The difference can be seen from the observation that the Bhabha tube is not tangent to the tube generated by a sphere at rest in the laboratory.)

One might argue that if the mass renormalization terms are ignored in the calculation of Section 4.4, then the corresponding term in (1) ought to be ignored also, and this is often done in practice. (Sometimes, this can be alternatively justified by observing that the $1/r$ in the second term makes it small at large distances from the particle, which is often where actual observations are made.)

To see an interesting consequence of (1), consider two concentric retarded spheres $S_{r_1}(0)$, $S_{r_2}(0)$ with radii $r_1 < r_2$, and suppose for simplicity that the original acceleration $a = a(0)$ persisted only for a small coordinate time Δt. Then to first order in Δt, the total radiation which passes through a sphere $S_r(0)$ is

$$\frac{2}{3}q^2\Delta t[-<a, a>u - \frac{1}{r}a] \ ,$$

and the difference in radiation passing through the two spheres is

radiation through $S_{r_1}(0)$ - radiation through $S_{r_2}(0)$

$$= \frac{2}{3}q^2\Delta t(\frac{1}{r_1} - \frac{1}{r_2})a + O(\Delta t^2) \ .$$

In other words, not all of the energy-momentum which enters the inner sphere exits through the outer sphere. Taking r_2 very large shows that none of the energy-momentum corresponding to the second term of (1) gets out to spatial infinity. Some authors argue (not altogether convincingly) that for this reason, the second term of (1) should not be considered as radiation.

(b) Many authors have attempted to treat the electron as a small spherically symmetric (relative, say, to the rest frame of the center) charge distribution. Though the electron has no known extension, the "classical electron radius"

$$r_e := \frac{e^2}{m} \approx 3 \times 10^{-13} \text{ cm} ,$$

where e is the charge and m the mass of the electron, is often used. The name arises from the fact that a spherically symmetric classical charge distribution of radius e^2/m will have electromagnetic energy (obtained by integrating $T^0{}_0 = (|\vec{E}|^2 + |\vec{B}|^2)/8\pi$ over all space) on the order of the electron mass m. (The exact energy depends on the nature of the charge distribution; for a uniform surface charge it is $m/2$.) Thus if the electron had this radius, all of its mass could be considered as of electromagnetic origin, arising solely from its field.

Proceeding in the spirit of exploring this idea, it is natural to take $r := r_e$ in (1) and interpret (1) as the total energy-momentum radiated by the electron. Generalizing slightly, take the rate of radiation relative to the rest frame of the center of the electron to be

(2) $$\frac{2}{3}e^2(-\langle a, a \rangle u + ka) ,$$

where k is a constant, so that taking $k := 0$ we can simultaneously treat the case of a point electron in which the mass renormalization term in (1) is ignored.

Consider an electron in arbitrary motion relative to the laboratory frame, and compute the laboratory-frame rate of radiation of pure energy predicted by (2). Specialize to the case $k = 0$ and show that it is the same as the rest-frame energy radiation rate

(3) $$-\frac{2}{3}e^2 \langle a, a \rangle .$$

In other words, under the assumption that pure energy is radiated in the rest frame (i.e. that the radiated energy-momentum is proportional to u), the radiated power is independent of the Lorentz frame in which it is computed.

(c) In Section 5.6 we shall discuss an experiment in which radiation is observed from electrons moving in a circle of radius R at nearly the velocity of light relative to the laboratory. The experiment observes loss of kinetic energy relative to the laboratory frame. (Of course, this must not be confused with energy loss relative to an electron's rest frame.) The experiment is usually analyzed by considering the electron as a point particle which gives off energy-momentum P to the field at the proper-time rate

(4) $$\frac{dP}{d\tau} = -\frac{2}{3}e^2 \langle a, a \rangle u .$$

Notice that (4) is a considerably stronger statement than the Larmor law. The latter merely asserts that the rest-frame rate of radiation of pure energy is

(5) $$\frac{dP^0}{d\tau} = -\frac{2}{3}e^2 \langle a, a \rangle \; ,$$

while (4) goes further to state that not only is the pure energy radiation given by (5), but that there is also no radiation of spatial momentum in the rest frame. The origin of (4) is, of course, the basic calculation of Section 4.4 culminating in 4.4(38), with the mass renormalization terms ignored.

When exploring the consequences of an electron of finite radius, it is not clear that it is appropriate to ignore the mass renormalization terms, which are finite in that context. Assume that the rest-frame energy-momentum radiation rate is given by (2) and write down an expression for the laboratory-frame rate of pure energy radiation. Specialize to the case in which the electron moves in a circle of radius R with constant speed v, and show that for all values of k, the lab-frame rate of energy radiation dE/dt is

(6) $$\frac{dE}{dt} = \frac{2}{3}e^2 \cdot \frac{v^4}{R^2} \cdot \gamma(v)^4 \; .$$

Chapter 5

Further Difficulties and Alternate Approaches

This chapter examines some interesting results and viewpoints which shed light on the problem of constructing a theory of electrodynamics which is both logically consistent and in agreement with experiment. The previous chapters presented the most widely accepted viewpoint, which culminates in the Lorentz-Dirac equation. There are many motivations and "derivations" of this equation, and the one presented in Chapter 4 is the one which I find most nearly rigorous and convincing. I tried to present the best possible case for the Lorentz-Dirac theory, but it is easy to see why many have serious doubts about that equation. (Others believe it is absolutely fundamental.)

In this last chapter we examine further difficulties with the Lorentz-Dirac equation and the fluid model and explore alternate approaches. No attempt has been made to survey all historically or potentially important theories, and the selection has been made solely according to taste. For example, the Kaluza-Klein model is not included because I feel that it contributes little insight into the problems of classical electrodynamics even though it is currently fashionable in related areas of physics and is well worth studying for its own sake.

The chapter is rather loosely organized, and the sections are largely independent. Sections 5.1 through 5.3 explore the standard fluid model and are not needed for the sequel except that the result of Section 5.1 is used peripherally in 5.4, and a simple lemma from 5.2 is used in 5.5. Section 5.2 questions the physical reasonableness of the Maxwell-Lorentz system without the retardation condition, while 5.3 questions the the very compatibility of the system and the condition! Section 5.4 presents an alternate fluid model proposed (and later abandoned) by Dirac. Although clearly unsatisfactory in its present form, it is hard to believe that ideas so natural and elegant will not find application somewhere. Section 5.5 presents relatively little-known work of Eliezer, later refined and extended by Huschilt and Baylis and by Hsing and Driver, which casts further doubt on the Lorentz-Dirac equation. Readers interested primarily in point electrodynamics may want to read this section first. Section 5.6 is an expository review of evidence for the usual energy-momentum tensor. Section 5.7 discusses the relation between energy-momentum tensors and equations of motion and examines alternate energy-momentum tensors.

5.1 The Cauchy problem for the Maxwell-Lorentz system.

In this section we consider the problem of formulating suitable initial conditions which, under appropriate analytic hypotheses, guarantee existence and uniqueness of local solutions of the Maxwell-Lorentz equations 3.6(5) and 3.6(6) for a charged dust on Minkowski space in which all the charges are of the same sign. This problem is closely related to the Cauchy problem for Maxwell's equations, which is a special case of the analysis of Section 2.9. The new Cauchy problem for the Maxwell-Lorentz equations is easily solved on the algebraic level, but, unfortunately, the resulting system of partial differential equations is not as tractable as that for the Maxwell equations alone. The new system is only quasilinear, rather than linear, and an appropriate existence theorem to guarantee C^∞ solutions from C^∞ initial data does not seem to be available. For this reason, we shall have to be content with an algebraic analysis ending in an application of the Cauchy-Kowalevski Theorem to obtain real analytic solutions from real analytic initial data. In the following, we make the blanket

5.1 Cauchy problem

hypothesis that all quantities (functions, vector fields, etc.) are real analytic.

The problem is probably most transparent from a three-dimensional point of view, so we shall work in a fixed Lorentz coordinatization on Minkowski space **M**. In terms of this coordinatization, write the four-velocity u of a charged dust as

$$u = \gamma(\vec{v})(1, \vec{v}) ,$$

and let $\vec{E}$ and $\vec{B}$ denote the electric and magnetic fields, respectively. Then the three-dimensional form of the Lorentz equation 3.6(6) (cf. 3.4(5)) is

(1) $$\frac{d(\gamma \vec{v})}{dt} = \frac{q}{m}(\vec{E} + \vec{v} \times \vec{B}) ,$$

and Maxwell's equations are

(2) $$\nabla \cdot \vec{E} = 4\pi \gamma \rho$$

(3) $$\frac{\partial \vec{E}}{\partial t} = \nabla \times \vec{B} - 4\pi \gamma \rho \vec{v}$$

(4) $$\nabla \cdot \vec{B} = 0$$

(5) $$\frac{\partial \vec{B}}{\partial t} = -\nabla \times \vec{E} .$$

Assume that $\rho, \vec{v}, \vec{E},$ and $\vec{B}$ are given at coordinate time $t = 0$, and consider the problem of calculating their time partial derivatives

$$\frac{\partial \rho}{\partial t}, \frac{\partial \vec{v}}{\partial t}, \frac{\partial \vec{E}}{\partial t}, \frac{\partial \vec{B}}{\partial t}$$

at $t = 0$ in terms of the given initial data.

The solution is almost immediate. Equation (3) gives $\partial \vec{E}/\partial t$ in terms of the initial data and (5) gives $\partial \vec{B}/\partial t$. Taking the inner product of both sides of (1) with $\vec{v}$ gives

$$(\gamma^3 \vec{v} \cdot \vec{v} + \gamma) \langle \frac{d\vec{v}}{dt}, \vec{v} \rangle = \frac{q}{m} \vec{E} \cdot \vec{v} ,$$

which can be solved for the component of $d\vec{v}/dt$ in the direction of $\vec{v}$ in terms of the initial data. The component of $d\vec{v}/dt$ orthogonal to $\vec{v}$ can be read off immediately from (1) as

$$\frac{q}{m\gamma}(\vec{E}_\perp + \vec{v} \times \vec{B}) ,$$

where $E_\perp$ denotes the component of $\vec{E}$ orthogonal to $\vec{v}$. Simple linear algebra now gives $d\vec{v}/dt$ in terms of $\vec{v}, \vec{E},$ and $\vec{B}$ at time 0, and then the easily verified relation

(6) $$\frac{d\vec{v}}{dt} = \frac{\partial \vec{v}}{\partial t} + (\vec{v} \cdot \nabla)\vec{v}$$

yields $\partial \vec{v}/\partial t$. (Here $d\vec{v}/dt$ represents the coordinate velocity of a single fluid particle, while $\partial \vec{v}/\partial t$ is the partial derivative with respect to coordinate time of $\vec{v}$ considered as a time and space dependent vector field.) Finally, $\frac{\partial \rho}{\partial t}$ can be obtained by differentiating (2), since $\frac{\partial \vec{E}}{\partial t}$ and $\frac{\partial \vec{v}}{\partial t}$ are already determined in terms of the initial data.

Thus it is routine to write a computer program which, given a network of values of $\rho, \vec{v}, \vec{E},$ and $\vec{B}$ at some given time t will compute approximations to the corresponding values at $t + \Delta t$. What makes the integration not quite trivial is that the initial values for these four quantities cannot be arbitrarily specified because the original equations (1)-(5) also impose constraints on the initial data. For example, (4) requires $\vec{B}$ to be divergenceless, and (2) relates $\vec{E}, \rho,$ and $\vec{v}$. A less obvious constraint is the condition 2.8(25)

(7) $$0 = \delta(\rho u) = \frac{\partial(\rho u^0)}{\partial t} + \sum_{\alpha=1}^{3} \frac{\partial(\rho u^\alpha)}{\partial x_\alpha}$$

for conservation of charge, which follows automatically from the Maxwell equation

$$\delta F = 4\pi \rho u^* \; .$$

In three-dimensional notation, (7) reads

(8) $$\frac{\partial(\gamma\rho)}{\partial t} + (\vec{v}\cdot\nabla)(\gamma\rho) + \gamma\rho \nabla\cdot\vec{v} = 0 ,$$

which, recalling from Figure 2-6 that $\gamma\rho$ represents the charge density as measured in the coordinate frame, is seen to be just the classical "continuity" equation. Because of these constraints (and the possibility of others, as yet unnoticed), it is not obvious that the formally integrated functions will, in fact, satisfy the full system. To assure this, we shall formulate the initial value problem in such a way that it is clear that all constraints will be conserved by the integration.

First observe that there need be only three unknown quantities because ρ is completely determined by $\vec{E}$ and $\vec{v}$ via (2). Therefore, let us take $\vec{v}, \vec{E},$ and $\vec{B}$ as unknowns, *define* ρ by (2) and change the initial value problem to that of specifying $\vec{v}, \vec{E},$ and $\vec{B}$ at time 0 subject to the single constraint

(4) $$\nabla\cdot\vec{B} = 0 \qquad \text{at } t=0 .$$

(Exercise 2.24.) Now notice that if $\vec{B}$ is chosen to satisfy (5) for all time, then (4) automatically holds for all time if it holds at time 0 because

$$\frac{\partial(\nabla\cdot\vec{B})}{\partial t} = \nabla\cdot\frac{\partial\vec{B}}{\partial t} = -\nabla\cdot(\nabla\times\vec{E}) = 0 ,$$

the last equality being a familiar vector identity (which is really just $d(d(E^*)) = 0$ in disguise). Once the problem is set up in this way, it should be apparent that the problem of integrating the Maxwell-Lorentz system in terms of given initial data satisfying (4) is algebraically solved if we define ρ by (2) and obtain $\vec{B}$ by integrating (5), $\vec{E}$ by integrating (3), and $\vec{v}$ by integrating (6). (By integrating (6) we mean integrating

$$\frac{\partial\vec{v}}{\partial t} = \frac{d\vec{v}}{dt} - (\vec{v}\cdot\nabla)\vec{v}$$

with $\frac{d\vec{v}}{dt}$ obtained from (1) by linear algebra as described above.) The very method of determining $\vec{B}, \vec{E},$ and $\vec{v}$ guarantees that (5), (3), and (1) hold for all time. Equation (2) holds tautologically from the definition of ρ, and we have already remarked that (4) for all time follows from (5) for all time and (4) at time 0. For completeness, we write down the resulting system:

(9) (i) $$\frac{\partial \vec{B}}{\partial t} = -\nabla \times \vec{E}$$

(ii) $$\frac{\partial \vec{E}}{\partial t} = \nabla \times \vec{B} - 4\pi\gamma\rho\vec{v}$$

(iii) $$\frac{\partial \vec{v}}{\partial t} = \frac{q\vec{E}\cdot\vec{v}}{m\gamma^3 \vec{v}\cdot\vec{v}}\vec{v} - (\vec{v}\cdot\nabla)\vec{v} + \frac{q}{m\gamma}(\vec{E}_\perp + \vec{v}\times\vec{B}) \; .$$

The verification that (9) is equivalent to (1), (3), and (5) is left as an exercise.

From a physical point of view it might seem more natural to specify ρ and $\vec{v}$ as initial data rather than $\vec{E}$, $\vec{B}$, and $\vec{v}$. If we think of the fluid as consisting of all the charges in the universe, then we might expect that the fluid itself would generate the $\vec{E}$ and $\vec{B}$ fields, so specifying the initial state of the fluid should be enough. This conclusion is correct as far as it goes. We *can* obtain a solution with arbitrary initial values for ρ and $\vec{v}$, but it will not necessarily be unique. There still remains some freedom to specify $\vec{E}$ and $\vec{B}$. Given ρ and $\vec{v}$ initially, we can always find a vector field $\vec{E}$ whose divergence is $4\pi\gamma\rho$ at time 0. This $\vec{E}$ is not unique because its curl at time 0 may be arbitrarily specified subject to the condition

$$\nabla\cdot(\nabla\times\vec{E}) = 0 \qquad \text{at } t=0 \; .$$

(This is just the result of Section 2.9 applied to the one-form E^*.) In this way, the present case reduces to the previous one.

The above analysis together with the Cauchy-Kowalevski Theorem [John, p.74] applied to the system (9) constitute a proof of the following theorem, which summarizes the results of this section. (A similar analysis of the analogous Cauchy problem in general relativity, can be found in [Synge, Chapter 10].)

Theorem.
Let D be an open subset of R^3. Let $\vec{x} \longmapsto \vec{E}_0(\vec{x})$, $\vec{x} \longmapsto \vec{B}_0(\vec{x})$, and $\vec{x} \longmapsto \vec{v}_0(\vec{x})$ be given vector-valued functions on D. Assume that $\vec{B}_0$ satisfies

$$\nabla\cdot\vec{B}_0 = 0 \; .$$

Assume also that $\vec{E}_0, \vec{B}_0$, and $\vec{v}_0$ are real analytic on D. Then there exists an open subset $\tilde{D}$ of R^4 containing D and real analytic vector functions $t, \vec{x} \longmapsto \vec{E}(t,\vec{x}), \vec{B}(t,\vec{x}), \vec{v}(t,\vec{x})$, defined for $t, \vec{x} \in \tilde{D}$, which satisfy the Maxwell-Lorentz equations (1) through (5) (with ρ defined by (2)) and the initial conditions

$$\vec{E}(0,\vec{x}) = \vec{E}_0(\vec{x}), \qquad \vec{B}(0,\vec{x}) = \vec{B}_0(\vec{x}), \qquad \vec{v}(0,\vec{x}) = \vec{v}_0(\vec{x}) \qquad \text{for all } \vec{x}\in D \; .$$

The solution is unique (among real analytic solutions).

Alternatively, ρ and $\vec{v}$, may be specified arbitrarily on $\{0\}\times D$ and then $\vec{B}$ and $\vec{E}$ may be specified arbitrarily subject to the consistency conditions

$$\nabla\cdot\vec{B} = 0 \quad \text{and} \quad \nabla\cdot\vec{E} = 4\pi\gamma\rho \qquad \text{on } \{0\}\times D \; .$$

5.2 Spherically symmetric solutions of the Maxwell-Lorentz system.

To make sense of spherical symmetry, we must work in a particular coordinatization of Minkowski space **M**. We choose as coordinates t, r, θ, ϕ, with t time and r, θ, ϕ, the usual spherical polar coordinates in Euclidean space, so that the metric tensor is

$$dt \otimes dt - dr \otimes dr - r^2(d\theta \otimes d\theta + \sin^2\theta \, d\phi \otimes d\phi) .$$

By a *spherically symmetric* solution of the Maxwell-Lorentz equations for a charged dust in which only one sign of charge is present (cf. Section 3.6),

(1) $$\frac{m}{q}\frac{du}{d\tau} = \hat{F}(u)$$

(2) $$dF = 0$$

(3) $$\delta F = 4\pi \rho u^* ,$$

we mean a solution of the form

(4) $$u = \gamma(v)\frac{\partial}{\partial t} + v\gamma(v)\frac{\partial}{\partial r} ,$$

(5) $$F = E(t,r) dt \wedge dr + B(t,r) \sin\theta \, d\theta \wedge d\phi ,$$

where v, E, and B are functions on **M** which do not depend on θ and ϕ. This is the same as requiring that u and F be invariant under the natural action of the spatial rotation group. Physically, (4) and (5) say that the fluid particles move radially inward or outward and that in the given coordinatization, the electric field $\vec{E}$ in the coordinate frame is radially outward with magnitude E, and the magnetic field is also radially outward. The solution is assumed to be defined on some open subset of **M** of the form $0 < |r - r_0| < \epsilon$, $|t - t_0| < \epsilon$, θ, ϕ, arbitrary. (The singularity of the spherical coordinates at $r = 0$ makes the spatial origin an annoying special case which we ignore for simplicity. We assume the solutions extended to the spatial origin when possible.) As explained in Section 3.6, we may view (3) as defining ρ.

Straightforward computation shows that the Maxwell equation $dF = 0$ is equivalent to

$$B(t, r) = \text{constant} .$$

Further computation shows that

$$\delta(\sin\theta \, d\theta \wedge d\phi) = 0 ,$$

and that the term in $\hat{F}$ corresponding to the magnetic field $B \sin\theta \, d\theta \wedge d\phi$ annihilates any u of the form (4). This shows that if u, F satisfies the Maxwell-Lorentz equations, then so does $u, E \, dt \wedge dr$, and hence we may as well assume that

(6) $$F = E(t,r) dt \wedge dr .$$

At each point of Minkowski space, define a vector w by

(7) $$w = v\gamma(v)\frac{\partial}{\partial t} + \gamma(v)\frac{\partial}{\partial r} .$$

5.2 Spherically symmetric solutions

Then w is a unit vector orthogonal to u, and the pair u, w spans the same subspace as $\frac{\partial}{\partial t}, \frac{\partial}{\partial r}$. From this one easily checks that for any function g,

(8) $$dg = u(g)u^* - w(g)w^* + \frac{\partial g}{\partial \theta} d\theta + \frac{\partial g}{\partial \phi} d\phi.$$

(Recall that $w(g)$ denotes the derivative of g in the direction of w. The anomalous minus sign comes from the fact that w has negative norm.)

For later convenience, set $k := m/q$. Since

$$F = E\, dt \wedge dr = E\, w^* \wedge u^*,$$

we have

(9) $$k \frac{du}{d\tau} = \hat{F}(u) = Ew,$$

so E is proportional to the magnitude of the proper acceleration $\frac{du}{d\tau}$.

It will be convenient to calculate δF by the formula 2.8(12):

$$\delta F = -\perp d \perp F.$$

First verify that

(10) $$\perp(w^* \wedge u^*) = \perp(dt \wedge dr) = \pm r^2 \sin\theta\, d\theta \wedge d\phi,$$

where the sign depends on the orientation chosen for Minkowski space. For definiteness we choose the orientation so that the sign in (10) is "+". From (10) and (8),

(11) $$\delta F = -\perp d(E\, r^2 \sin\theta\, d\theta \wedge d\phi)$$
$$= -\perp(u(Er^2)\sin\theta\, u^* \wedge d\theta \wedge d\phi - w(Er^2)\sin\theta\, w^* \wedge d\theta \wedge d\phi)$$
$$= -\frac{u(Er^2)}{r^2} w^* + \frac{w(Er^2)}{r^2} u^*,$$

so that the Maxwell equation

$$\delta F = 4\pi \rho u^*$$

is equivalent to

(12) $$u(Er^2) = 0,$$

This says that Er^2 is constant on the worldline of any fluid particle, say

(13) $$Er^2 = Q,$$

where Q is a constant which depends on the particle. Put differently, Q is a function on spacetime which is constant on the worldlines.

Also, from (11) and the Maxwell equation $\delta F = 4\pi \rho u^*$,

(14) $$4\pi \rho = \frac{w(Er^2)}{r^2} = \frac{w(Q)}{r^2}.$$

Solving (4) and (7) for $\frac{\partial}{\partial r}$ yields

$$\frac{\partial}{\partial r} = -v\gamma u + \gamma w ,$$

so

$$\frac{\partial Q}{\partial r} = \gamma w(Q) = 4\pi \gamma r^2 \rho ,$$

and hence at any fixed coordinate time t,

(15) $\quad Q(t, r) = Q(t, r_0) + \int_{r_0}^{r} \gamma(v(t, s))\rho(t, s) 4\pi s^2 \, ds ,$

where r_0 is arbitrary. Recall from Figure 2-6 that $\gamma \rho$ is the charge density as measured in the coordinate frame, and so for $r > r_0$,

$$Q(t, r) - Q(t, r_0)$$

is the total charge between radial coordinates r_0 and r. If we make the formal choice $r_0 = 0$ and assume that E is bounded, then (13) implies that $Q(t, 0) = 0$, and $Q(t, r)$ becomes the total charge in a ball of radius r centered at the origin at time t. Once Q is identified as charge, (13) is recogized as the form of the classical Coulomb law which states that the electric field at a distance r from the center of a ball of charge is Q/r^2, where Q is the total charge inside the sphere of radius r about the center.* However, (12) is more general than the classical Coulomb law in that the particles in the ball may be moving in the coordinate frame, whereas the classical Coulomb law is usually stated only for stationary charges.

From (12) it is easy to derive a second-order ordinary differential equation which characterizes the trajectories of the fluid particles. This is most conveniently done by parametrizing the radial position $r(t)$ of a fluid particle by coordinate time t rather than proper time τ. We have already noted that (9) shows that E is a constant multiple of the scalar proper acceleration $-\langle du/d\tau, du/d\tau \rangle^{1/2}$. To obtain the relation between the latter and the coordinate acceleration d^2r/dt^2, differentiate (4) with respect to τ (more precisely, compute $D_u u$, which, since $D_u(\partial/\partial t) = 0 = D_u(\partial/\partial r)$, amounts to differentiating the components of u with respect to τ):

$$\frac{du}{d\tau} = \frac{d\gamma}{d\tau}\frac{\partial}{\partial t} + \frac{d(v\gamma)}{d\tau}\frac{\partial}{\partial r}$$

$$= \gamma^3 \frac{dv}{d\tau}[v\frac{\partial}{\partial t} + \frac{\partial}{\partial r}] = \gamma^3 \frac{dv}{dt} w .$$

We restate this result as a lemma, which will be needed in a later section.

---

* Incidentally, if the 4π had originally been omitted from the Maxwell equation $\delta F = 4\pi \rho u^*$, it would have popped up in the denominator of the Coulomb law, and the latter convention is used in some texts. Our convention was chosen because it simplifies the retarded potentials and the Lorentz-Dirac equation. Unfortunately, there seems to be no normalization which eliminates factors of π everywhere.

Lemma.

Let a particle move on a line in Euclidean 3-space. Denote the Euclidean coordinate of the line as r, so that the coordinate velocity is $v = dr/dt$ and the four-velocity is $u = \gamma(v)\partial/\partial t + v\gamma(v)\partial/\partial r$. Define the *scalar proper acceleration* A by

$$\frac{du}{d\tau} = A\left(v\gamma(v)\frac{\partial}{\partial t} + \gamma(v)\frac{\partial}{\partial r}\right),$$

so that $A^2 = -\langle du/d\tau, du/d\tau\rangle$. Then the relation between A and the coordinate acceleration dv/dt is:

$$A = \gamma^3 \frac{dv}{dt} = (1-v^2)^{-3/2}\frac{dv}{dt}.$$

Comparing the relation just established with (9) gives

(16) $$E = kA = k\gamma^3\frac{dv}{dt} = k\gamma^3\frac{d^2r}{dt^2}.$$

Hence (13) is equivalent to the second-order ordinary differential equation:

(17) $$\frac{d^2r}{dt^2} = k^{-1}\left[1 - \left(\frac{dr}{dt}\right)^2\right]^{3/2}\frac{Q}{r^2},$$

where Q is a function on $\mathbf{M}$ which is constant on each worldline. The solutions are uniquely determined by specifying $r(t_0)$ and $v(t_0) := dr/dt(t_0)$ at some initial time t_0. Thus if for each radial coordinate r', we are given the initial velocity $v_0(r')$ of the fluid particle with $r(t_0) = r'$, and if we choose $Q(t_0, r')$ as an arbitrary function $f(r')$, then from the standard existence and uniqueness theorem for ordinary differential equations we obtain a unique local solution to (17) with

$$r(t_0) = r', \qquad \frac{dr}{dt}(r_0) = v_0(r'), \text{ and } \quad Q(t_0, r') = f(r').$$

Note that (15) shows that once the initial velocities $v_0(r')$ are chosen, specifying the function

$$r' \longmapsto Q(t_0, r') = f(r')$$

is, up to a constant of integration, the same as specifying the initial proper charge density

$$r' \longmapsto \rho(t_0, r').$$

Specifying the constant of integration $Q(t_0, r_0)$ is, by (13), the same as specifying $E(t_0, r_0)$. Given these initial specifications, (17) determines u, and (13) and (15) determine E.

The following theorem summarizes the above analysis. The proof, which is left to the reader, consists of the above outline interspersed with standard results from the theory of ordinary differential equations to check that everything makes sense. (For example, that a collection of worldlines constructed via solutions to (17) satisfying given initial conditions at $t = t_0$ fill an open subset of $\mathbf{M}$. This is necessary to ensure that the function $t, r \longmapsto Q(t, r)$, is well-defined by the specification of its initial values $Q(t_0, r') = f(r')$ and the condition that it be constant on the worldlines.)

Theorem.

Let an initial time t_0 and a radial coordinate r_0 be given. Suppose that also given are:

a C^∞ initial velocity function $r \longmapsto v_0(r) < 1$,

a C^∞ initial charge density function $r \longmapsto \rho_0(r) > 0$, and

a value E_0 for the electric field at time t_0 and radial coordinate r_0.

Then there exist $\epsilon > 0$ and a unique spherically symmetric C^∞ solution u, F for the Maxwell-Lorentz system (1), (2), (3), which is defined on the open subset of **M** for which $|t-t_0| < \epsilon$, $|r-r_0| < \epsilon$, and which satisfies (in the notation above):

(i) $\qquad F = E\, w^* \wedge u^* = E\, dt \wedge dr$,

(ii) $\qquad v(t_0, r) = v_0(r) \quad$ and $\quad E(t_0, r) = E_0 \quad$ for $\; |r-r_0| < \epsilon$,

(iii) $\qquad \rho(t_0, r) = \rho_0(r) \quad$ for $\; |r-r_0| < \epsilon$.

The solution also satisfies:

(iv) $\qquad Er^2$ is constant on the worldline of any fluid particle; that is,

$$u(Er^2) = 0 \;,$$

and

(v) $\qquad$ for any t, r with $|t-t_0| < \epsilon$, $|r-r_0| < \epsilon$,

$$E(t, r) = E(t, r_0)\frac{r_0^2}{r^2} + \frac{1}{r^2} \int_{r_0}^{r} \gamma(v(t,s))\rho(t,s)\, 4\pi s^2\, ds \;.$$

We have seen that specifying the initial proper charge density ρ, the initial velocities $v_0(r)$, and a constant of integration $Q(t_0, r_0)$ uniquely determines the solution up to a free magnetic field $B \sin\theta\, d\theta \wedge d\phi$. To understand the significance of the constant of integration, suppose that given r_0, we choose $Q(t_0, r_0) = 0$. Then, since the magnitude of the proper acceleration is proportional to $E = Q/r^2$ by (9) and (13), the acceleration on the sphere $r = r_0$ vanishes at $t = t_0$. Since Q is constant on the particle worldlines, a "zero-acceleration sphere" which coincides with the sphere $r = r_0$ at $t = t_0$ expands outwards (or contracts inwards) with the fluid particles. From (15) and (17), particles outside this sphere are accelerated outwards, while particles inside are accelerated inwards. †

The fact that particles inside the sphere are accelerated inwards is contrary to the intuition that charges of like sign repel each other. To see this clearly, suppose that all the fluid particles are initially inside the sphere, and assume for simplicity that all initial velocities are zero, so that the zero-acceleration sphere remains fixed at $r = r_0$ for all time. Then one sees from (15) and (17) that the ball of charge will contract toward the spatial origin, and the worldline of each particle will end at the origin (assuming that the local solutions of (17) are defined as far as the origin). One can object to this solution on the grounds that E is singular at the spatial origin. However, in general relativity, similar singularities ("black holes") are

---

† Physically realistic solutions in a universe without point charges would presumably have $r_0 = 0$ for all time, so that no particles could be inside the sphere. However, the solutions with $r_0 > 0$ still hold some mathematical and conceptual interest.

common. The intuition that like charges repel and unlike charges attract can be saved by interpreting the singularity at the origin as a point charge of variable strength which sucks in the fluid charge, but the mathematics is obviously independent of any such picture.

It seems that no explicit solutions to equation (17) are known. However, its analog in two-dimensional Minkowski space (one space and one time dimension) is easily solved, and the solutions illustrate other features of these equations which seem to violate ordinary intuition. The above analysis (with omission of the angle variables, of course) goes through unchanged for two-dimensional Minkowski space up to equation (10). In two dimensions, the dual $\perp(w^* \wedge u^*)$ of the 2-form $w^* \wedge u^*$ is a 0-form; that is, a scalar function. The analog of (10) is

(10') $$\perp(w^* \wedge u^*) = \pm 1 ,$$

and that of (11) is:

(11') $$\delta F = -u(E)w^* + w(E)u^* .$$

Hence

(12') $$u(E) = 0 ,$$

which says that the electric field, and hence from (9) the proper acceleration, is constant on the particle worldlines. Characterization of worldlines of particles moving in one space dimension with constant proper acceleration is a standard exercise in special relativity (Exercise 1.22). The solution in terms of Lorentz coordinates t (= time), r (= space) is

(18) $$t = A_p^{-1} \sinh(A_p \tau)$$

$$r = A_p^{-1}(\cosh(A_p \tau) - 1 + pA_p) ,$$

where p is a real parameter which labels the particles, chosen so that $r(0) = p$ and $t(0) = 0$, and A_p is the scalar proper acceleration of particle p. This is known as "hyperbolic motion" because the worldlines are hyperbolas in Minkowski space. The analogs of (13), (16), and (15) are

(13') $$E = Q ,$$

(16') $$E = k\gamma^3 \frac{d^2r}{dt^2} = k A_p ,$$

where p is the label of the worldline through the point at which both sides are evaluated, and

(15') $$Q(t, r) = Q(t, r_0) + \int_{r_0}^{r} \gamma(v(t, s))\rho(t, s)\, ds .$$

Hence for any r, the proper acceleration of the particle at space coordinate r at time 0 is

(19) $$A_r = k^{-1} Q(0, r_0) + k^{-1} \int_{r_0}^{r} \gamma(v(0, s))\rho(0, s)\, ds .$$

The above relations are valid in two-dimensional Minkowski space *without* any assumption of spatial symmetry, and in the absence of symmetry, there is no natural choice for the constant of integration $Q(0, r_0)$. If, however, we introduce symmetry about the origin via the assumption

(20) $\quad\quad\quad \rho(t, -r) = \rho(t, r) \quad$ for all t, r,

then the obvious choice would be $r_0 = 0$ and

(21) $\quad\quad\quad Q(0, 0) = 0$.

Then

(22) $\quad\quad\quad A_r = k^{-1} \int_0^r \gamma(v(0, s)) \rho(0, s) \, ds$,

and the worldlines appear as in Figure 1, which seems physically reasonable except, perhaps, for the fact that the velocity of each fluid particle approaches the speed of light as $t \to \pm\infty$.

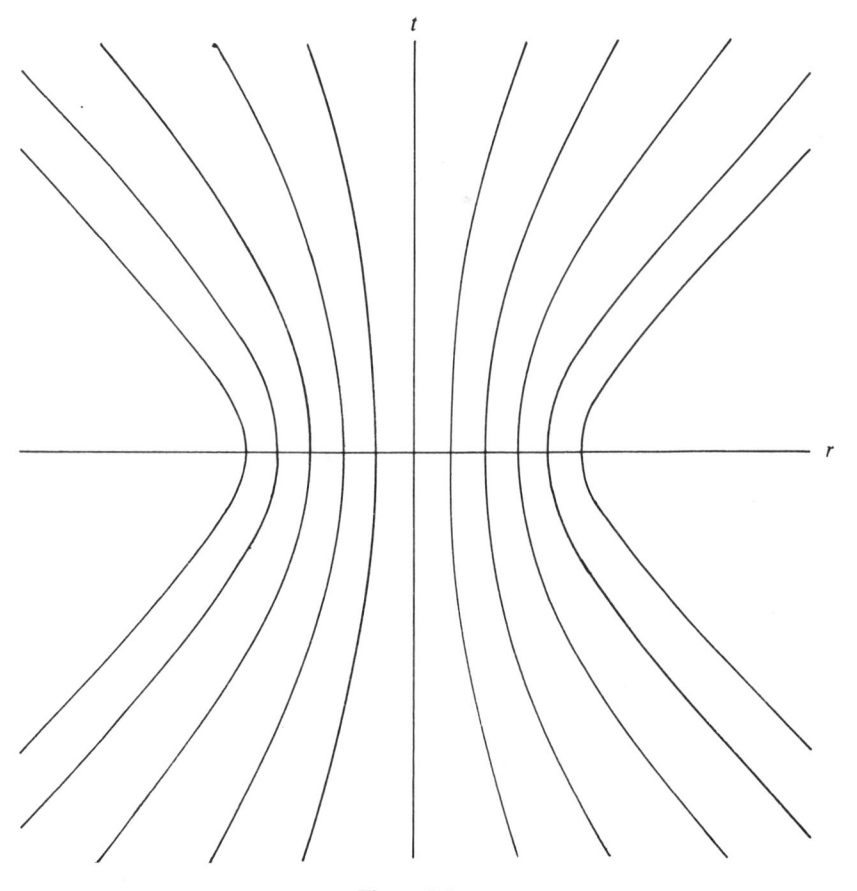

Figure 5-1

5.2 Spherically symmetric solutions

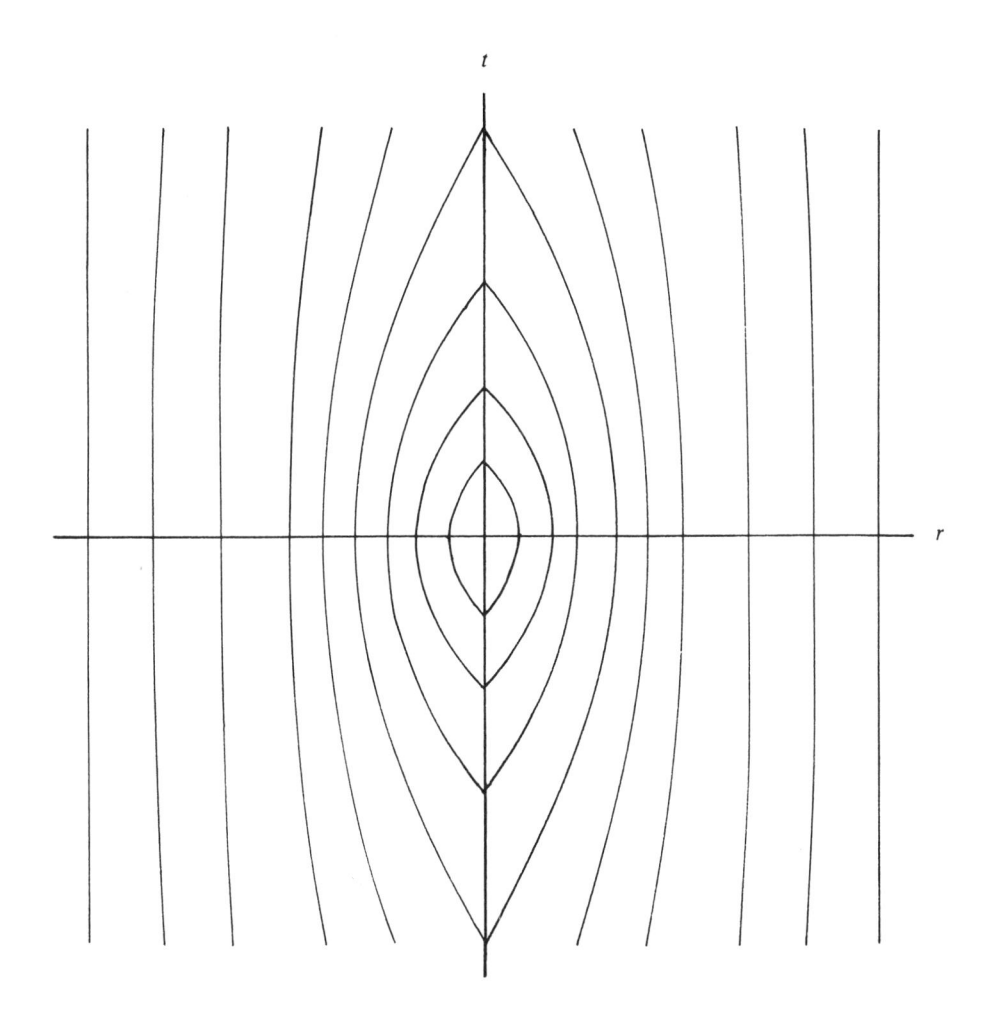

Figure 5-2

Up to now we have assumed that the sign of all the charges is the same. Let us now relax this condition to see what happens when charges of both signs are present. In this case, it is best to write (1) as

(1')
$$\rho_m \frac{du}{d\tau} = \rho \hat{F}(u) ,$$

where

$$\rho_m := \frac{m}{q} \rho = k\rho$$

is the proper mass density, and we set $k := m/q$. If the fluid particles are identical except for sign, then $|k|$ is constant and ρ_m is locally proportional to ρ. For simplicity, let us think of this case, though our analysis will not require that $|k|$ be constant, but only the weaker assumption that mass is conserved. (Mass conservation in this context means that $\delta(\rho_m u) = 0$.)

First we observe that k is constant on the particle worldlines. This is seen as follows. It is an easy exercise that the equation of conservation of charge,

$$\delta(\rho u) = 0 ,$$

can also be written as

$$0 = \delta(\rho u) = u(\rho) + \rho\delta(u) \quad , \text{ or}$$

$$u(\log \rho) = -\delta(u) .$$

If also $\delta(\rho_m u) = 0$, then

$$u(\log \rho_m) = -\delta(u) = u(\log \rho) ,$$

and so

$$u(\log(\frac{\rho_m}{\rho})) = u(\log \rho_m) - u(\log \rho) = 0 ,$$

which says that $\log(\rho_m/\rho)$, and hence ρ_m/ρ, is constant on the worldlines. Thus in solving (17), k may be treated as constant, and (18) and (19) are still valid.

Now suppose we take the previous solution with $Q(0, 0) = 0$ (in which all the charges had the same sign) and reverse the sign of all the charge to the left of the origin. This corresponds to replacing assumption (20) with

(23) $\rho(t, -r) = -\rho(t, r)$ for all t, r.

Intuition that like charges repel and unlike charges attract suggests that the charges of opposite sign on opposite sides of the origin should attract each other, and so it would be reasonable to expect worldlines of the general form shown in Figure 2. However, this is not what happens if we maintain the "natural" choice (21) for the constant of integration. Inspection of (22) shows that, owing to the fact that $k = \frac{\rho_m}{\rho}$ and ρ change sign simultaneously, the worldlines are the same as before (shown in Figure 1). Thus reversing the sign of all the charge to the left of the origin has no effect on the motion of the fluid! The same general phenomenon occurs in the three-dimensional case also; reversing the sign of all the charge inside a zero-acceleration sphere does not affect the motion of the fluid.

It should be noted, however, that in the one-dimensional case when both signs of charge are present, (21) is no longer the only plausible choice for the constant of integration. If instead we require

(21') $\quad Q(0, -\infty) = 0 = Q(0, \infty)$,

then existing symmetry is preserved, the analog of (22) is

(22') $\quad A_r = k^{-1} \int_{-\infty}^{r} \gamma(v(0, s)) \rho(0, s) \, ds$,

and the worldlines do appear as in Figure 2. However, the four-velocity field u is not continuous at the origin in this case, so we really have two disjoint solutions (one for $r < 0$ and one for $r > 0$) rather than one.

We have presented two situations (the collapsing ball of like charges in three space dimensions and the example just discussed) which call into question the intuition that like charges repel and unlike charges attract. In these two cases the cause of the anomalies was relatively transparent, and ordinary intuition could be saved by simple adjustments. However, these examples may leave the reader with a feeling of unease. In more complicated situations it might not be obvious how to choose the arbitrary constants or reinterpret the solutions in order to avoid counterintuitive behavior. Is there some general principle which tells how to do this? If not, is it possible that the seemingly counterintuitive behavior might actually be physically realistic? Where does our intuition that like charges repel and unlike charges attract come from in the first place? The answer to the first question is that no such general principle is known. Whether apparently counterintuitive behavior (such as a ball of like charges contracting) is physically possible is also unknown, partly because very few explicit solutions to the Maxwell-Lorentz equations are known, and partly because charged dusts are not readily available for observation.* Finally, our intuition concerning the respective repulsion or attraction of like or unlike charges can be traced back to the Coulomb field

$$E = \frac{q}{r^2}$$

of high school electrostatics, which is part of the retarded field. Since our analysis of the Maxwell-Lorentz equations did not use the retarded fields at all, it is not surprising that solutions may fail to conform to intuition derived from latter.

This suggests that it might be necessary to add the retardation condition to the Maxwell-Lorentz equations in order to guarantee physically realistic solutions. However, the extent to which the retardation condition is compatible with the Maxwell-Lorentz equations is not known. Not only is there no theorem which guarantees the existence of retarded solutions to the latter, but also, to the best of my knowledge, no rigorous retarded solution to the Maxwell-Lorentz equations with a nonzero charge density has ever been constructed. One obstacle to such constructions is that the domain on which retarded solutions are defined must, by definition, contain at least all backward light cones with vertices at each of its points, and physically realistic solutions would presumably be globally defined on all of spacetime. However, we shall see in the next section that for some physically reasonable initial conditions, there are no such solutions. This shows that retarded solutions for the Maxwell-Lorentz equations are not *always* available in physically realistic situations, and it is even barely conceivable that retarded solutions may *never* exist with nonzero charge density.

---

* It would be probably be hard to find anyone who would bet on observing a contracting ball of charge. However, even stranger effects in which people do believe (such as black holes) abound in relativity theory.

However, this last possibility seems unlikely. In fact, any everywhere-defined spherically symmetric solution which is a pure electric field in the coordinate frame is *both* retarded and advanced. To see why, suppose we can find a spherically symmetric C^∞ solution of the Maxwell-Lorentz equations which is defined on all of Minkowski space. As explained earlier, we may assume that the magnetic field vanishes. Moreover, the electric field must be radially outward, and it follows that if the solution is defined everywhere and continuous, then the electric field must be zero at the spatial origin. From the expression

$$\nabla \cdot \vec{E} = \frac{1}{r^2}\frac{\partial}{\partial r}(r^2 E_r) + \frac{1}{r\sin\theta}\frac{\partial}{\partial \theta}(\sin\theta E_\theta) + \frac{1}{r\sin\theta}\frac{\partial E_\phi}{\partial \phi}$$

for the divergence in polar coordinates, where E_r, E_θ, E_ϕ denote the components of $\vec{E}$ along unit vectors in the r, θ, ϕ directions, one sees immediately that any such solution of the Maxwell equation

$$\nabla \cdot \vec{E} = 4\pi\gamma\rho$$

is of the form

(24) $$E(t,r) = \frac{Q(t,r)}{r^2},$$

where E is the magnitude of the electric field and $Q(t,r)$ is given by (15) with $r_0 = 0$ and $Q(t,0) = 0$. Given the original global solution of the Maxwell-Lorentz equations, we can construct the retarded field F^{ret} corresponding to the given current ρu. The retarded field will also be everywhere defined and spherically symmetric and so has electric field satisfying (24). One can see directly from the definition 3.7(9) of the retarded potential A^{ret} that in the spherically symmetric situation it is of the form

$$A^{ret} = \phi_0 dt + \phi_1 dr,$$

where ϕ_0 and ϕ_1 depend only on r and t. Hence

$$F^{ret} = dA^{ret} = \left(\frac{\partial \phi_0}{\partial r} - \frac{\partial \phi_1}{\partial t}\right) dr \wedge dt,$$

so F^{ret} is a pure electric field in the coordinate frame and must also be given by (24); in other words, F^{ret} is the original field. Of course, similar reasoning shows that the original solution was advanced as well. Thus the problem of constructing spherically symmetric retarded solutions which are everywhere defined reduces to the problem of constructing globally defined spherically symmetric solutions. We shall see in the next section that for some initial conditions there is an obvious obstruction which prevents such global solutions. However, for other initial conditions there is no such obstruction apparent, and it seems likely that some global solutions with nonzero charge densities do exist.

5.3 Nonexistence of global solutions of the Maxwell-Lorentz equations.

In this section we shall apply the information obtained about spherically symmetric solutions of the Maxwell-Lorentz equations to show that solutions globally defined on all of Minkowski space need not exist even for physically reasonable initial conditions. Standard existence theorems applied to the ordinary differential equation 5.2(17) guarantee *local* solutions; i.e. solutions

$$u = u(t, r), \qquad F = F(t, r)$$

defined on some open set

$$t_1 < t < t_2, \qquad r_1 < r < r_2.$$

However, we shall see that for certain initial conditions, there is a finite minimum time t_1 and a finite maximum t_2 beyond which the solution cannot be extended. Very roughly speaking, this happens when the initial fluid density thins out sufficiently rapidly as $r \to \infty$. This implies that these solutions are not retarded (since retarded solutions, by definition, must be defined on backward light cones).

We shall see that the nonexistence of global solutions is not a manifestation of some unlikely pathology, but rather is due to a simple phenomenon well-known in relativity theory as "shell-crossing". Though we shall only prove the occurence of shell-crossing for certain spherically symmetric solutions, it will be apparent that it is to be expected in a wide variety of situations. This calls into question the extent to which the retardation condition is consistent with the Maxwell-Lorentz equations.

Consider a spherically symmetric solution with all initial ($t = 0$) velocities zero and zero acceleration at the spatial origin. For notational simplicity we assume that $m/q = 1$, but what we do will work for arbitrary m and q. According to 5.2(9) and 5.2(13), a particle at radial coordinate r at time t will experience a proper acceleration Q/r^2, where Q is the total charge inside the sphere of radius r (centered at the origin) at time t. Moreover, Q is constant on the worldline of the particle. Thinking physically, the constancy of Q suggests that if we fix our attention on a sphere, or "shell", of fluid particles, say the particles with radial coordinate r at time 0, and watch this shell expand in time, then particles initially inside the shell should always remain inside, and particles initially outside should remain outside. If the charge density were physically obtained by counting discrete charged particles per unit volume, then these conditions would ensure the constancy of Q and conversely. A little more reflection should convince the reader that *independently* of the constancy of Q, particles would not normally be expected to cross the shell because that would entail two distinct four-velocities (one for the particle on the shell and one for the crossing particle) at the same point in spacetime except in the improbable event that the crossing occurred in such a way that both particles had precisely the same coordinate velocity at the moment of crossing. We shall use the constancy of Q to show that certain initial conditions imply that a crossing with different velocities must occur. When this happens, the solution obviously breaks down at the crossing because the fluid four-velocity cannot be well-defined there.

Let $v := \dfrac{dr}{dt}$ denote the coordinate velocity of a particular fluid particle whose radial coordinate is described by the function $t \longmapsto r(t)$. By 5.2(17), the coordinate acceleration $\dfrac{dv}{dt}$ is given by

(1) $$\frac{dv}{dt} = \gamma^{-3} \frac{Q}{r^2},$$

where the charge Q is given by 5.2(15) with $r_0 = 0$:

$$Q(t, r) = \int_0^r \gamma(v(t, s)) \rho(t, s) 4\pi s^2 \, ds \ .$$

(Although 5.2(15) was not proved for $r_0 = 0$, it follows from 5.2(13) that $\lim_{r \to 0} Q(t, r) = 0$ when the acceleration vanishes at the spatial origin, and this implies 5.2(15) with $r_0 = 0$.)

Differentiating γ with respect to coordinate time gives

$$\frac{d\gamma}{dt} = \gamma^3 v \frac{dv}{dt} = \frac{Q}{r^2} v = \frac{Q}{r^2} \frac{dr}{dt} \ ,$$

and so

$$\gamma(v(t)) = \gamma(v(0)) + \int_0^t \frac{Q}{r^2} \frac{dr}{dt} \, dt \ .$$

Since Q is constant on the worldline, it may be taken outside the integral, and so we have

(2) $\qquad \gamma(v(t)) = \gamma(v(0)) - \dfrac{Q}{r(t)} + \dfrac{Q}{r(0)} \ .$

This equation has a simple physical interpretation. Recall that $m\gamma(v)$ is the relativistic energy of a particle with rest mass m and coordinate velocity v. If by analogy we think of $\gamma(v(t))$ as the "kinetic energy" of an infinitesimal fluid particle, then (2) is similar to the Newtonian law of conservation of energy in which Q/r plays the role of potential energy.

The assumption of zero initial velocities and equation (1) imply that all particles move outwards for $t > 0$ with ever-increasing velocities and that $r(\infty) = \infty$. Hence

(3) $\qquad \gamma(v(\infty)) = 1 + \dfrac{Q}{r(0)} \ .$

Now consider a second fluid particle whose radial coordinate at time t is denoted $\bar{r}(t)$ and velocity denoted $\bar{v}(t)$. The charge inside the sphere with radius $\bar{r}(t)$ is independent of t (constant on the worldline) and will be denoted $\bar{Q}$. We cannot denote the corresponding charge for the original particle with trajectory $t \longmapsto r(t)$ as Q without abuse of notation (the problem being how to distinguish this Q, which is a constant, from the original function $t, r \longmapsto Q(t, r)$, so we call it $\tilde{Q}$. Suppose that both

(4) $\qquad \bar{r}(0) < r(0) \quad$, and

(5) $\qquad \gamma(\bar{v}(\infty)) = 1 + \dfrac{\bar{Q}}{\bar{r}(0)} > 1 + \dfrac{\tilde{Q}}{r(0)} = \gamma(v(\infty)) \ .$

In other words, suppose that the barred particle starts inside the sphere of the original particle but has greater potential energy. Since we still have the freedom to specify the initial charge density $\rho(0, \cdot)$, given (4) we can always arrange (5) by making the charge density between $\bar{r}(0)$ and $r(0)$ very small so that $\bar{Q} = Q(0, \bar{r}(0))$ is only slightly smaller than $\tilde{Q} = Q(0, r(0))$. It follows from (5) that

(6) $\qquad \bar{v}(\infty) > v(\infty).$

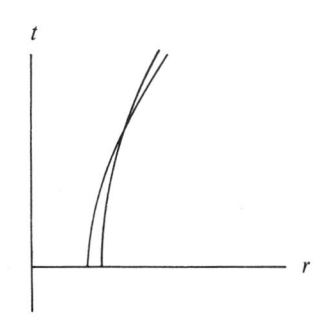

Figure 5-3. Two worldlines of fluid particles intersect non-tangentially. The fluid four-velocity is not well-defined at the point of intersection.

Now (6) says that the terminal velocity of the inner particle is larger than that of the outer; hence at some time t_1 we must have

$$\bar{r}(t_1) = r(t_1) \ .$$

That is, the worldlines of the two particles (with the same angular coordinates) intersect (see Figure 3), which contradicts (5), since $\bar{Q}$ and $\tilde{Q}$ are both constant on the worldlines. (It is also easy to see that the velocities $\bar{v}(t_1)$ and $v(t_1)$ must be different at the point of intersection; if they were the same, then the worldlines would be the same because initial positions and velocities uniquely determine the solution to (1).) This shows that a solution with all initial velocities zero and initial charge density such that (4) and (5) hold cannot be extended beyond some time $t_1 > 0$. Since (1) is symmetric under time reversal, it follows that the solution cannot have existed before time $-t_1$. We summarize:

Theorem.

Let $0 < r_1 < r_2$ be given radial coordinates, and suppose the initial charge density function $r \longmapsto \rho(0, r)$ is chosen so that

$$(7) \qquad \frac{1}{r_1} \int_0^{r_1} \rho(0, r) 4\pi r^2 \, dr \ > \ \frac{1}{r_2} \int_0^{r_2} \rho(0, r) 4\pi r^2 \, dr \ .$$

Then there is no solution to the Maxwell-Lorentz equations which is defined on all of Minkowski space, and has initial charge density function $\rho(0, \cdot)$ and all initial (i.e. $t=0$) velocities zero. Moreover, there is some time $t_{\max} < \infty$ such that any solution with these initial conditions can be defined only for $|t| \leq t_{\max}$.

This result states that for a large class of initial positions and velocities, the Maxwell-Lorentz equations have no global solution. Even worse, an examination of what went wrong (different worldlines intersecting with different velocities) suggests that this is probably typical, and that the retardation condition may well be incompatible with the Maxwell-Lorentz system.

What may be failing in this example is not so much the general ideas of continuum electrodynamics as the simple charged dust model in which it is envisioned that nearby particles have nearly the same velocity and that the fluid four-velocity at a spacetime point coincides with the velocity of an infinitesimal particle at that point. A model in which the charged fluid were treated as a gas with many different particle velocities at a given point (it being assumed in this idealization that the particles would not collide) might be able to handle the retardation condition. If only one sign of charge is present (positive, say), it would be natural to try to define a "fluid four-velocity u" and a "charge density ρ" via the inhomogeneous Maxwell equation (cf. 3.6(7)):

$$\rho := \frac{<\delta F, \delta F>^{1/2}}{4\pi} ,$$

$$u := \frac{\delta F}{4\pi\rho} ,$$

where F would be taken as the superposed retarded fields of all the particles. Then one might hope to identify u with some average of the four-velocities of the individual particles and ρ with the charge density in the rest frame defined by u. However, the theorem still shows that if the fluid is initially at rest (i.e. $u(0, r) = (1, 0, 0, 0)$ and (7) holds, then u could not satisfy the Lorentz equation for all time, though the individual particles might always satisfy the Lorentz equation. It would be interesting to carry out this program sufficiently explicitly that one could see if an equation different from the Lorentz equation for the average four-velocity u would emerge.

5.4 An alternate fluid model.

In this section we present a very suggestive model for continuum electrodynamics. Though it is not equivalent to the standard formulation given in Chapter 3 and is physically unsatisfactory in its present form, it is natural and elegant and may well contain the germ of a successful future theory. The mathematical structure of the theory is virtually identical to one proposed by P. Dirac [1951] and revised in two later papers [1952], [1954]. However, his motivation and presentation were quite different from that given below.

As always, let u and ρ denote, respectively, the four-velocity and proper charge density of a charged fluid (cf. 3.2). What we shall do below will work in an arbitrary spacetime M, and taking M to be Minkowski space does not simplify anything.

Let u be the four-velocity field and ρ the proper charge density of a charged fluid. For simplicity of exposition we assume that there is only one sign of charge present and take the charge/mass ratio q/m as unity, but these assumptions are not essential. The Lorentz equation is now

$$D_u u = \hat{F}(u) .$$

(Recall from Section 3.2 that $du/d\tau = D_u u$.) The key insight of this approach is that the differential $d(-u^*)$ of the one-form field $-u^*$ *automatically* satisfies the Lorentz equation with $d(-u^*)$ in place of F. This is not immediately obvious but follows from an short calculation which we shall now do.

Let us compute the differential du^* of the one-form u^* at a point $p \in M$. For this purpose, it will be convenient to choose orthonormal vector fields near p of the form $e_0 := u, e_1, e_2, e_3$, where e_1 is chosen as a multiple of $D_u u$ if the latter does not vanish at p. Now du^* (like any one-form field) can be written

$$du^* = \sum_{i<j} f_{ij} e_i^* \wedge e_j^* ,$$

with the f_{ij} functions. (This is one of the rare places where the upper-lower index summation convention doesn't work.) Collecting terms with e_0^* and changing notation for later convenience, we see that du^* can be written

(1) $$du^* = y^* \wedge u^* + \beta ,$$

with

$$<y,u> = 0 \quad \text{and} \quad \beta =: \sum_{1 \leq i < j \leq 3} \beta_{ij} e_i^* \wedge e_j^* .$$

In the rest frame of the particle at the point in question, the term $y^* \wedge u^*$ represents a pure electric field in the direction of y, and β is a pure magnetic field.

The following is a well-known and very useful formula for the differential of an arbitrary one-form α applied to two vector fields $y, z \in M_p$ [Warner, p. 70]:

(2) $$(d\alpha)(y,z) = y(\alpha(z)) - z(\alpha(y)) - \alpha([y,z]) .$$

(Recall that vectors are considered as directional derivatives, so, for example, $y(\alpha(z))$ denotes the directional derivative of the function $p \longmapsto \alpha_p(z_p)$ in the direction of y_p. The formula may appear a bit suspect because the left side only depends on the values y_p, z_p of y and z at p, while the right requires them to be vector fields defined in a neighborhood of p. However, the right side actually only depends on y_p, z_p, as one easily sees by checking linearity over the ring of C^∞ functions.)

From (2), we find that for any vector field z which is orthogonal to u,

(3) $$du^*(u, z) = u(<u, z>) - z(<u, u>) - <u, [u, z]>$$
$$= -<u, D_u z - D_z u> = <D_u u, z> ,$$

where we have used several times that fact that $<u, u>$ and $<u, z>$ are constant. From (1),

$$du^*(u, z) = (y^* \wedge u^*)(u, z) + \beta(u, z) = -<y, z> .$$

Comparing with (3), we see that $y = -D_u u$ and

(4) $$du^* = -D_u u^* \wedge u^* + \beta .$$

From (4) it is immediate that

(5) $$-d\hat{u}^*(u) = D_u u .$$

That is, (5) says that $-\hat{du}^*$ satisfies the Lorentz equation if we define

(6) $$F := -du^* .$$

In Chapter 3 we saw that the three fundamental assumptions of continuum electrodynamics are:

(i) the linearity of $\hat{F}$ in the Lorentz equation,

(ii) the Maxwell equation $dF = 0$, and

(iii) the Maxwell equation $\delta F = 4\pi \rho u^*$.

(A fourth assumption, which is not always made and may even be inconsistent with the first three is the retardation condition discussed in 3.7.) What is exciting about the definition (6) is that it tautologically guarantees (i) and (ii), leaving only (iii) to be separately postulated. Moreover, (iii) is easy to motivate. We want the equation

(7) $\qquad \delta(\rho u) = 0$

of conservation of charge which implies that, at least locally, ρu^* is the divergence of some two-form, and it is economical and elegant to postulate that ρu^* is the divergence of the naturally ocurring 2-form $-du^*$. Thus, given the definition (6), we can recover the whole Maxwell-Lorentz system 3.6(5), 3.6(6) from the single postulate

(8) $\qquad \delta du^* = -\rho u^*$.

Since (8) may be viewed as defining ρ in terms of u (cf. Section 3.6), there is really only the one unknown u in the system and (8) just asserts that δdu^* is a multiple (varying from point to point) of u^*.

The elegance of postulating (8) as the single non-tautological equation of electrodynamics is enhanced by the fact that in Minkowski space, the "space" part β of du^* has a simple physical interpretation as the intrinsic rotation ("curl") of the fluid as seen from the rest frame of a fluid particle. To see this, choose a particular spacetime point p, and let u, e_1, e_2, e_3 be an orthonormal basis at this point. Coordinatize Minkowski space relative to this basis, so that $u_p = (1, 0, 0, 0)$, $e_1 = (0, 1, 0, 0)$, etc. Let $\vec{v} = (v^1, v^2, v^3)$ denote the spatial fluid velocity field in this frame. Of course $\vec{v}$ vanishes at p itself, but is in general nonzero in a neighborhood of p. Writing $v := |\vec{v}|$, we have

$$u = \gamma(v) \cdot (1, \vec{v}).$$

Consider the e_i, $i = 1, 2, 3$, as constant vector fields on **M**. (Of course, u, e_1, e_2, e_3 is not generally orthonormal at points other than p.) From (1) and (2), the coefficients β_{ij} of the magnetic part of du^* are given by

(9) $\qquad \beta_{ij} = du^*(e_i, e_j) = e_i(<u, e_j>) - e_j(<u, e_i>) - <u, [e_i, e_j]>$

$\qquad\qquad = <D_{e_i} u, e_j> - <D_{e_j} u, e_i>$.

Now at p itself (where $\vec{v} = 0$ and $\gamma = 1$),

$$<D_{e_i} u, e_j> = e_i(\gamma)<(1, \vec{0}), e_j> + \gamma \cdot <(0, e_i(v^1), e_i(v^2), e_i(v^3)), e_j> = -e_i(v_j).$$

This shows that the right side of (9) is (up to a sign) one of the components of the curl $\vec{\nabla} \times \vec{v}$ computed in the rest frame at p. In fact, if we define a 3-vector $\vec{R}$ by $\vec{R}^* := \perp_3 \beta$, where $\perp_3$ denotes the Hodge duality operation within the three-dimensional space spanned by e_1, e_2, e_3 (with the orientation taken so that e_1, e_2, e_3 is positively oriented in this space when u, e_1, e_2, e_3 is a positively oriented basis for **M**), then one easily checks that the signs in (9) are such as to give

(10) $\qquad \vec{R} = -\nabla \times \vec{v}$

and

(11) $\qquad -du^* = (D_u u^*) \wedge u^* + \perp_3 \vec{R}^*$.

(A reader who is fanatical enough to check signs should keep in mind that $\perp_3 \perp_3 \beta = -\beta$, rather than "+" as in usual Euclidean 3-space, due to the fact that the metric on the space spanned by e_1, e_2, e_3 is negative definite rather than positive definite.)

Thus the "space" part of du^* describes the local rotation of the fluid in that it is the dual of the curl of the velocity field relative to the rest frame at the point at which (11) is evaluated. Now the space part of the field tensor F is $-\perp_3 \vec{B}^*$, where $\vec{B}$ is the magnetic field, and so if $F = -du^*$, then

$$\vec{B} = -\vec{R} = \nabla \times \vec{v}.$$

That is, in a theory based on (6) and (8), the magnetic field describes the local rotation of the fluid. Besides being a very attractive interpretation, this provides, in principle at least, an operational method for measuring $\vec{B}$ by observing the motion of the fluid. In contrast, in the standard theory one conceives of measuring the electric and magnetic fields $\vec{E}$ and $\vec{B}$ by observing hypothetical point particles with all possible velocities at a given spacetime point. In a pure fluid theory, there is (in any given frame) only one velocity at any point, and so the usual method of measuring the fields really doesn't make sense. It is certainly aesthetically desirable that one should be able to measure the fields solely by observing the fluid, and this is done in a very nice way in the present framework.

It is natural to wonder if the above theory might actually be the standard continuum theory in disguise. That is, it is conceivable, *a priori*, that every solution u, F of the Maxwell-Lorentz equations might satisfy $F = -(m/q)du^*$. Unfortunately, this turns out to be false. A physical example similar to one given by Dirac [1951] is a cloud of electrons at rest at a particular coordinate time in an external magnetic field. At that time, $\nabla \times \vec{v}$ would vanish, and couldn't be the magnetic field. That there actually exist solutions to the Maxwell-Lorentz equations of this type is a consequence of the solution of the Cauchy problem in Section 1; at any particular coordinate time one can arbitrarily specify the fluid velocity field and the electric and magnetic fields subject only to the constraint $\nabla \circ \vec{B} = 0$. After Dirac published [1951], which in essence proposed a theory based on (6) and (8), the fact that it was not equivalent to the usual theory was pointed out by D. Gabor. Apparently, Dirac found the above example so convincing that he then published [1952], in which he modified the original theory so as to include the example as a solution. Insofar as the fluid equations of motion are concerned, the modified theory is the same as the usual one based on the Maxwell-Lorentz equations (without the retardation condition), and the attractive features of the original theory noted above are unfortunately lost. The novelty of the modified theory derives from Dirac's motivation of it via several interesting variational principles.

Note that the above example might be impossible if the retardation condition were imposed, and it is just barely conceivable that imposing this condition could save the original Dirac theory. However, I know of no reason to think of this as a realistic possibility, especially since the retardation condition is difficult to reconcile with the Maxwell-Lorentz equations, as we saw in Section 5.3.

There is one interesting positive result regarding the possible equality of F and $-du^*$: if they are equal at any given coordinate time in **M**, then they are always equal. Thus we can obtain the original Dirac model from an initial condition rather than the stronger global assumption that $F = -du^*$ always. For instance, if one could think of a plausible physical reason why F should become asymptotic to $-du^*$ in the infinite past, say, the original theory might be resurrected.

The algebraic part of the proof is very simple. We assume the reader is familiar with the concept of the Lie derivative $L_v G$ of a 2-form G along a vector field v, and we recall the formula [Warner, p. 70] (which is rather ugly in the present notation):

(12) $$L_v G = -d(\hat{G}(v)^*) + v \lrcorner dG,$$

where $v \lrcorner dG$ denotes the two-form obtained by inserting v in the 3-form dG:

$$(v \lrcorner dG)(w, z) := dG(v, w, z) \qquad \text{for all vectors } w, z.$$

Applying this with $G = F + du^*$ and $v = u$ (so that $dG = 0$ and also $\hat{G}(u) = 0$ because $-du^*$ satisfies the Lorentz equation) shows that

(13) $\qquad L_u(F + du^*) = 0$.

The proof is completed by a routine lemma, to be proved below, stating that (13) and the initial condition $F + du^* = 0$ at coordinate time $t = 0$ imply that $F + du^*$ vanishes identically.

Lemma.

Let u be the four-velocity field of a fluid on **M** (so that every point of **M** lies on some integral curve of u which cuts the hyperplane $t = 0$), and let G be any 2-form field on **M** . Suppose that

(i) $\quad G = 0$ at coordinate time $t = 0$,

and

(ii) $\quad L_u G = 0$ at all points of **M** .

Then $G = 0$ on **M** .

Proof.

Choose an orthonormal basis e_0, e_1, e_2, e_3 of vector fields (which might as well be constant), and write

$$G = \sum_{i<j} G_{ij}\, e_i^* \wedge e_j^*$$

with the G_{ij} functions. Recall the following formula for $L_u G$ [Warner, p.70]:

(14) $\qquad u(G_{ij}) = u(G(e_i, e_j)) = (L_u G)(e_i, e_j) + G([u, e_i], e_j) + G(e_i, [u, e_j])$

$\qquad\qquad\qquad = G([u, e_i], e_j) + G(e_i, [u, e_j])$.

The right side may be expanded in the form $\sum_{k<l} f_{ij}^{kl} G_{kl}$, where the f_{ij}^{kl} are functions. Let us restrict our attention to (14) on one particular worldline (integral curve of u), parametrized by proper time τ , where $\tau = 0$ corresponds to $t = 0$. On this worldline, (14) becomes the system of ordinary differential equations:

$$\frac{dG_{ij}}{d\tau} = \sum_{k<l} f_{ij}^{kl} G_{kl}, \qquad 0 \leq i < j < 3 .$$

By the standard existence and uniqueness theorem for such systems, the solution is uniquely determined by the initial values of G_{ij} at $\tau = 0$, so if G vanishes at $\tau = 0$ then it vanishes everywhere on the worldline. Since this holds for all worldlines, the lemma is proved.

The lemma and the remarks just before it establish the following theorem. Although stated for Minkowski space **M** , it is valid in any spacetime with the hyperplane $t = 0$ replaced by a spacelike hypersurface.

Theorem.

Let F, u be a solution of the Maxwell-Lorentz equations 3.6(5), 3.6(6) defined on **M**, and suppose $F = -du*$ at coordinate time $t = 0$. Then $F = -du*$ on all of **M**.

5.5 Peculiar Solutions of the Lorentz-Dirac Equation.

In this section we shall examine unusual properties of special solutions of the Lorentz-Dirac equation 4.3(15):

(1) $$ma = q\hat{F}_{ext}(u) + \frac{2}{3}q^2[\frac{da}{d\tau} + <a, a>u] .$$

For simplicity of notation we use units which make $m = \frac{2}{3}q^2$ (see Exercise 3.14), yielding

(1)' $$\frac{da}{d\tau} - a + <a, a>u = -\frac{3}{2q}\hat{F}_{ext}(u) .$$

Not all solutions of (1) are expected to describe physical particles. The simplest case is when $F_{ext} = 0$. Then (1)' can be viewed as the following second-order equation for u (when $F_{ext} \neq 0$, the worldline $z(\cdot)$ enters implicitly through F_{ext}):

(2) $$\frac{d^2u}{d\tau^2} - \frac{du}{d\tau} + <\frac{du}{d\tau}, \frac{du}{d\tau}>u = 0 .$$

Obviously,

$$u = \text{constant}$$

are solutions corresponding to a particle of constant velocity, and these are presumably the only physically meaningful solutions. However, there are obviously many others, since *both* $u(0)$ and $a(0) = du/d\tau(0)$ may be arbitrarily specified.

For the case of one-dimensional motion, (2) is easily solved. Assume that the motion takes place along the x-axis, and suppress the other two space directions from the notation. Write $u = \gamma(v)(1, v)$. Define

$$w := \gamma(v)(v, 1) ,$$

so that

$$<w, u> = 0 \quad \text{and} \quad <w, w> = -1 .$$

Since $du/d\tau$ must be orthogonal to u, and hence in the direction of w, we may write

$$\frac{du}{d\tau} = Aw ,$$

where this defines the scalar proper acceleration A.

Note that since w is a unit vector, $\frac{dw}{d\tau}$ is orthogonal to w, and hence

$$\frac{dw}{d\tau} = Bu \quad \text{for some function } B .$$

The relation

$$B = \left\langle \frac{dw}{d\tau}, u \right\rangle = -\left\langle w, \frac{du}{d\tau} \right\rangle = -\langle w, Aw \rangle = A$$

shows that the u-components of the first and third terms of (2) cancel, and hence (2) may be written

(3) $\qquad \left(\dfrac{dA}{d\tau} - A\right)w = 0$.

Thus a solution of (2) for one-dimensional motion requires that

$$A = Ce^\tau$$

for some constant C. The solutions with $C \neq 0$, known for obvious reasons as *runaway* solutions, are clearly unphysical. They describe a particle which starts at rest and, in the absence of any external fields, accelerates exponentially and reaches a velocity arbitrarily close to that of light. Energy-momentum is not conserved (in the sense described in Section 4.3) for such solutions. This is not surprising because, even assuming mass renormalization and ignoring the integrals over the caps, the Lorentz-Dirac equation only guarantees conservation of energy-momentum under the additional hypothesis that

(4) $\qquad a(-\infty) = a(\infty)$. †

In solving the Lorentz-Dirac equation for nonzero F_{ext}, if one wants conservation of energy-momentum, one must take care to impose appropriate auxiliary conditions so that the solutions satisfy (4). Unfortunately, it is not clear that this is always possible.

In fact, there are physically reasonable initial value problems whose solutions are so startling as to cast doubt on the validity of the equation. A simple example is two particles of equal mass and opposite charge moving symmetrically with respect to the origin in one dimension.* Suppose that the initial velocities are toward each other or zero. Every high school physics student "knows" that the opposite charges will attract each other and eventually collide. Nevertheless, we shall show that the Lorentz-Dirac equation implies that they can *never* collide, but instead will eventually run away from each other at velocities which approach the speed of light! Moreover, $a(\infty) \neq 0$.

More specifically, consider two particles of equal mass 2/3 and charges -1 and $+1$ with respective initial positions $(-x_0, 0, 0)$ and $(x_0, 0, 0)$ ($x_0 > 0$) in R^3, initial velocities $(-v_0, 0, 0)$ and $(v_0, 0, 0)$, and initial accelerations $(-c_0, 0, 0)$ and $(c_0, 0, 0)$ (see Figure 4). We assume that each particle obeys the Lorentz-Dirac equation (1)' with the external field taken as the retarded field of the *other* particle. Physically, we expect the motion to be along the x-axis and symmetric about the origin, and we restrict our attention to solutions of this type. This means that if the right-hand particle has position $x(t)$ at coordinate time t, then the left-hand particle has position $-x(t)$ at coordinate time t.

In our analysis of the motion we shall focus our attention on the *right-hand* particle which we name "particle R", the other being "particle L". We suppress the two extraneous

---

† Most authors also demand that $a(-\infty) = 0 = a(\infty)$. This is necessary for the usual justification of ignoring the integrals over the caps presented in Section 4.3. However, if the caps are taken as light cones, the integrals over the caps can be computed exactly and are found to merely contribute a mass renormalization even when the retarded acceleration is nonzero (Exercise 4.2).

* Although the motion takes place in one dimension, the solutions we shall analyze are genuine three-dimensional solutions which might reasonably be expected to describe the motion of real particles (unlike the one-dimensional solutions of the Maxwell-Lorentz equations considered in Section 5.2).

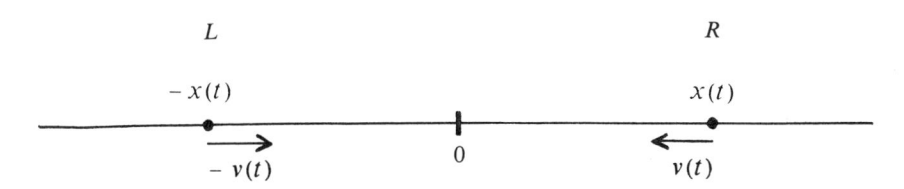

5.5 Peculiar solutions of Lorentz-Dirac equation

```
         L                              R
      -x(t)                           x(t)
   ●────→─────────────┼─────────────←───────●
        -v(t)         0              v(t)
```

Figure 5-4. The particles move symmetrically with respect to the origin.

space dimensions in the notation and work in two-dimensional Minkowski space. Let

$$v := \frac{dx}{dt}$$

denote its coordinate velocity and

$$u = \gamma(v)(1, v)$$

its four-velocity. We shall use the same notation as in the analysis of runaway solutions above, so that

$$\frac{du}{d\tau} = Aw,$$

where τ is proper time for the right-hand particle.

The Lorentz-Dirac equation (1)′ for the right-hand particle can be written (cf. (1)′, (3))

$$(5) \qquad \left(\frac{dA}{d\tau} - A\right)w = -\frac{3}{2}F_{ext}(u),$$

where F_{ext} is the retarded field produced by particle L, which is given by 4.2(16) as

$$(6) \qquad F_{ext}(u) = -\frac{1}{r_{ret}^2}w,$$

and r_{ret} is the "retarded distance" (the distance between the particles in the rest frame of particle L at the retarded time). To avoid possible confusion we point out that the "w" in (6) is not the same as the "retarded" w in 4.2(16), but (6) holds anyway. The reason is that if v and y are *any* orthogonal unit vectors in two-dimensional Minkowski space, then $v^* \wedge y^* = \pm dt \wedge dx$, so that the $w_{ret}^* \wedge u_{ret}^*$ of 4.2(16) is equal to the present $w^* \wedge u^*$, the sign being determined by noting that both w's point in the positive x-direction. (The minus sign in (6) is because the left-hand particle has negative charge. The signs are such that (6) represents a force to the left.) An exact expression for r_{ret} is given in equation (10) below, but we do not need it to see the main idea, which we explain first.

We rewrite (5) as

$$\frac{dA}{d\tau} - A = \frac{3}{2}\frac{1}{r_{ret}^2}$$

which, after multiplication by the integrating factor $e^{-\tau}$ becomes

$$(7) \qquad \frac{d(e^{-\tau}A)}{d\tau} = \frac{3}{2}\frac{e^{-\tau}}{r_{ret}^2}.$$

Hence

$$(8) \quad A = e^\tau \left[\frac{3}{2} \int_0^\tau \frac{e^{-s}}{r_{ret}(s)^2} \, ds + A_0 \right],$$

where A_0 is the initial proper acceleration. Mere inspection of (8) shows peculiarities. The integral on the right is always positive, and when the particles get close, there is a large contribution to the integrand in the direction of repulsion. Of course, the initial acceleration A_0 is expected to be negative (i.e. to the left), and it is certainly not obvious, *a priori,* that the integral will get big enough before the assumed collision to cause the acceleration to change sign. Nevertheless, it is clear that the integral is working to turn the particles around. An observation which will be important in the proof of the theorem below is that A is obviously monotonically increasing when it is positive, and so if it ever becomes positive, it remains positive thereafter.

It turns out no matter what the initial velocity and acceleration, a collision *cannot* occur. Moreover, if the right-hand particle is ever at zero velocity or moving to the left, then that particle eventually turns around and accelerates at an ever-increasing rate toward $+\infty$. In other words, *all* solutions in which the particles are not always moving away from each other are "runaway" in the opposite direction that would be expected from the dictum that "unlike charges attract".

This situation was first analyzed by Eliezer [1943], who came to conclusions similar to the above, but some of his proofs were apparently hard to read and have been questioned in the literature. † However, his conclusions concerning one-dimensional motion are correct. Proofs which are quite readable are given in [Hsing and Driver, 1975], which treats more general one-dimensional motion and unfortunately was never published, and [Hsing and Driver, 1977]. Related and overlapping results which are also of interest appear in [Huschilt and Baylis, 1976]. The proofs below were inspired by [Hsing and Driver, 1977].

Now we calculate r_{ret}. Recall that the field F_{ret} which affects particle R at a point $(t, x(t))$ on its worldline may be thought of as being "emitted" by particle L at a point $(t', -x(t'))$ on *its* worldline and then traveling at the velocity of light to reach particle R at $(t, x(t))$. We think of the "retarded time" t' as a function of t, and in the proofs below, priming a symbol representing a time always denotes the corresponding retarded time. Now t' and t are related by the equation

$$(9) \quad t - t' = x(t) + x(t'),$$

and r_{ret} is the spatial distance from $(t', -x(t'))$ to $(t, x(t))$ as measured in the rest frame of particle L at coordinate time t'. Since the two events are connected by a light ray, the spatial distance in the given frame is the same as the time between them as measured in that frame. Thus, if we denote by $v' := -v(t')$ the velocity of the left-hand particle at the retarded time t' and $u' := \gamma(v')(1, v')$ the corresponding four-velocity then we have

$$(10) \quad r_{ret} = \langle u', (t, x(t)) - (t', -x(t')) \rangle$$

$$= \gamma(v')(t - t') - v' \gamma(v')(x(t) + x(t'))$$

$$= \left(\frac{1-v'}{1+v'} \right)^{1/2} (x(t) + x(t')).$$

---

† Usually, and perhaps always, incorrectly. There are a good many unsubstantiated statements and conflicting claims in various papers analyzing this and related problems.

The following lemma is the key which opens (8) to rigorous analysis. It states that if the right-hand particle is always moving left on some finite coordinate-time interval, then its velocity is bounded away from the speed of light on that interval. If the particle were released at rest in the coordinate frame and were attracted toward a collision at the origin in a finite time, as intuition suggests, then the lemma says that it could not acquire unbounded relativistic energy before the collision. This is quite different from a classical Newtonian particle attracted by a Coulomb force $-1/r^2$. The classical particle would have potential energy $-1/r$, which becomes negatively unbounded as it approaches the center, so by conservation of energy, its kinetic energy would become positively unbounded. In this sense the lemma is counterintuitive unless one already knows that the particles cannot get arbitrarily close.

In the following proofs we always assume as given a solution of (5) defined on some semi-infinite coordinate time interval $-\infty < t < c$. ‡ Of course, this is what is expected physically, unless one is willing to consider creation of particles. Also, the very definition of the retarded interaction precludes a solution only defined on a finite time interval.

Lemma 1.

Assume that a solution to (5) satisfies $v(t) \leq 0$ on some finite coordinate time interval $a \leq t < c < \infty$. Then $|v|$ is bounded away from the speed of light on $[a, c)$.

Proof. Let

$$V := v\gamma(v) = \frac{v}{(1-v^2)^{1/2}}.$$

Since v is negative on $[a, c)$, it is enough to show that V is bounded below there. Recall the lemma of Section 5.2 relating the coordinate acceleration dv/dt to the scalar proper acceleration A for motion on a line:

(11) $$A = \gamma^3 \frac{dv}{dt} = (1-v^2)^{-3/2} \frac{dv}{dt}.$$

Using this, we have

(12) $$\frac{dV}{dt} = \frac{dv}{dt}\gamma + v^2 \gamma^3 \frac{dv}{dt} = \gamma^3 \frac{dv}{dt} = A.$$

Integration of (12) shows that V is bounded below on any finite interval $[a, s)$ on which A is bounded below, and it is obvious from (8) that A is bounded below on any finite coordinate time interval. (If (8) is written with the initial time at a, then for all subsequent times the right side of (8) consists of a positive quantity plus $e^\tau A_0$, and it is obvious that $e^\tau A_0$ is bounded below on any finite coordinate time interval because the proper time τ cannot exceed the coordinate time t by Exercise 1.17.)

Lemma 2.

If, as above, a' and b' denote the retarded times corresponding to times a and b, respectively, then

$$a' \leq b' \quad \text{if and only if} \quad a \leq b.$$

Proof. This says that if two light signals are sent by particle L at times a' and b', then

---

‡ Essentially the same proof applies also to worldlines which are symmetric but otherwise arbitrary (driven by external forces) up to some given coordinate time and thereafter conform to the retarded interaction.

they are received by particle R in the same temporal order. Consider a stationary observer at the spatial origin $x = 0$, and let a'' and b'' denote the respective times at which the signals are received by this observer. We shall show that $a'' \leq b''$ is equivalent to $a' \leq b'$, and then essentially the same argument (which does not depend on the direction of time) will show that $a'' \leq b''$ is also equivalent to $a \leq b$.

By elementary geometry, forward light rays (45° lines) sent from $(a', -x(a'))$ and $(b', -x(b'))$ will intersect the time axis at $t = a'' = a' + x(a')$ and $t = b'' = b' + x(b')$, respectively, and hence

$$b'' - a'' = b' - a' + x(b') - x(a') \ .$$

Since the velocity of the particle is always less than 1,

$$|x(b') - x(a')| \leq b' - a' \ ,$$

so

$$b'' - a'' \geq b' - a' - |x(b') - x(a')| \geq 0$$

when $b' - a' \geq 0$.

Lemma 3.

Suppose v is bounded away from 1 on some coordinate-time interval $a \leq t < c$. Then there exists a constant M such that

(13) $\qquad r_{ret} \leq Mx(t) \qquad$ for $a \leq t < c$.

Proof. Since $|v| < 1$ wherever the solution is defined, it follows trivially from continuity that if v is bounded away from 1 on $[a, c)$, then it is also bounded away from 1 on any larger interval $[\bar{a}, c)$, $-\infty < \bar{a} \leq a$. By Lemma 2, the interval $[a', c)$ includes all retarded times t' corresponding to times t with $a \leq t \leq c$, and so we can conclude that the velocities $|v'|$ at all retarded times t' corresponding to $a \leq t < c$ are also bounded away from 1. Given this, a glance at the expression (10) for r_{ret} makes clear that it is enough to show the existence of a constant K such that

(14) $\qquad x(t') \leq Kx(t) \qquad$ for $a \leq t < c$.

That there must exist such a constant can be seen intuitively as follows. Between the retarded time t' and the given time t, a light signal emitted by the left-hand particle at time t' must travel a distance $x(t') + x(t)$, while the left-hand particle itself travels a distance $|x(t') - x(t)|$. If $x(t')$ could be arbitrarily large relative to $x(t)$, then the ratio of these two distances would be arbitrarily close to 1, which would imply that the particle velocity would be arbitrarily close to the speed of light.

For a formal proof, let v_{max} denote the supremum of $|v|$ on $[a', c)$. The average speed of the left-hand particle between times t' and t is

$$\frac{|x(t') - x(t)|}{t - t'} \leq v_{max} \ .$$

If there were no constant K for which (14) holds, then there would exist a sequence of points $t_n \in [a, c)$ such that

$$x(t_n') > nx(t_n), \qquad n = 1, 2, , 3 \cdots .$$

Then, using (9), we would have

$$1 > v_{max} \geq \frac{x(t_n') - x(t_n)}{t_n - t_n'} = \frac{x(t_n') - x(t_n)}{x(t_n) + x(t_n')}$$

$$= \frac{1 - \frac{x(t_n)}{x(t_n')}}{1 + \frac{x(t_n)}{x(t_n')}} \geq \frac{1 - \frac{1}{n}}{1 + \frac{1}{n}} \longrightarrow 1 ,$$

as $n \to \infty$, a contradiction.

Theorem.

Consider a solution to the Lorentz-Dirac equation for which the motion is on a line and symmetric about the spatial origin as described above and which for some $b < \infty$ is defined for all coordinate times t satisfying $-\infty < t < b$. Then

(i) $x(t)$ is bounded away from 0 for $0 \leq t < b$ (i.e. for some $\delta > 0$, $x(t) > \delta$ for all $0 \leq t < b$). Moreover, if the solution is defined for all time, then $x(t) \to \infty$ as $t \to \infty$. In particular, the particles cannot approach arbitrarily closely for $t > 0$.

(ii) If the solution is defined for all times $0 \leq t < \infty$, and if there is any time t_0 at which $v(t_0) \leq 0$, then $v(t) \to 1$ as $t \to \infty$, and $A(t)$ is positive and monotonically increasing for large t. In other words, if the particles are ever stationary or moving toward each other, then after sufficient time they will be moving away from each other at velocities asymptotic to the speed of light and ever-increasing (repulsive) proper acceleration.

Proof. The heart of the matter is to show that $x(t)$ is bounded away from 0 on any finite time interval; given this, the rest is a mere exercise. First we shall show that if $v(t) \leq 0$ on any finite interval $a \leq t < c$, then $x(t)$ is bounded away from 0 on that interval; in particular, a collision cannot occur at the end of a finite time interval on which the particle is always moving left. From (7) and recalling that $\gamma = dt/d\tau$, we have

$$\frac{d(e^{-\tau}A)}{dt} = \gamma^{-1} \frac{3}{2} \frac{e^{-\tau}}{r_{ret}^2} .$$

By Lemma 1, $|v|$ is bounded away from 1 on $[a, c)$. Given this, Lemma 3 implies that there exists a positive constant K such that

(15) $$\frac{d(e^{-\tau}A)}{dt} \geq \frac{K}{x(t)^2} \quad \text{for } a \leq t < c .$$

Since $-1 < v := dx/dt \leq 0$ on $[a, c)$, it follows that

$$\frac{d(e^{-\tau}A)}{dt} \geq \frac{K}{x(t)^2} \geq -\frac{K}{x(t)^2} \frac{dx}{dt} \quad \text{for } a \leq t < c .$$

Integration from a to t yields

(16) $$A \geq \left[\frac{K}{x(t)} + B\right] e^\tau \quad \text{for } a \leq t < c ,$$

where B is a constant. Suppose that $x(t)$ is not bounded away from zero on $[a, c)$. Then (16) shows that A is positive somewhere on $[a, c)$, and since A is monotonically increasing when it is positive from (8), it follows that A is positive on some smaller interval $[\bar{a}, c)$, $a < \bar{a} < c$. From (11),

$$(17) \qquad \frac{dv}{dt} = (1 - v^2)^{3/2} A \geq -(1 - v^2)^{3/2} A \frac{dx}{dt} \qquad \text{on } [\bar{a}, c),$$

where the inequality follows as previously from $A > 0$ and $0 \leq -dx/dt < 1$. Given that v is bounded away from 1 on $[a, c)$, it follows from (16) and (17) that for some *positive* constant L,

$$\frac{dv}{dt} \geq -L \left[\frac{K}{x(t)} + B \right] \frac{dx}{dt} \qquad \text{on } [\bar{a}, c).$$

A second integration shows that

$$v(t) \geq -LK \log x(t) - LB x(t) + C \qquad \text{on } [\bar{a}, c),$$

where C is another constant. Since v is bounded by the speed of light and L is positive, this last equation implies that $x(t)$ is bounded away from 0 on $[\bar{a}, c)$. It is obvious that $x(t)$ is bounded away from zero on the closed interval $[a, \bar{a}]$ (because it is continuous and can't be zero by (15)), and it follows that $x(t)$ is bounded away from zero on $[a, c)$.

We just showed that $x(t)$ is bounded away from 0 on any finite time interval on which v is nonpositive. To extend the proof to show that $x(t)$ is bounded away from 0 on any finite interval on which the solution is defined, note that it is trivial that $x(t)$ is bounded away from 0 on any half-closed interval $[a, c)$ on which v is positive (i.e. any interval on which the right-hand particle is always moving right). As noted above, it is also trivial that $x(t)$ is bounded away from 0 on any finite *closed* interval $[0, a]$ on which the solution is defined. Putting these two trivial facts together with what was just proved, we see that if there is a finite interval $[0, c)$ on which $x(t)$ is not bounded away from zero, then v must take on both signs in any smaller interval of the form $[a, c)$, and $V := v \gamma(v)$ has the same property. Given this, the Intermediate Value Theorem of elementary calculus implies that V has infinitely many zeros on any interval $[a, c)$. Recalling from (12) that $dV/dt = A$, we can conclude from the Mean Value Theorem that A has infinitely many zeros on any interval $[a, c)$. On the other hand, we see directly from the positivity of the integrand in (8) that if A ever vanishes, then it is strictly positive after that time, and this contradiction completes the proof that $x(t)$ is bounded away from 0 on any finite interval.

Now we show that if the solution is defined for all times $-\infty \leq t < \infty$, and if $v(t_0) \leq 0$, then $v(t) \to 1$ as $t \to \infty$. If the proper acceleration A were always negative on $[t_0, \infty)$, the coordinate acceleration

$$(18) \qquad \frac{dv}{dt} = (1-v^2)^{3/2} A$$

would have the same property, and the velocity would be strictly monotonically decreasing. Under the assumption that $v(t_0) \leq 0$, a collision would occur in a finite time, but we just showed that this cannot occur. Thus A must eventually become positive, (8) shows that it remains positive from that time on, and we conclude that dv/dt is positive for large t. This implies that v is monotonically increasing for large t and hence either approaches 1 as $t \to \infty$ or is bounded away from 1 as $t \to \infty$. If v were bounded away from 1, then (18) would show that dv/dt would be bounded away from 0 for large t, which would imply that $v(t) \to \infty$ as $t \to \infty$, an impossibility. Hence $v(t) \to 1$ as $t \to \infty$.

The final conclusion of (i), that $x(t) \to \infty$ as $t \to \infty$, follows trivially from what has just been proved. If v is always positive, it is obvious; if v is ever negative or zero, it follows from $v(t) \to 1$ as $t \to \infty$. This completes the proof of the theorem.

One consequence of this theorem is that if the Lorentz-Dirac equation holds, one can in principle extract an arbitrarily large amount of kinetic energy from a pair of oppositely charged point particles of equal mass by releasing them at rest and waiting until they shoot off in opposite directions with sufficiently high velocities. Presumably, no one believes that this is actually possible, and many interpret this result as saying that the Lorentz-Dirac equation has no physically realistic solutions for one-dimensional symmetric motion of opposite charges. (For an opposite view, see [Rohrlich, p.186].) The reader may well wonder how this can be reconciled with the apparent widespread belief in the Lorentz-Dirac equation. First of all, the theorem is not widely known, and is not even mentioned in most texts. One gets the impression that up until the 1970's its correctness was in doubt, though the research literature in this area is often so murky that it is hard to tell. Apart from this, there are ways to try to explain away the difficulty.

For example, recall that in setting up the equation we chose $m = 2q^2/3$ to avoid having to carry a constant which does not essentially affect the analysis, and later we chose $q = \pm 1$. However, if we put the units back in, it turns out that the Lorentz-Dirac equation differs from the Lorentz equation by a term so tiny that no one has been able to devise a practical experiment to measure it. Some argue that the solutions probably behave more or less like solutions of the Lorentz equation until the particles get very close, so close in fact that either they bump into each other (if they are not true point particles) or quantum effects set in and we are out of the domain of the classical theory. The "Compton wavelength" h/mc (see Exercise 1.23) is often mentioned as the distance at which quantum effects might become important. However, it has not been proved, to my knowledge, that solutions with masses and charges found in nature and physically reasonable initial accelerations actually must get this close before turning around.

Another possible way out is to take the point of view that true 1-dimensional motion is impossible because there is always a slight uncertainty in the angle between the respective initial velocities. Some speculate that for three-dimensional motion, the generic case (e.g. an open set of small but nonzero angles) might behave more reasonably. Since very little is known about three-dimensional solutions, this possibility cannot be completely ruled out, but neither is there any evidence for it.

5.6. Evidence for the usual energy-momentum tensor.

The weakest point in the traditional formulation of electrodynamics would seem to be the choice of the usual energy-momentum tensor 3.8(20):

(1) $$T_{usual} = \hat{F}^2 - \frac{1}{4}\mathrm{tr}(\hat{F}^2)I \ .$$

From the point of view developed in Chapter 3, it is chosen because it makes the Lorentz equation equivalent to conservation of energy-momentum in the fluid model. Since the Lorentz equation is eventually abandoned in point electrodynamics anyway, this motivation is reasonable *a priori*, but not *a posteriori*. From the point of view of the engineer, it is used because it "works" (more or less). However, there might be other tensors with the same properties. In this section we examine mathematical and physical evidence bearing on the usual energy-momentum tensor. In the next we shall look at some proposals for alternate energy-momentum tensors and alternate equations of motion.

A. Mathematical evidence.

The question of the uniqueness of T_{usual} obviously depends on the kind of constructions which one is willing to permit. For example, an energy-momentum tensor at a space-time point z might conceivably be constructed from the field F_z at z (as T_{usual} is), or the field together with its derivatives up to a certain order at z, or the retarded potential at z produced by all charges in the universe, etc.

For some of the less exotic contructions, there are good uniqueness theorems. For example, suppose we are given the field tensor F on Minkowski space $\mathbf{M}$ and seek a tensor $T = (T^{ij})$ with the following three properties (a), (b), and (c):

(a) T is constructed in a Lorentz-invariant way from F. This means that with respect to any Lorentz frame L, the components T^{ij} of T are defined as given functions ϕ^{ij} of the components of F,

$$T^{ij} := \phi^{ij}((F^{kl})) = \phi^{ij}(F^{00}, F^{01}, \cdots, F^{33}) \ ,$$

and that if we pass to a new frame $\bar{L}$ related to the old by a Lorentz transformation Λ, denoting the components of tensors in the new frame by bars, e.g.

$$\bar{F}^{ij} := \Lambda^i{}_\alpha \Lambda^j{}_\beta F^{\alpha\beta} \ ,$$

then

(2) $$\bar{T}^{ij} := \phi^{ij}((\bar{F}^{kl}))$$
$$= \Lambda^i{}_\alpha \Lambda^j{}_\beta T^{\alpha\beta} = \Lambda^i{}_\alpha \Lambda^j{}_\beta \phi^{\alpha\beta}((F^{kl})) \ .$$

For those who are repelled by such transformation formulae, one way to think about (2) on a higher algebraic level is to view T_z, with $z \in \mathbf{M}$ fixed, as a tensor-valued function

$$T = \phi(\cdot) : \mathbf{M}_0^2 \longrightarrow \mathbf{M}_0^2$$

on the space $\mathbf{M}_0^2$ of all 2-contravariant tensors on $\mathbf{M}$. The domain of ϕ is thought of as the space of all F_z's, and the codomain as the space of all T_z's. † The group of Lorentz transformations acts in the natural way on both the domain and codomain, and (2) just says that ϕ intertwines these natural actions.

---

† Although the domain and codomain happen to coincide in this simple example, it's better to think of them as separate. For instance, we might consider a T constructed from several given tensors, such as F

(b) T is symmetric: $T^{ij} = T^{ji}$ for all i, j.

(c) If F satifies the free-space Maxwell equations

$$dF = 0 \; ; \quad \delta F = 0 \; ,$$

then T has vanishing divergence: $\delta T = 0$.

Condition (c), of course, is what is needed to obtain the usual integral law of conservation of energy-momentum for a free electromagnetic field on Minkowski space.

In this situation, the following uniqueness theorem for T, which is a slight variant of a result of [Kerrighan, 1982], can be proved.

Theorem. Let g denote the metric tensor on $\mathbf{M}$. If (a), (b), and (c) hold, then there exist constants a and b such that

(3) $\qquad T^{ij} = aT^{ij}_{usual} - bg^{ij}$.

Surprisingly, this attractive and fundamental result is quite recent (1982). Actually, Kerrighan considered a tensor T defined on an arbitrary spacetime and constructed from both F and the metric tensor, and proved (3) under these conditions. However, the same ideas work in the above context. For the proof, which is elementary but quite detailed, the reader will have to consult Kerrighan's original paper.

However, the theorem does not cover all plausible constructions. For example, given any function ϕ on Minkowski space $\mathbf{M}$, the following tensor, written with components with respect to a Lorentz coordinatization, has identically vanishing divergence:

(4) $\qquad T^{ij} := \dfrac{\partial^2 \phi}{\partial x_i \partial x_j} - g^{ij} \Box \phi$.

If ϕ is any function constructed invariantly from F, for example:

$$\phi(x) := \psi(<F_x, F_x>)$$

where ψ is any real-valued function of a real variable and $<F_x, F_x> := (F_x)_{\alpha\beta} F_x^{\alpha\beta}$, then (4) satisfies (b), and (c) above (but not (a), of course, since it does not depend just on F, but also on its derivatives). Note, however, that the indicial expression (4) does *not* define a tensor in general spacetimes because it does not transform as the components of a tensor must under general coordinate transformations. It is only because Minkowski space has the additional structure of Lorentz coordinatizations (as distinguished from general coordinatizations) that (4) works there.

Another relevant result [Kerrighan, 1981] states, roughly, that any symmetric tensor $T = (T^{ij})$ on a general spacetime whose divergence vanishes identically and which is constructed invariantly from the metric tensor together with a potential 1-form A and their first and second partial derivatives is actually independent of A (and its derivatives). This precludes obtaining a new energy-momentum tensor by adding something analogous to (4) satisfying the hypotheses of the theorem to T_{usual}.

---

and the potential, or F and the metric tensor on a general spacetime, and the same general viewpoint would apply.

Since the retarded potentials A_{ret}, which are arguably the physically relevant ones, satisfy

$$\delta(A_{ret}) = 0 \quad,$$

a natural question is whether one can construct a (symmetric) 2-contravariant tensor with identically vanishing divergence when $\delta dA = 0$ from a 1-form A with $\delta A = 0$ which is not of the obvious form (3) above. This appears to be an open problem.

Another approach is to try to contruct zero-divergence tensors from given particle worldlines. We shall see in the next section that this can be done in natural ways, but time will have to judge the physical relevance of such constructions. Since in many practical applications one has better knowledge of the fields than the particle worldlines, a tensor constructed just from the fields would seem preferable, *a priori*.

Finally, we should at least mention a completely different and very influential point of view. One can "derive" Maxwell's equations from a variational principle based on an appropriate Lagrangian. This prescription is described in virtually all advanced texts on electomagnetism,‡ and we shall not repeat it here. The Lagrangian itself and the variational principle seem to have no obvious physical or intuitive significance, and it can be argued that writing down the Lagrangian and applying the variational principle differs little from simply writing down the equations. However, this approach has proved a powerful heuristic tool in both classical and quantum mechanics. One reason is the ease with which laws of symmetry are established; if the Lagrangian has certain kinds of symmetry, then the dynamical equations derived from it are guaranteed to have similar symmetry. Also, there is a well-known procedure for obtaining a zero-divergence 2-contravariant tensor, called the "canonical energy-momentum tensor" $T = T^{ij}$, from the Lagrangian. If the Lagrangian is constructed from some "fields" (such as F or A), then so is the tensor, and it has zero divergence whenever the fields satisfy the equations obtained from the variational principle. In general, the canonical energy-momentum tensor is not symmetric, and its symmetrization does not necessarily have vanishing divergence. However, T_{usual} may be viewed as the "gauge-independent" part of the particular canonical energy-momentum tensor obtained from the usual Lagrangian which yields Maxwell's equations.

The Lagrangian approach has proved enormously fruitful in quantum mechanics, where one often wants to guess the equations of motion for a system whose physical properties are unknown. It is sometimes easier to write down Lagrangians with desired symmetry properties than to directly guess equations with those properties. Moreover, once one has the Lagrangian, one gets for free the canonical energy-momentum tensor and other useful general constructs. For these reasons, some virtually identify constructing a physical theory with writing down a Lagrangian.

If one does adopt this point of view, then the first question one asks is what Lagrangian will yield Maxwell's equations. Again, the answer may depend on what constructions one allows (e.g. is the Lagrangian to be constructed from F, or the potential, or something else?) but it seems unlikely that there are any plausibly simple possibilities other than trivial variants of the usual one. Thus in this or some similar sense, some may view T_{usual} as "natural", or even "unique".

B. Physical evidence

Before beginning the physical discussion, the reader should be aware that I am a mathematician, not a physicist, and at this point am venturing beyond my domain of primary competence. Moreover, even physicists often disagree violently in this area. One can find

---

‡ The treatment of [Jackson] is particularly clear.

arguments for almost any point of view in the physics literature, but clean, definitive experimental evidence is hard to come by. The following should be viewed simply as the conclusions of one mathematician who has seriously tried to find out why the usual energy-momentum tensor for the electromagnetic field is so widely accepted.

To begin on a very concrete level, no one doubts that electromagnetic radiation exists and that mechanical energy can disappear at one point of spacetime and reappear at another point connected to the first by a light ray. For instance, one can crank a generator which powers a transmitter and later and farther away recover mechanical energy via, say, vibrations of the diaphragm of an earphone connected to a a few passive electronic components (a "crystal radio" for those old enough to remember; the point is that no battery or other external power source is needed at the receiving end). The conventional view, which is essentially universally accepted, is that mechanical energy is changed into field energy by accelerating charges and conversely, the quantitative relation being given by 4.2(24) or 4.4(38)(ii), which state that the rate dE/dt of radiation of energy in the rest frame of the particle is proportional to the square of the proper acceleration a:

$$(5) \qquad \frac{dE}{dt} = -\frac{2}{3} q^2 <a, a> .$$

Equation (5) was essentially known before the advent of relativity and is customarily derived nonrelativistically by integrating the Poynting vector $\vec{E} \times \vec{B}/4\pi$ for the retarded field over a small sphere with the particle instantaneously at rest at the center. In this calculation as done in many practically oriented texts, the particle is considered as "slowly moving" and no distinction is made between coordinate time and proper time nor between retarded spheres and laboratory spheres. If one interprets the coordinate frame used as the rest frame of the particle at the emission time, and the integration as over a retarded sphere $S_r(\tau)$ (see Section 4.4), then the calculation is relativistically exact, and we saw in 4.2(23) that the result is the Larmor radiation law:

$$(6) \qquad \frac{dE}{dt} := \lim_{r \to 0} \int_{S_r(\tau)} \frac{\vec{E} \times \vec{B}}{4\pi} \circ d\vec{S} = \frac{2}{3} q^2 \vec{a} \circ \vec{a} ,$$

where $\vec{a} := d\vec{v}/dt$ is the vector acceleration (in the rest frame at the emission time, exactly, and in the coordinate frame, approximately). The integral in (6) represents the rate at which energy is passing through a sphere of radius r at time $t+r$ relative to the rest frame of the particle at proper time t, given that the particle was at the center of the sphere at time t, and the limit $r \to 0$ is actually unneccessary because the integral happens to be independent of r.

One criticism of this approach is that it only counts pure energy relative to the rest frame; purely spatial momentum which passes through the sphere is ignored. The latter can be computed exactly (Exercise 4.10), and the result is that the proper-time rate at which purely spatial momentum (relative to the rest frame) passes through the sphere is

$$(7) \qquad \frac{2}{3} q^2 \frac{1}{r} a ,$$

which becomes infinite as $r \to 0$. Of course, this is just a mass renormalization term, but the rather shaky arguments for abandoning such in the discussion of the Lorentz-Dirac equation in Chapter 4 do not really apply in the present context. If we believe in the usual energy-momentum tensor as describing physically measurable quantities, then we ought in principle to be able to "see" that spatial momentum coming out of the sphere just by measuring the fields, independently of any observations of the particle. Even if we are only interested in measuring pure energy in the laboratory frame, a large component of spatial momentum in the rest frame can be Lorentz-transformed into a large component of pure energy in the laboratory frame. The "slowly-moving" approximation does not help here; if the particle's velocity is given as nonzero, then this fixes the Lorentz transformation which converts from the rest frame to the laboratory frame, and if the flux of rest-frame spatial momentum through the sphere becomes infinite as as the retarded sphere shrinks, then the laboratory-frame rate of pure energy radiation can also become infinite.

Of course, in special circumstances better justifications for ignoring the infinite (in the limit $r \to 0$) flux of spatial momentum might be given. At any rate, it *is* usually ignored in analyses of actual experiments. This is often tacitly done by replacing the classical Larmor law (5) with the generalization:

(8)   Rest-frame energy-momentum radiation rate $= -\frac{2}{3}q^2 <a, a> u$.

Assumption (8) implies (5) but not conversely: (8) states that not only is the pure energy radiation in the rest frame given by (5), but also that there is no radiation of purely spatial momentum in that frame. (For a typical argument for (8), see [Rohrlich, Section 5-1].) Although it is not clear that evidence for (8) is the same as evidence for T_{usual}, for the rest of the Section we shall follow custom and assume (8). Most of the physical measurements will involve measurements of energy in the laboratory frame.

Recall from Section 4.2 that the component of the Poynting vector (relative to the rest frame at the emission time) which is orthogonal to the retarded sphere (and hence the only part which contributes to the energy flow through the sphere) is

(9) $\vec{P}_{rad} := \frac{q^2}{4\pi r^2} (\vec{a}_\perp \cdot \vec{a}_\perp) \vec{w}$,

where $\vec{w}$ is the radially outward vector from the emission point relative to the rest frame at the emission time. Using (9) as an approximation to the laboratory-frame Poynting vector and the interpretation of the Poynting vector as the local energy flux, one can construct crude models of macroscopic physical systems which radiate, such as antennas. Most standard models involve unverifiable but not unreasonable assumptions (such as particular current distributions within an antenna) and predict measurable quantities such as intensity of radiation as a function of distance and orientation of the radiator. Agreement with experiment is good enough to make the models useful.

The most impressive verification of (8) seems to be measurements of synchrotron radiation. A *synchrotron* is a particular kind of particle accelerator. The exact details of their construction are not important in the present context; all we need to know is that they are one of a class of accelerators in which charged particles move in a circle. Electric fields accelerate electrons to nearly the speed of light in a horizontal plane, and a vertical magnetic field bends their paths into a horizontal circle. For the purpose of explaining the ideas of the measurements, we shall think of the magnetic field as constant, though time-varying magnetic fields are often used in practice. The explanation to follow is based on an early experiment of [Corson], but is highly idealized and simplified for pedagogical purposes and therefore does not accurately describe any actual experiment. (For example, that experiment used a time-varying magnetic field.)

When the accelerating electric fields are turned off, the electrons "coast" in the magnetic field. Think of the field as spatially uniform and of magnitude B in the vertical direction. If the motion were described by the Lorentz force, then we would have

(10) $\frac{d(M\vec{v})}{dt} = e\vec{v} \times \vec{B}$,

where e is the electron charge, $\vec{v}$ the coordinate velocity, and $M := m\gamma(\vec{v})$ the relativistic mass, with m the rest mass. From (10) one easily computes (Exercise 3.13) that the magnitude $v := |\vec{v}|$ of the velocity must be constant and that the electron will move in a circle of radius r given by

(11) $$r = \frac{Mv}{eB}.$$

which suggests that if the particle loses (kinetic) energy by radiation, then the speed will drop and the radius of the circle should decrease. It is not hard to show (Exercise 3.12) that if the speed *does* drop, then (10) cannot hold and (11) doesn't follow either, but it is assumed that dv/dt is sufficiently small that (10) and (11) are acceptable approximations. Note that even if dv/dt is small, dr/dt need not be small since

$$\frac{dM}{dt} = \frac{1}{(1-v^2)^{3/2}} \vec{v} \cdot \frac{d\vec{v}}{dt} = \frac{1}{(1-v^2)^{3/2}} \frac{1}{2} \frac{d(v^2)}{dt} = \frac{1}{(1-v^2)^{3/2}} v \frac{dv}{dt}.$$

At relativistic velocities ($v \approx 1$) a small change in dv/dt can easily be magnified several orders of magnitude by the factor $(1-v^2)^{-3/2}$.

The rate of decrease of r can be observed by placing targets in the path of the beam. First the electrons are driven in a circle by electric fields, and then the electric fields are turned off and the rate of orbit contraction is measured. From this one can deduce from (11) the actual rate of energy loss in the laboratory frame. The theoretical prediction is obtained by applying a Lorentz transformation to (8) (Exercise 4.10). The result is that the laboratory-frame rate of energy loss dE/dt is given by

(12) $$\frac{dE}{dt} = \frac{2}{3} \frac{q^2}{r^2} v^4 \gamma(v)^4.$$

At the velocities at which the orbit contraction becomes noticeable for electrons v may be taken as unity (in the experiment mentioned above, $v > .99999$ and $\gamma(v) \approx 400$), and we may think of $\frac{dE}{dt}$ as very nearly proportional to γ^4; i.e. to the fourth power of the relativistic mass-energy of the particle:

(13) $$\frac{dE}{dt} \sim \gamma(v)^4.$$

This predicts that radiation losses will increase dramatically with particle energy once relativistic velocities are attained; doubling the energy increases radiation losses by a factor of $2^4 = 16$.

The actual measurements are in nearly perfect agreement with (13). This is usually taken as a verification of (8). However, it is more accurate to say that (13) is *consistent* with (8), since there are reasonable expressions for radiation other than (8) which also imply (13). For example, we shall see in the next Section that if we use a particular energy-momentum tensor different from the usual one, then the negative of the term

(14) $$\frac{2}{3} q^2 \left[\frac{da}{d\tau} + <a,a>u \right]$$

which appears on the right of the Lorentz-Dirac equation may reasonably be interpreted as the rate of radiation of energy-momentum of the particle. In the particle's rest frame, (14) represents purely spatial momentum, so (14) is in a sense opposite to (8), which represents radiation of pure mass-energy. If (14) is assumed in place of (8), then (13) still follows (Exercise 4.9). This is a special to motion at constant speed (which, assuming the Lorentz equation, is the case whenever the field is a pure magnetic field in the laboratory frame) and is essentially due to the fact that in that case a and hence $da/d\tau$ happen to be purely spatial in the laboratory frame, so that the $da/d\tau$ does not contribute to the lab-frame energy radiation. For instance, assumptions (8) and (14) do not predict the same lab-frame energy radiation rate, for one-dimensional motion (Exercise 4.9).

The latter is a particular instance of radiation known as *bremsstrahlung* (braking radiation), which occurs when charged particles with relativistic velocities hit a stationary target. Although *bremsstrahlung* can easily be observed (for instance, this is a common way of producing x-rays), uncertainty about the precise nature of the forces producing the deacceleration makes it difficult to make quantitative predictions with confidence. Unfortunately, it seems that no experiment which can distinguish between energy-momentum radiation rates (8) and (14) has been performed.

Returning to synchrotron radiation, it is interesting to note that not only does the measured energy loss (13) of the particles conform to (8), but this radiation can actually be *seen* as a beam of visible light tangent to the particles' orbit. The spectral distribution can be calculated by a (nontrivial) Fourier analysis of the Poynting vector [Schwinger], [Jackson, Chapter 14], and the calculations are in good agreement with the observed spectrum.

It is interesting that though the γ^4 law [Schott, 1912]) and spectral calculations were well known in 1947 when synchrotron radiation was first seen, they were not universally believed relevant, and the first visual observation of such radiation was an unexpected consequence of a malfunction in the machine. Early circular accelerators did not produce energies high enough for the γ^4 losses to be noticed. As more powerful machines were built which approached energy levels at which these losses might be expected to be observed, there was considerable doubt as to whether they would actually occur. ‡ The basic reasoning, considerably oversimplified, was that if one models the particle beam as a continuous stream of infinitesimal particles, then it would be analogous to direct current flowing in a wire, and it was well known that "a direct current does not radiate", as one expert put it. More sophisticated models consider random fluctuations which cause the electrons to "bunch up" and lead ultimately back to the γ^4 law. That is, the relation of energy loss to velocity turns out to be the same as that predicted for a single particle. All this shows how difficult and sometimes ambiguous is the analysis of actual experiments.

To summarize, most consider the verification of the γ^4 law (13) for radiation losses in synchrotrons to be direct, quantitative evidence in favor of (8), and   in favor of T_{usual}. In addition there is much indirect evidence for T_{usual} in the form of qualitatively successful calculations of various physical aspects of radiation using the radiation Poynting vector. However, we should also mention that there are cases in which the interpretation of the Poynting vector as *local* energy flow seems physically questionable. Perhaps the simplest example is a point charge sitting next to a permanent magnet. This is a static situation; the only fields are the Coulomb electric field of the point charge and the magnetic field of the magnet, and these do not change with time. One might think that in such an electrostatic situation there would be no energy flow and that the Poynting vector would vanish identically. Yet in almost any configuration, at most points of space $\vec{E}$ and $\vec{B}$ will not be parallel, and the Poynting vector does *not* vanish. If one insists on interpreting the Poynting vector as describing the local flux of field energy, then this means that there is a constant flow of field energy. Of course, there is no mathematical contradiction; for example, the energy flow might just go around forever in closed curves. In an electrostatic situation (i.e. all fields constant in time and no work done on charges), the integral of the Poynting vector over any closed surface vanishes by 3.8(30), indicating no *net* outflow of energy from the volume enclosed by the surface, though there may be outflow through some parts of the surface

---

‡ [Baldwin] gives a very interesting first-hand account of some of the history of the discovery of synchrotron radiation. For another interesting perspective, see [Pollock].

compensated by inflow through others. Presumably, this perpetual energy flow is physically undetectable, but it does call into question calculations, such as those of antenna radiation patterns, in which the Poynting vector is interpreted as local energy flow and is considered to be observable (by, for example, a field-strength meter). Why can one observe the Poynting vector in one situation and not the other? (One possible answer is that it may be the radiation Poynting vector $\vec{P}_{rad}$ rather than $\vec{E} \times \vec{B}/4\pi$ which is actually observed.) The fascinating text of [Feynman, Leighton, and Sands] discusses other interesting examples of "unphysical" Poynting vectors.

5.7 Alternate energy-momentum tensors and equations of motion.

A. Introduction.

The difficulties with the Lorentz-Dirac equation noted in Sections 4.3 and 5.4 have caused many to seek other equations of motion. The problem is how to get away from the Lorentz-Dirac equation without losing conservation of energy-momentum. For example, the following have been proposed:

(1) (Lorentz-Dirac) $\qquad ma = q\hat{F}_{ext}(u) + \frac{2}{3}q^2[\frac{da}{d\tau} + <a,a>u]$

(2) [Eliezer, 1948]

$$ma = q\hat{F}_{ext}(u) + \frac{2q^3}{3m}[\frac{d(\hat{F}_{ext}(u))}{d\tau} + <\hat{F}_{ext}(u), a>u]$$

(3) [Mo and Papas]

$$ma = q\hat{F}_{ext}(u) + \frac{2q^3}{3m}[\hat{F}_{ext}(a) - <\hat{F}_{ext}(a), u>u]$$

(4) [Herrera, 1977]

$$ma = q\hat{F}_{ext}(u) + \frac{2q^4}{3m^2}[\hat{F}_{ext}^2(u) + <\hat{F}_{ext}(u), \hat{F}_{ext}(u)>u]$$

(5) [Bonnor]

$$\frac{d(mu)}{d\tau} = q\hat{F}_{ext}(u) + \frac{2}{3}q^2<a,a>u \quad .$$

These are equations for a particle whose worldline $\tau \longmapsto z(\tau)$ in Minkowski space **M** is parametrized by proper time τ and which is acted on by an external field F_{ext}. As usual, $u := dz/d\tau$ and $a := du/d\tau$.

The Bonnor equation (5) is just the equation 4.3(8) which emerges naturally from 4.3(5) as an expression of conservation of energy-momentum by integrating the usual field energy-momentum tensor,

(6) $\qquad T_{usual} = \hat{F}^2 - \frac{1}{4}\text{tr}(\hat{F}^2)I \quad ,$

over a Bhabha tube. Taking the inner product of both sides of (5) with u shows that it can hold only if the particle's rest mass is variable. The radiated energy is furnished by a decrease in rest mass. The numerical constants in (5) are such that under physically reasonable accelerations, an electron could radiate away all its mass in an observable time. To avoid this problem, Bonnor proposed the equation only for "macroscopic" charged particles.

We saw in Section 4.3 that the Lorentz-Dirac equation may be motivated as a replacement for the Bonnor equation 4.3(8). Assuming the validity of mass renormalization, it implies a law of conservation of energy-momentum for particles with asymptotically vanishing acceleration and guarantees constant rest mass.

The equations proposed by Eliezer, Mo and Papas, and Herrera are all variants of the Lorentz-Dirac equation, the idea being to approximate a in some sense by $(q/m)\hat{F}_{ext}(u)$ on the right to get rid of the term $da/d\tau$ associated with the "preacceleration" discussed in Section 4.3. Since the Lorentz equation

$$a = \frac{q}{m} \hat{F}_{ext}(u)$$

is very nearly exact, it is not unreasonable to hope that an equation obtained in this way might conserve energy-momentum to a close approximation. However, none of the proposers of these equations provided an exact conservation law.

Since the Bonnor equation and ultimately the Lorentz-Dirac equation come from reasonable formulations of conservation of energy-momentum in terms of T_{usual}, it is natural to try to obtain the other equations from different field tensors. We shall see that this can be done. In fact, *any* equation of the form

(7) $$\frac{d(mu)}{d\tau} = v(\tau),$$

where $v(\tau)$ is any vector function on the worldline (such as the right sides of (5.1)-(5.5)), can be obtained from an appropriate symmetric, zero-divergence energy-momentum tensor by essentially the procedure outlined in Section 4.3 except that mass renormalization is *not* necessary. From this point of view, an equation of motion of form (7) *is* a law of conservation of energy-momentum of *field* + *particle* on the worldline relative to some field "energy-momentum" tensor. † The conservation law is completed by requiring that the divergence of the tensor vanish off the particle worldlines.

The situation may be reviewed as follows. We want an equation of motion which is indistinguishable from the Lorentz equation for motions of individual particles directly observable with present (1985) technology. We also want it to be an expression of conservation of energy-momentum relative to a tensor which can reproduce the successes of T_{usual}. The latter requirement suggests that the energy flux vector of the new tensor should be closely related to the Poynting vector.

Of course, there are other requirements. For instance, we would like to rule out preacceleration and behavior such as that described in Section 5.5. The really objectionable feature of the latter is not so much that opposite charges repel (though that is bad enough!), but that energy is only conserved formally (i.e. assuming mass renormalization) and even then not in a particularly useful sense. For instance, we can start with two particles at rest and end up with arbitrarily high kinetic energies.

---

† Since this is not the usual tensor, some may prefer a name other than "energy-momentum" for the field quantity which is conserved. We choose this terminology because this quantity plays a role in a theory based on one of the alternate equations analogous to that of energy-momentum in the traditional theory. It should also be noted that opinions differ as to what properties should be required of an "energy-momentum tensor". For instance, some feel that it should be derived in the "usual" way (which we have not discussed) from a Lagrangian. The reader who shares these views can simply substitute another name.

B. The method.

The method we shall use to obtain an equation of motion as an expression of conservation of energy-momentum relative to a given energy-momentum tensor is similar to that of Sections 4.3 and 4.4, and we shall use the notation of those sections. We consider a three-dimensional Bhabha tube $S_r(\tau_1,\tau_2)$ of retarded spheres $S_r(\tau)$ surrounding the worldline of a particle along with caps $C_r(\tau_i)$, $i = 1, 2$, which cut the tube in retarded spheres $S_r(\tau_i)$. (See Figure 4-5). In Section 4.3 we used spacelike hyperplanes as caps, but in the present context it will be more convenient to use truncated light cones:

(8) $\qquad C_r(\tau_i) := \{ \lambda(u(\tau_i) + w) \mid 0 < \lambda < r, \ \langle w, u(\tau_i)\rangle = 0 \}$.

This is simpler because the integrals over such cones vanish for the particular tensors which we shall consider.

Let $T = (T^i{}_j)$ be a given energy-momentum tensor (not necessarily the usual one). We take as the condition for conservation of energy-momentum equation 4.3(5) in the limit $r \to 0$:

(9) $\qquad mu(\tau_2) - mu(\tau_1) = - \lim_{r \to 0} \left[\int_{S_r(\tau_1,\tau_2)} \perp T + \int_{C_r(\tau_2)} \perp T + \int_{C_r(\tau_1)} \perp T \right]$,

where the orientations for $C_r(\tau_2)$, $C_r(\tau_1)$, $S_r(\tau_1, \tau_2)$ are chosen as the boundary orientations for the four-dimensional manifold which they bound. In Section 4.3 the limit existed only after mass renormalization, but the limit will always exist for the class of tensors which we shall introduce, and no renormalizations will be necessary. Differentiating with respect to τ_2 gives the equation of motion associated with T:

(10) $\qquad \dfrac{d(mu)}{d\tau} = - \dfrac{d}{d\tau} \lim_{r \to 0} \left[\int_{S_r(\tau_1,\tau)} \perp T + \int_{C_r(\tau)} \perp T \right]$.

The second term on the right vanishes for the tensors which we shall use and is only included for completeness.

The physical motivation is similar to that of Section 4.3. It is not as easy to directly interpret integrals over a light cone as over a spacelike surface, but an interpretation can be obtained by using the fact that our T's will have zero divergence to relate $\int_{C_r(\tau_i)} \perp T$ to an integral over a smaller tube of radius $r' < r$ plus an integral over a spacelike hyperplane. (See Figure 5.) The details are left to the reader.

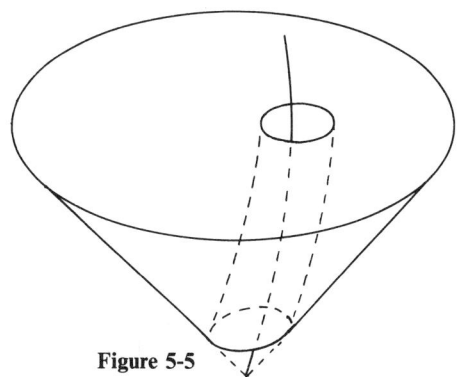

Figure 5-5

Before introducing the class of tensors which will yield equations of form (7), we review the notation of Sections 4.2 and 4.3. The situation is as follows. We have a worldline $\tau \longmapsto z(\tau)$ parametrized by proper time τ, with four-velocity denoted $u(\tau)$, and four-acceleration $a(\tau)$. Given any point $x \in \mathbf{M}$ not on the worldline, let $\tau_{ret}(x)$ denote the "retarded" proper time at which the backward light cone from x intersects the worldline. (See Figure 4-1.) The four-velocity u is initially only defined on the worldline but may be extended to a vector field $x \longmapsto u_x$ defined on all of $\mathbf{M}$ by defining

$$u_x := u(\tau_{ret}(x))$$

for x not on the worldline. Similar remarks apply to other vector and scalar functions on the worldline.

Recall from Section 2.4 that if v is a vector and f a scalar function, then $v(f)$ denoted the directional derivative of f in the direction of v. In this section we shall change the notation for this directional derivative to $v[f]$ in order to avoid a point of possible confusion. When $\tau \longmapsto v(\tau)$ is a vector function on the worldline as in (7), the notation $v(\tau_{ret})$ for the directional derivative would invite confusion of the *vector* field $x \longmapsto v(\tau_{ret}(x))$ with the *scalar* function $x \longmapsto v_x[f]$.

Let $x \in \mathbf{M}$ be the field point at which we are to evaluate the tensor. We assume that x is not on the worldline. (The tensors to be introduced are undefined on the worldline itself.) Decompose the vector from the source point $z(\tau_{ret}(x))$ to the field point x as

(11) $\qquad x - z(\tau_{ret}(x)) = (u(\tau_{ret}(x)) + w)r$,

where w is a spacelike unit vector which is orthogonal to u, and r is a scalar (the *retarded distance*). (See Figure 4. For brevity, we drop some of the subscripts "ret" used in Section 4.2.) Equation (11) defines both r and w, which we view as functions of the field point x:

(12) $\qquad r := \langle x - z(\tau_{ret}(x)), u(\tau_{ret}(x)) \rangle \quad$ and

$$w := \frac{x - z(\tau_{ret}(x))}{r} - u(\tau_{ret}(x)).$$

Also define a lightlike vector c pointing from the source point $z_{ret}(x)$ to the field point x by

(13) $\qquad c := u(\tau_{ret}(x)) + w$.

We also view c as a function of x; i.e. as a vector field defined off the worldline.

We can now introduce our class of energy-momentum tensors T: they are of the form

(14) $\qquad T := f \dfrac{c \otimes c^*}{r^2}$,

where f is a scalar function of the special form

$$f(x) = g(w, z(\tau_{ret}(x))),$$

with g an arbitrary scalar function. A more geometrical way to say this is that f can be any function which is constant on forward light rays emanating from the worldline. For example, we shall see that a tensor of this form which yields the Lorentz-Dirac equation (1) as (10) is

5.7 Alternate e-m tensors

(15) $$T_{L-D} := \frac{3}{4\pi} <k\frac{da}{d\tau} + q\cdot F_{e\hat{x}t}(u), w> \frac{c\otimes c^*}{r^2},$$

where $k := \frac{2}{3}q^2$. (Later we shall consider other choices for k.)

C. Zero divergence.

An acceptable energy-momentum tensor must have zero divergence off the particle worldlines in order to ensure conservation of the energy-momentum of the field. First we shall do the computation which establishes that the divergence of tensors of the form (14) does vanish. [†] We shall need expressions for

$$v[\tau_{ret}], \quad v[r], \quad D_v u, \text{ and } \quad D_v w,$$

where v is an arbitrary vector in $\mathbf{M}$. The first is provided by the expression 4.2(13) for the differential of the retarded time:

(16) $$d\tau_{ret} = c^*.$$

In other words (16) says that for any $v \in \mathbf{M}$,

$$v[\tau_{ret}] = <c, v> = <u+w, v>.$$

Since $u_x = dz/d\tau(\tau_{ret}(x))$, it follows from the chain rule and (16) that for any vector $v \in \mathbf{M}$,

(17) $$D_v u = \frac{du}{d\tau}(\tau_{ret}(x)) \cdot v[\tau_{ret}] = <c, v>a.$$

From 4.2(14) and (16),

(18) $$v[r] = -<v, w> + r<v, c><w, a>.$$

Finally, $D_v w$ can be obtained by applying D_v to (11) or (12). All the resulting derivatives except $D_v w$ have already been computed above, and solving for $D_v w$ gives

(19) $$D_v w = \frac{1}{r}\bigl[v - (<v, u> + r<v, c><w, a>)u$$
$$+ (<v, w> - r<v, c><w, a>)w - r<v, c>a\bigr].$$

Some particular cases of (19) which we shall use are:

(20) $$D_u w = -<w, a>c - a,$$

(21) $$D_w w = <w, a>c + a = -D_u w,$$

and if e is orthogonal to both u and w, then

(22) $$D_e w = \frac{e}{r}.$$

[†] This useful fact seems to have escaped wide notice; the only statement which I have seen in the literature is in [Rowe], where it is only mentioned in passing.

At any point $x \in M$, choose an orthonormal basis of the form u_x, w_x, e_2, e_3. Then from (17), we obtain

(23) $\quad \delta(u) = \langle D_u u, u \rangle - \langle D_w u, w \rangle - \langle D_{e_2} u, e_2 \rangle - \langle D_{e_3}, e_3 \rangle$

$\qquad\qquad = \langle a, w \rangle$,

and similarly from (20) and (21),

(24) $\quad \delta(w) = -\langle a, w \rangle + \dfrac{2}{r}$.

Hence

$$\delta(c) = \delta(u+w) = \frac{2}{r}.$$

From the component formula 2.8(6) for the covariant divergence of a vector field, it follows that for any vector field v and scalar function f,

$$\delta(fv) = \frac{\partial(fv^i)}{\partial x_i} = v^i \frac{\partial f}{\partial x_i} + f \cdot \frac{\partial v^i}{\partial x_i} = v[f] + f\delta(v).$$

Hence

(25) $\quad \delta(\dfrac{c}{r^2}) = c[\dfrac{1}{r^2}] + \dfrac{1}{r^2} \cdot \dfrac{2}{r} = \dfrac{-2c(r)}{r^3} + \dfrac{2}{r^3} = 0$,

where the relation $c[r] = 1$ follows from (18). Moreover, for any function f,

$$\delta(f\frac{c}{r^2}) = \frac{c[f]}{r^2} + f\delta(\frac{c}{r^2}) = \frac{c[f]}{r^2}.$$

If f is constant on forward light rays emanating from the worldline (which are lines of the form $\lambda \mapsto \lambda c$, $\lambda > 0$), then $c[f] = 0$, and so

$$\delta(f\frac{c}{r^2}) = 0.$$

Similarly, Exercise 3.21(b) shows that when $c(f) = 0$,

(26) $\quad \delta(f\dfrac{c \otimes c}{r^2}) = \delta(f\dfrac{c}{r^2})c + D_c(\dfrac{f}{r^2}c) = 0 + \dfrac{f}{r^2}D_c c = 0$.

This completes the proof that tensors of the form (14) have zero divergence on the worldline.

D. Special integrals over the tube.

Now we compute

$$\int_{C_r(\tau)} \perp T \qquad \text{and} \qquad \int_{S_r(\tau_1,\tau_2)} \perp T$$

for certain tensors of the form (14) using the method and notation of Section 4.4. First we show that the integrals over the cone caps vanish; in fact, $\perp T$ vanishes when restricted to the tangent space of the cone at any point. This last fact makes it eminently reasonable to simply ignore the singularity in (14) at the vertex of the cone. (If the integral over the cone vanished by virtue of some complicated cancellation, one would want to investigate more carefully what was happening around the vertex.)

Let $r>0$ be given. Recall from Section 4.4 that for each proper time τ we have a map $\phi(\tau, \cdot): S_r^{std} \longrightarrow S_r(\tau)$, which anti-isometrically identifies the "standard" sphere S_r^{std} of radius r in R^3 with the retarded sphere $S_r(\tau)$, and that for each τ, s we write $\phi(\tau, s) = z(\tau) + r(u(\tau) + w(\tau, s))$ with $w(\tau, s)$ a spacelike unit vector orthogonal to $u(\tau)$. The cone $C_r(\tau)$ is defined as the image of the map

$$\psi : (0, r) \times S_r^{std} \longmapsto \mathbf{M}$$

defined by

$$\psi(\lambda, s) = z(\tau) + \lambda \cdot (u(\tau) + w(\tau, s)), \qquad 0 < \lambda < r, \quad s \in S_r^{std}.$$

We could follow the method of Section 4.4 and integrate $\perp T$ over the cone by integrating its pullback under ψ over $(0, r) \times S_r^{std}$ (cf. Exercise 4.2). However, there is a shortcut for tensors of the special form $fc \otimes \perp c^*$ because the restriction of $\perp c^*$ to the tangent space of the cone happens to vanish identically, and of course its integral over any subset of the cone also vanishes. To see this, recall from 2.2(21) that for any vectors v_1, v_2, v_3,

$$\perp c^*(v_1, v_2, v_3) = \Omega(c, v_1, v_2, v_3) .$$

This clearly vanishes when all the v_i are tangent to the three-dimensional cone because then, since c is also tangent to the cone, the set c, v_1, v_2, v_3 must be linearly dependent.

Now we compute the integral over the tube of certain tensors of form (14). Recall from 4.4(19) that $\frac{\perp c^*}{r^2}$ pulls back under $\phi^\dagger$ to

$$(27) \qquad \phi^\dagger\left(\frac{\perp c^*}{r^2}\right) = \Omega_P ,$$

where Ω_P denotes the volume 3-form on the product $(\tau_1, \tau_2) \times S_r^{std}$.

Hence

$$(28) \qquad \int_{S_r(\tau_1,\tau_2)} f \cdot \frac{(u+w) \otimes \perp(u^*+w^*)}{r^2}$$

$$= \int_{\tau_1}^{\tau_2} d\tau \int_{S_r^{std}} \tilde{f}(\tau, s) \frac{u(\tau) + w(\tau, s)}{r^2} d\sigma_r(s) ,$$

where $\tilde{f}(\tau, s)$ stands for the pullback of f to the product: $\tilde{f}(\tau, s) := f(\phi(\tau, s))$.

Of course, (28) is to be viewed as the four scalar equations obtained by inserting indices on u and w. Note that the term involving $u(\tau)$ is independent of the sphere variable and hence can be moved outside the integral over the sphere:

$$(29) \quad \int_{\tau_1}^{\tau_2} d\tau \int_{S_r^{std}} \tilde{f}(\tau,s) \frac{u(\tau)}{r^2} d\sigma_r(s)$$

$$= \int_{\tau_1}^{\tau_2} d\tau \, u(\tau) \int_{S_r^{std}} \frac{\tilde{f}(\tau,s)}{r^2} d\sigma_r(s).$$

The two cases which will be of interest are

(i) when $\tilde{f}$ is odd in s (i.e. $\tilde{f}(\tau,-s) = -\tilde{f}(\tau,s)$), in which case (29) vanishes, and

(ii) when $\tilde{f}$ is independent of s, in which case the inner integral collapses to $4\pi \tilde{f}(\tau)$.

The part of the inner integral of (28) involving w, with indices inserted, is

$$(30) \quad \int_{S_r^{std}} \frac{\tilde{f}(\tau,s)}{r^2} w^i(\tau,s) d\sigma_r(s).$$

For fixed τ, since $w(\tau, s)$ is orthogonal to $u(\tau)$ and the map $s \longmapsto rw(\tau, s)$ is an anti-isometry, we can choose a Lorentz frame in which $u(\tau) = (1, 0, 0, 0)$ and $w(\tau, s) = (0, \vec{s})/r$, with $\vec{s} \in R^3$. In this frame (30) is identified (cf. 4.4(25)) with the integral written in three-dimensional notation as

$$(31) \quad \int_{S_r^{std}} \frac{\tilde{f}(\tau,\vec{s})}{r^2} \frac{\vec{s}}{r} d\sigma_r(\vec{s}).$$

In one of the two cases of interest, that in which f is independent of s, (31) vanishes because the integrand is odd in s. The other case of interest is as follows. Suppose $v = v(\tau) \in M$ is orthogonal to $u(\tau)$, so that in the above τ-dependent frame, $v = (0, \vec{v})$, and suppose f is of the form

$$\tilde{f}(\tau, s) = \langle v(\tau), w(\tau, s) \rangle = -\frac{\vec{v} \cdot \vec{s}}{r}.$$

Then by symmetry under spatial rotations leaving $\vec{v}$ fixed, we see immediately that (31) must be a multiple of $\vec{v}$, and when the elementary integration is carried out, it is found to be

$$\int_{S_r^{std}} \frac{\tilde{f}(\tau,\vec{s})}{r^2} \frac{\vec{s}}{r} d\sigma_r(\vec{s}) = \int_{S_r^{std}} -(\vec{v} \cdot \frac{\vec{s}}{r}) \frac{\vec{s}}{r} \frac{d\sigma_r(\vec{s})}{r^2} = -\frac{4\pi}{3} \vec{v}.$$

Moreover, for such an f, which is odd in s, (29) vanishes. Putting all this together shows that if we are given any vector function $\tau \longmapsto v(\tau) \in M$ such that $v(\tau)$ is always orthogonal to $u(\tau)$, and if we extend v to a vector field on M by $v_x := v(\tau_{ret}(x))$ for all $x \in M$, then

$$(32) \quad \int_{S_r(\tau_1,\tau_2)} \frac{3\langle v, w \rangle}{4\pi r^2} (u + w) \otimes_\perp (u^* + w^*) = -\int_{\tau_1}^{\tau_2} v(\tau) \, d\tau.$$

This implies that *any* equation of motion of the form

(33) $$m\frac{du}{d\tau} = v(\tau)$$

with $v(\tau)$ orthogonal to $u(\tau)$ can be obtained without renormalization as (10) from an energy-momentum tensor of the form

(34) $$\frac{3<v, w>}{4\pi r^2}(u + w) \otimes (u^* + w^*) .$$

Of course, all of the equations (1)-(5) are of the form (33). We shall see in a moment that with a different choice of f one can obtain (33) without any restriction on $v(\tau)$, so the Bonnor equation

(5) $$\frac{d(mu)}{d\tau} = q\hat{F}(u) + k<a, a>u ,$$

where $k := \frac{2}{3}q^2$, can also be obtained from this kind of tensor.

Recall from Section 4.4 that if mass renormalization is allowed, then the usual energy-momentum tensor T_{usual} also yields the Bonnor equation. This suggests an alternate method for "obtaining" from an energy-momentum tensor any equation of the form

(35) $$\frac{d(mu)}{d\tau} = q\hat{F}(u) + p(\tau)$$

as an expression of conservation of energy-momentum. Simply add to T_{usual} a term of the form (14) which gives (33) with $v(\tau) := p(\tau) - k<a, a>u$. Though it is not clear the extent to which T_{usual} is fundamental, some may feel more comfortable adding to it a term which in most circumstances is unobservably small rather than abandoning it altogether.

E. The main result.

We summarize the above discussion in the following theorem.

Theorem.

Let $\tau \longmapsto z(\tau)$ be a particle worldline parametrized by proper time τ and $\tau \longmapsto v(\tau) \in \mathbf{M}$ a vector function. View v as a vector field on $\mathbf{M}$ via the "backward light cone" construction described above:

$$v_x := v(\tau_{ret}(x)) \quad \text{for all } x \text{ not on the worldline.}$$

Then the following symmetric 2-contravariant tensor field

(36) $$x \longmapsto S_x := [-<v_x, u_x> + 3<v_x, w_x>]\frac{c_x \otimes c_x}{4\pi r^2} ,$$

defined for x not on the worldline, has the following properties:

(i) $\delta(S) = 0$;

(ii) For any $\tau_1 \leq \tau_2$, and any $r > 0$,

$$\int_{S_r(\tau_1, \tau_2)} \lrcorner \hat{S} = -\int_{\tau_1}^{\tau_2} v(\tau) \, d\tau ,$$

where $\hat{S} = (S^i{}_j)$ denotes the mixed form of S, and

(iii) the integral of $\lrcorner \hat{S}$ over any open subset of a forward light cone $C_r(\tau)$ emanating from the worldline vanishes.

Proof. We proved above that (36) has the stated properties when $v(\tau)$ is always orthogonal to $u(\tau)$, so by linearity it is enough to prove it when $v(\tau)$ is a multiple of $u(\tau)$, say

$$v(\tau) = g(\tau)u(\tau)$$

for some function $\tau \longmapsto g(\tau)$. View g as a function on $\mathbf{M}$ via the backward light cone construction. Now (36) reduces to

$$\hat{S} = \frac{-g}{4\pi r^2}(u+w) \otimes (u^* + w^*) \ ,$$

and

$$\int_{S_r(\tau_1,\tau_2)} \lrcorner \hat{S} = \int_{\tau_1}^{\tau_2} d\tau \int_{S_r^{std}} \frac{-g(\tau)}{4\pi r^2} (u(\tau) + w(\tau,s)) \, d\sigma_r(s) \ .$$

Since $w(\tau,s)$ is odd in s, its integral over S_r^{std} vanishes, and

$$\int_{S_r(\tau_1,\tau_2)} \lrcorner \hat{S} = -\int_{\tau_1}^{\tau_2} d\tau \, g(\tau)u(\tau) = -\int_{\tau_1}^{\tau_2} v(\tau) \, d\tau \ .$$

This completes the proof. Note that (ii) is valid for any nonzero r, and taking the limit $r \to 0$ in (10) is actually unnecessary for tensors of the form (36).

The first result of this type was obtained by Pryce [1938]. Pryce's tensor is

$$(37) \qquad T_{pryce} := \frac{q^2}{4\pi}[-9\!<\!a,w\!>^2 + 2\!<\!\frac{da}{d\tau},c\!> - 3\!<\!a,a\!>]\frac{c \otimes c^*}{r^2} \ ,$$

plus terms involving F_{ext} which are the same as those of T_{usual}, and it yields the Lorentz-Dirac equation as (10). However, this tensor has not gained wide acceptance, possibly because of its complexity and lack of physical motivation. A result which is formally similar to the theorem above was obtained by [Bhabha and Harish-Chandra], but their tensors are more complicated than (36) and are not symmetric except in unusual special cases.

F. General properties of the tensors given by the theorem.

One possible objection to energy-momentum tensors of the form

(38) $$T = f \frac{c \otimes c^*}{r^2}$$

is that the local energy density at $x \in M$ relative to a Lorentz frame with unit timelike vector e_0 is

(39) $$\langle T_x e_0, e_0 \rangle = \frac{f(x)}{r^2} \langle u+w, e_0 \rangle^2 ,$$

which is not positive where $f(x)$ is not positive. On the other hand, it is not clear that one needs to demand such local positivity. For example, the Poynting vector $\vec{E} \times \vec{B}$ is supposed to represent local energy flux, but when so interpreted can give strange results which appear as "unphysical" as negative energy densities (cf. the end of Section 5.6). Some feel that what is important is not the interpretation of $\vec{E} \times \vec{B}$ as local energy flow, but rather the interpretation of its integral over any closed surface as the rate of radiation leaving the volume bounded by the surface. In the same way, it may be that one can live with negative local energy densities so long as the integral of such over any constant-time hyperplane is finite and bounded below (by a bound independent of the hyperplane.) This would prevent particles from extracting infinite energy from the field. (Even this condition may be too strong. For instance, T_{usual} does not prohibit this, since the usual field energy is infinite.)

The integral of an energy-momentum tensor of the form (38) over a spacelike hyperplane $V(\tau)$ which is orthogonal to the worldline at $z(\tau)$ has a transparent physical interpretation; it is the total energy-momentum radiated by the particle up to proper time τ. To see this, consider the four-dimensional region R in Figure 5 bounded by:

(i) a Bhabha tube $S_r(\tau_1, \tau)$,

(ii) the part $L_r(\tau_1)$ of a forward light cone with vertex at $z(\tau_1)$ which lies outside the tube, and

(iii) the part $V_r(\tau_1, \tau)$ of the spacelike hyperplane $V(\tau) + ru(\tau)$ which lies outside the tube and within the cone $L(\tau_1)$. (Thus $V_r(\tau_1, \tau)$ intersects the tube in the retarded sphere $S_r(\tau)$).

Since the divergence of T vanishes on R,

$$0 = \int_R \delta(T) = \int_{\partial R} \bot T = \int_{S_r(\tau_1,\tau)} \bot T + \int_{L_r(\tau_1)} \bot T + \int_{V_r(\tau_1,\tau)} \bot T .$$

We have already observed in Part D, that

$$\int_{L_r(\tau_1)} \bot T = 0 ,$$

and so

(40) $$\int_{V_r(\tau_1,\tau)} \bot T = - \int_{S_r(\tau_1,\tau)} \bot T .$$

In the limit $r \to 0$, this says that

$$\int_{V_0(\tau_1,\tau)} \perp T$$

represents the total energy-momentum radiated by the particle between proper times τ_1 and τ. In the limit $\tau_1 \to -\infty$,

$$\int_{V_0(\tau_1,\tau)} \perp T \longrightarrow \int_{V(\tau)} \perp T ,$$

and so the integral of $\perp T$ over $V(\tau)$ is the total energy-momentum radiated by the particle up to proper time τ.

This pushes the question as to whether unbounded energy (relative to some given Lorentz frame) can be extracted from the field back to the question of whether whatever equation of motion is postulated allows solutions with arbitrarily large kinetic energy. Of course, for *some* (unphysical) F_{ext}, such as a uniform electric field throughout all space and time, this would be expected to happen, so to answer this, one must replace F_{ext} with the field of the other particles in the system and solve all the equations of motion simultaneously. This is, unfortunately, an extremely difficult mathematical problem even for just two particles. There is no question that it would be more convenient to have an energy-momentum tensor in which the total energy in a spacelike hyperplane were, *a priori*, both finite and bounded below.

Finally, one possible advantage of tensors like (15) over T_{usual} is that the constant k which multiplies the radiation term in the Lorentz-Dirac equation (and similar constants in the other equations) does not have to be $\frac{2}{3} q^2$, but can be chosen arbitrarily. The nominal choice just given comes from "derivations" of the Lorentz-Dirac equation starting from T_{usual} similar to that of Sections 4.3 and 4.4. There seems to be no experimental evidence bearing on the correct value for k, and if T_{usual} were abandoned, there would be no reason to use the traditional value. *A priori*, it would seem more natural to guess that k would be proportional to q instead of q^2. For example, if one could glue together two electrons, would the resulting particle really emit $2^2 = 4$ times the radiation of a single electron with the same worldline, as the conventional choice for k predicts?

G. Further requirements for equations of motion and energy-momentum tensors.

Now we shall critically examine equations (1)-(5) and the corresponding energy-momentum tensors defined by the theorem. Absolutely minimum conditions for a candidate for an electromagnetic energy-momentum tensor would seem to be that it yield the Lorentz force and predict the same laboratory-frame energy loss for uniform circular motion as the classical Larmor radiation law.

Some object [Huschilt and Baylis, 1974] to the Mo-Papas and Herrera equations because the radiation reaction term in each drops out for one-dimensional motion, in which case they reduce to the Lorentz equation. [Huschilt and Baylis, 1973] have investigated certain numerical solutions for this one-dimensional equation in the special case of symmetric motion of two particles of like charge and consider them unphysical. Apart from these objections, the main problem with these equations seems to be that no one has worked out a good way to relate them to traditional pictures of conservation of energy-momentum. For instance, how does one explain *bremsstrahlung* (cf. Section 5.6) in the absence of a radiation reaction term?

I know of no compelling objection to the Eliezer (1.3) equation, and in fact it would seem to be an attractive alternative to the Lorentz-Dirac equation. However, this equation has not undergone the intense scrutiny which the Lorentz-Dirac equation has, and difficulties might well surface.

Many, perhaps most, practical applications of the usual energy-momentum tensor involve analyzing the Poynting vector $\vec{E} \times \vec{B}/4\pi$, which is the space part of the first column of the matrix for $T_{usual} = (T^i{}_j)$, and is physically interpreted as the energy flux vector. A typical example might be an estimation of the spatial radiation pattern of an antenna by assuming a particular current flow and computing the magnitude of the "radiation" part of $\vec{E} \times \vec{B}$ at a large distance as a function of the angular coordinates. Such calculations often give good qualitative agreement with experiment, and an energy-momentum tensor is hardly likely to be taken seriously as a replacement for T_{usual} unless it can reproduce these successes. Therefore, it is important to look at the corresponding energy flux vectors for these alternate tensors.

First we look more closely at the Poynting vector given by the usual energy-momentum tensor for a field

$$F = F_{ret} + F_{ext}$$

consisting of the retarded field for a single particle plus an external field F_{ext}. If we indicate the dependence of T on F by writing

(41) $$\hat{T}(F) := \hat{F}^2 - \frac{1}{4}\text{tr}(\hat{F}^2)I \quad,$$

then we have

(42) $$\hat{T}(F) = \hat{T}(F_{ret}) + \hat{T}(F_{ext}) + \hat{T}_{mix}$$

with

(43) $$\hat{T}_{mix} := \frac{1}{4\pi}[\hat{F}_{ret}\hat{F}_{ext} + \hat{F}_{ext}\hat{F}_{ret} - \frac{1}{4}\text{tr}(\hat{F}_{ret}\hat{F}_{ext} + \hat{F}_{ext}\hat{F}_{ret})]$$

Note that the covariant divergences of all three terms on the right of (42) vanish separately.

Recall from 3.8(28) that the Poynting vector $\vec{E} \times \vec{B}/4\pi$ appeared in the first column of the matrix $(T^i{}_j)$ for the usual energy-momentum tensor. In the same way, one obtains the "energy flux vector" $\vec{P}$ for an arbitrary symmetric energy-momentum tensor $T = (T^{ij})$ relative to a given Lorentz frame by writing the matrix for the mixed form $\hat{T} = (T^i{}_j)$ with respect to that frame; the energy flux vector is then the space part of the first column. † The physical interpretation of $\vec{P}$ is that for any purely spatial unit vector $\vec{n}$, $\vec{P}\cdot\vec{n}$ is the coordinate-time rate per unit area at which energy is crossing an infinitesimal surface to which $\vec{n}$ is normal in the direction of $\vec{n}$. In other words, if e_0, e_1, e_2, e_3 is an orthonormal basis of coordinate vectors which defines the frame, with e_0 timelike, then

(44) Energy flux vector for T = space part of $\hat{T}e_0 = \hat{T}e_0 - <\hat{T}e_0, e_0>e_0$.

In this way, from the expression 4.4(20) for $T(F_{ret})$ one obtains the energy flux vector $\vec{P}_{ret}$ relative to the rest frame of the particle at the retarded time:

(45) $$\vec{P}_{ret} := \hat{T}(F_{ret})(u) - <\hat{T}(F_{ret})(u), u>u = \frac{q^2}{4\pi r^3}a_\perp - \frac{q^2}{4\pi r^2}<a_\perp, a_\perp>w \quad.$$

---

† Actually, the energy flux vector for any $T = (T^i{}_j)$, symmetric or not, is the negative of the space part of the first row of the matrix, but it is easier to speak of the first column. This is another instance in which the "natural" guess as to sign is wrong; the first row seems to have the wrong sign.

Of course, this is just the usual Poynting vector $\vec{E} \times \vec{B}/4\pi$ for the retarded fields; the part

(46) $$P_{rad} := -\frac{q^2}{4\pi r^2} <a_\perp, a_\perp> w$$

was called the "radiation Poynting vector" $\vec{P}_{rad}$ in 4.2(22). As pointed out there, the first term

(47) $$\frac{q^2}{4\pi r^3} a_\perp$$

makes no contribution to the energy flux through a retarded sphere, since $a_\perp$ is tangent to the sphere.

Of course, actual measurements are usually carried out in a fixed "laboratory" Lorentz frame rather than an instantaneous rest frame for the particle. However, if the particle has sufficiently low velocity in the lab frame, as we assume henceforth, then its four-velocity u will approximate $(1, \vec{0})$ in the lab frame, and it is apparent from (45) that the right side can be taken as an approximation to the lab-frame Poynting vector at a fixed spatial point in the lab frame.

In the so-called "far-field" approximation, one considers the limit $r \to \infty$. In this limit, the $O(r^{-3})$ terms are unimportant, but of course we cannot drop the $O(r^{-1})$ terms in $\hat{T}_{mix}$ as we did when considering the limit $r \to 0$ in Section 4.4. We shall assume in our analysis of the Poynting vector for $\hat{T}(F)$ that we can ignore the $O(r^{-3})$ term (47) in the Poynting vector for F_{ret}, either because we are a long way from the source or because we are only interested in radially outward radiation. ‡

Straightforward calculation gives $(\hat{T}_{mix})_x$ at an arbitrary field point $x \in \mathbf{M}$ as:

(48) $$(\hat{T}_{mix})_x := \frac{q}{4\pi r^2}[(\hat{F}_{ext})_x(w) \otimes u^* + u \otimes (\hat{F}_{ext})_x(w)^*$$

$$- (\hat{F}_{ext})_x(u) \otimes w^* - w \otimes (\hat{F}_{ext})_x(u)^* + <(\hat{F}_{ext})_x(u), w>I]$$

$$+ \frac{q}{4\pi r}[(\hat{F}_{ext})_x(c) \otimes a_\perp^* - (\hat{F}_{ext})_x(a_\perp) \otimes c^*)$$

$$- c \otimes (\hat{F}_{ext})_x(a_\perp)^* + a_\perp \otimes (\hat{F}_{ext})_x(c)^* - <\hat{F}_{ext}(c), a_\perp>I] \ .$$

where $c := u + w$. We write $(\hat{F}_{ext})_x$ explicitly as a reminder that $\hat{F}_{ext}$ is evaluated at the field point rather than the retarded point $z(\tau_{ret}(x))$. In the limit of low velocity, the Poynting vector $\vec{P}$ for $\hat{T}(F)$ at $x \in \mathbf{M}$ may be approximated ($\approx$) by the rather complicated expression

(49) $$\vec{P} \approx \vec{P}_{ret} + (\hat{T}_{mix} + \hat{T}(F_{ext}))_x(u) - <(\hat{T}_{mix} + \hat{T}(F_{ext}))_x(u), u>u$$

$$= \vec{P}_{ret} + \frac{q}{4\pi r^2}[(F_{ext})_x(w) - <(F_{ext})_x(w), u>u]$$

$$+ \frac{q}{4\pi r}[-(\hat{F}_{ext})_x(a_\perp) + <(\hat{F}_{ext})_x(a_\perp), u>u + <(\hat{F}_{ext})_x(u), a_\perp>w$$

---

‡ It should also be noted that the process of mass-renormalization effectively throws out the term $(q^2/4\pi r^3)a_\perp \otimes u^*$ in the energy-momentum tensor 4.4(20) which gives rise to (47), so believers in this prescription might well argue that one should throw out (47) even when it does not vanish in the lab frame.

$$- <(\hat{F}_{ext})_x(u), w>a_\perp] + (\hat{F}_{ext}^2)_x(u) - <(\hat{F}_{ext}^2)_x(u), u>u \ .$$

This expression is exact in the rest frame at the retarded time and approximate in the lab frame.

The radially outward component $-<\vec{P}, w>w$ of (49), which is the only part which will contribute to radiation through a retarded sphere with the particle at the center, is considerably simpler:

$$(50) \qquad -<\vec{P}, w>w \approx P_{rad} + \frac{q}{4\pi r}<(\hat{F}_{ext})_x(u-w), a_\perp>w - <(\hat{F}_{ext}^2)_x(u), w>w \ .$$

In general, we cannot ignore the terms involving F_{ext} in the limit $r \to \infty$, and we can proceed no further without more information about F_{ext}. In analyses of antenna radiation patterns, the assumption $F_{ext} = 0$ is usually made. For example, one can think of a transmitter driven by cranking a generator inside a grounded metallic building. If the transmitter is not connected to an antenna, no fields will be observed outside the building. If an external antenna is connected, then radiation can be observed and its intensity as a function of position measured. In the conventional picture, turning the crank moves a wire through a magnetic field, which causes the electrons inside the wire to move, and, after a long chain of intervening events, eventually causes the electrons in the antenna to accelerate and thus radiate. Although there are no accepted relativistically valid models to describe the action of "mechanical" forces on electrons, in some way whose details are obscure, the mechanical energy fed the generator gets translated into motion of electrons. The situation can be abstracted by imagining a single charge driven by forces whose nature is not specified except to say that they are nonelectromagnetic in the sense that they contribute nothing to F_{ext}.

A criticism of this abstraction from a fundamental point of view is that no such forces are generally recognized. Of the four so-called "fundamental forces" (gravitational, electromagnetic, weak, and strong), one (the gravitational force) is not really a "force" within the framework of general relativity. Two (the weak and strong) are essentially quantum in nature and therefore, besides being only incompletely understood, lie outside the domain of classical physics. This leaves only the electromagnetic force as classically describable. From this point of view, it could be argued that considering accelerated electrons while taking $F_{ext} = 0$ is inadmissible. Nevertheless, it is customary in analyses of antenna radiation patterns, and we shall follow this custom and ignore the above philosophical objections. The fact is that a completely coherent and satisfying mathematical description of nature has not yet been invented, and sometimes one has to bypass questions of principle in order to effectively use tools which have proven productive in the past.

When analyzing the tensors similar to those mentioned above for the Eliezer, Mo-Papas, and Herrera equations within this framework, we take $F_{ext} = 0$ at the field point but not at the retarded point, the hope being that whatever fields cause the acceleration may be ignored in this context at large distances. It would certainly be hard to analyze the radiation pattern of the transmitter in the example above without some such assumption. For another example on a more fundamental level, imagine a light particle such as an electron acted on by a much heavier nearby particle such as a proton with the same absolute charge. The proton's acceleration will be smaller than the electron's by a factor on the order of the ratio of their masses (very roughly 10^3), and, assuming the Larmor law, the proton's radiation field will be smaller by a factor of the square, or about 10^6. Thus it is plausible that the proton's field may be ignored relative to the electron's at large distances from the pair, since the part of (50) due to the proton's Coulomb field falls off like r^{-3}.

The effect of this ignoring of $(F_{ext})_x$ at the field point x is that we may use T_{mix} for the part of the tensor which yields the Lorentz force without worrying about its influence on the radiation pattern. This same procedure is customarily employed (usually without explicit mention) in analyzing the Poynting vector for T_{usual} via the assumption $(F_{ext})_x = 0$. The reason that we belabor the point is that when using a tensor like

$$(51) \quad (\hat{T}_{EZR})_x := \frac{2q^3}{4\pi r^2 m} < \frac{d(\hat{F}_{ext})_{z(\tau_{ret}(x))}(u)}{d\tau}, w > (u+w) \otimes (u+w)^* + (\hat{T}_{mix})_x$$

to obtain the Eliezer equation, at first sight it looks a little peculiar to set $F_{ext} := 0$ in the last term but not the first.

The analysis for $T(F_{ret})$ under the above assumptions is almost immediate. The only term which is not negligible at large distances gives the radiation Poynting vector (46), which is approximately a vector pointing from the emission point to the field point in the lab frame. Note that

$$(52) \quad -<a_\perp, a_\perp> = -<a, a>\sin^2\theta \ ,$$

where θ denotes the angle between the acceleration a and the radially outward vector w (exact in the rest frame, approximate in the lab frame). Many features of actual antenna radiation patterns are qualitatively consistent with the $\sin^2\theta$ dependence of (52).

The tensor given by the theorem for the Lorentz-Dirac equation is, at a field point x:

$$\frac{2q^2}{4\pi r^2} < \frac{da}{d\tau} + (\hat{F}_{ext})_{z(\tau_{ret}(x))}(u), w > (u+w) \otimes (u+w)^* \ .$$

Note that $\hat{F}_{ext}$ is evaluated at the retarded point $z(\tau_{ret}(x))$ rather than x, unlike (48). Since the term involving $\hat{F}_{ext}(u)$ has magnitude approximately proportional to the magnitude of the acceleration, rather than its square, it seems more promising to look instead at

$$(53) \quad (\hat{T}_{L-D})_x := \frac{2q^2}{4\pi r^2} < \frac{da}{d\tau}, w > (u+w) \otimes (u+w)^* + T_{mix} \ .$$

This yields the Lorentz-Dirac equation as (10); the Theorem shows that the first term on the right gives the radiation reaction term, and we saw in Section 4.4 that T_{mix} gives the Lorentz force.

The energy flux vector for this tensor relative to the particle's rest frame at the retarded time (taking $F_{ext} = 0$ as discussed above) is

$$(54) \quad \text{Energy flux vector for } \hat{T}_{L-D} \text{ (with } F_{ext} = 0\text{)} = \frac{2q^2}{4\pi r^2} < \frac{da}{d\tau}, w > w \ ,$$

which initially appears to be of quite a different form than (46). However, if we average (54) over time, assuming periodic motion of the source, then we can hope to integrate (54) by parts and get rid of the $da/d\tau$. This averaging is physically reasonable because it is very difficult to observe *instantaneous* radiated power; actual test equipment usually measures average power of a periodic wave over many frequency cycles. The integration by parts is surprisingly delicate and lengthy and is therefore relegated to an appendix. The result, however, is easy to describe: the analog of (46) and (52) for T_{L-D} in the limit of low velocity and amplitude of motion is

(55)   Time-averaged outward energy radiation per unit area for $\hat{T}_{L-D}$

at an angle of θ with a in the lab frame

$$\approx \frac{q^2}{m}\frac{1}{r^2}(8\cos^2\theta - 2) \cdot \frac{1}{\tau_2-\tau_1} \int_{\tau_1}^{\tau_2} -<a(\tau), a(\tau)> d\tau \ ,$$

which does not have the "correct" $\sin^2\theta$ angular dependence. (See Figure 6. The approximation symbol " $\approx$ " is simply a reminder that the result is exact only in the limit described.) However, the analog of (55) for Pryce's tensor (37) has the factor $(8\cos^2 - 2)$ replaced by the desired $\sin^2\theta$. It is easy to see how this happens. Pryce's tensor (37) has additional terms with an angular dependence of $3 - 9\cos^2\theta$. These additional terms contribute nothing to the equation of motion because they integrate to zero over the sphere; however, they are just what is needed to convert $8\cos^2\theta - 2$ into $\sin^2\theta$. It is interesting that the rather complicated Pryce tensor gives the correct angular dependence in this particular limit, since Pryce did not consider this question in his paper.

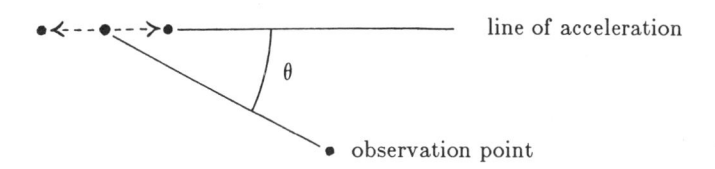

Figure 5-6

It is clear that this kind of device can be used to convert any tensor with an angular dependence of

(56) $$C_1\cos^2\theta + C_2 ,$$

with C_1, C_2 constants into one with a $\sin^2\theta$ dependence, supposing only that (56) is not a multiple of $1 - 3\cos^2\theta$. (If the latter assumption is not valid, then there is no average energy radiation in the limit considered, as defined by that tensor.) For example, the analysis given in the appendix for T_{L-D} goes through without essential change for T_{EZR}, and, assuming that we can approximate $\hat{F}(u)$ by a, we again obtain (55). The incorrect $8\cos^2\theta - 2$ angular dependence can be converted to the desired $\sin^2\theta$ by adding to (51) the terms

(57) $$\frac{q^2}{4\pi r^2}[-9<a,w>^2 - 3<a,a>](u+w)\otimes(u+w)^* ,$$

which does not change the equation of motion for the reason just described. Alternatively, all the a's in (57) could be replaced by $\hat{F}(u)$'s.

Finally, we mention another possible energy-momentum tensor for the Eliezer equation. Recall that the term in the usual tensor which is responsible for the Larmor radiation is given by (46). Since the Lorentz equation is very nearly exact the substitutions

(58) $$a \longrightarrow \frac{q}{m}\hat{F}_{ext}(u) ,$$

and

(59) $$a_\perp = a + <a,w>w \longrightarrow \frac{q}{m}[\hat{F}_{ext}(u) + <\hat{F}_{ext}(u),w>w]$$

should not alter (46) very much. This and the observation that $<a_\perp,a_\perp> = <a_\perp,a>$ partially motivate the following definition of a second energy-momentum tensor T_{EZR2}:

$$\text{(60)} \quad \hat{T}_{EZR2} = -\frac{q^3}{m} [<\hat{F}_{ext}(u), a> + <\hat{F}_{ext}(u), w><w, a>] \frac{(u+w) \otimes (u+w)^*}{4\pi r^2}$$

$$+ \ T_{mix} \ .$$

The reason for the seemingly inconsistent use of the substitution (59) for $a_\perp$ in $<a_\perp, a>$ but not (58) for a will become clear later. This tensor has the correct $\sin^2\theta$ dependence when (58) is a valid approximation. The tensor T_{EZR2} is easily integrated over a Bhabha tube using the machinery developed in Section 4.4, and the result is

$$\text{(61)} \quad \int_{S_r(\tau_1,\tau_2)} \perp T_{EZR2} = -\int_{\tau_1}^{\tau_2} [q\hat{F}_{ext}(u(\tau)) + \frac{2}{3}\frac{q^3}{m} <\hat{F}_{ext}(u(\tau)), a(\tau)> u(\tau)] \, d\tau \ .$$

Hence the equation of motion which arises from (10) using T_{EZR2} is

$$\text{(62)} \quad \frac{d(mu)}{d\tau} = q\hat{F}_{ext}(u) + \frac{2q^3}{3m} <\hat{F}_{ext}(u), a> u \ ,$$

which is similar to the Bonnor equation and has the same difficulty of nonconstant rest mass. However, if we mimic the reasoning of Section 4.3 which led to the replacement of the Bonnor equation by the Lorentz-Dirac equation, we obtain the Eliezer equation. Of course, we lose conservation of energy-momentum relative to T_{EZR2} in the strict sense of (10), but energy-momentum conservation still holds for scattered particles between infinite past and future for the same reason that the Lorentz-Dirac equation guarantees it relative to T_{usual}. More explicitly,

$$\text{(63)} \quad -\int_{\tau_1}^{\tau_2} [q\hat{F}_{ext}(u) + \frac{2}{3}\frac{q^3}{m} <\hat{F}_{ext}(u), a> u] \, d\tau =$$

$$= -\int_{\tau_1}^{\tau_2} [q\hat{F}_{ext}(u) + \frac{2}{3}\frac{q^3}{m} (\frac{d(\hat{F}_{ext}(u))}{d\tau} + <\hat{F}_{ext}(u), a> u)] \, d\tau$$

$$+ \hat{F}_{ext}(u(\tau_2)) - \hat{F}_{ext}(u(\tau_1)) \ .$$

If $\hat{F}_{ext}$ vanishes asymptotically in the infinite past and future, then it is apparent from (63) that the Eliezer equation guarantees conservation of energy-momentum (relative to T_{EZR2}) between infinite past and future.

The electrodynamical theory just outlined, in which T_{EZR2} is taken as the energy-momentum tensor and the Eliezer equation as the equation of motion, seems to retain most (perhaps all) of the experimentally verifiable elements of the usual classical electrodynamics and yet sidesteps the difficulties associated with the Lorentz-Dirac equation. (But it is certainly not clear that new ones will not arise!) Since it is very close in several senses to (46), which seems to be the important radiation term in T_{usual}, it is somewhat more conservative than a theory based on (51) and (57). However, it is a shame to lose the exact conservation of energy-momentum in the sense of (10) which (51) guarantees.

The suggestions above are not offered as finished products, but rather in the spirit of pointing out the kinds of ways in which it might be possible to modify classical electrodynamics to obtain a theory more nearly consistent and intellectually satisfying than that given by the Lorentz-Dirac equation. Given that the latter has never been experimentally tested, it would be surprising if the above theories based on the Eliezer equation and T_{EZR} or T_{EZR2} could be disproved by experiment. All seem physically sterile at present. For this reason, it would be premature to take them seriously. Moreover, no theory of classical electrodynamics is likely to be considered successful unless it can be quantized and reproduce the remarkable successes of the present quantum electrodynamics.

Many, perhaps most, will feel that this goal is hopeless, or at least too risky to pursue, but there have always been voyagers willing to embark in small boats across vast unknown seas. The only frontiers accessible to the common person in this age are those of the mind. To those who have a good boat and the inclination to risk it, may your journey reward your daring.

References

Little of the material in this chapter appears in books, and the reader who wants to learn more will have to consult the original papers cited in the text. Perhaps a word of warning may be useful here. This is a difficult and controversial area of physics, and it is not uncommon to encounter questionable claims and even serious disagreement among different authors of good reputation. In my experience, it is not safe to take anything for granted, and the only way to be sure of a proof or calculation is to check it oneself.

Two interesting papers which present worthwhile approaches not discussed in this chapter are [Teitelboim, 1970] and [Rowe, 1978]. A long chapter by P. Pearle in [Teplitz] discussing electron models is also of interest.

Appendix to Section 5.7. Calculation of an energy flux vector.

We shall consider a particle of charge q executing periodic motion on a line in the limit of low amplitude and velocity of oscillation and shall compute the time-averaged radiation intensity as a function of direction. We work in a fixed "laboratory" Lorentz frame. For definiteness, assume that the line of oscillation is the x-axis and that the x-coordinate of the particle as a function of coordinate time t is given by the simple harmonic motion

(1) $\qquad x(t) = b + b \sin \omega t$,

where b is a small constant. (The displacement by b is for later convenience. The precise nature of the motion is not critical to our considerations so long as it is periodic and satisfies some simple conditions to be described below.) The coordinate velocity $v(t) := dx/dt$ and acceleration $A(t) := dv/dt$ are then

(2) $\qquad v(t) = b\omega \cos \omega t$,

and

(3) $\qquad A(t) = -b\omega^2 \sin \omega t$,

and we shall suppose that b and $b\omega$ are suitably small, while maintaining constant the maximum acceleration K :

$$b\omega^2 = K \quad \text{(constant)} .$$

Of course, this implies that the angular frequency ω may be arbitrarily large and that

(4) $\qquad \dfrac{dA}{dt} = -b\omega^3 \cos \omega t$

can be arbitrarily large. Since (4) appears in the energy-momentum tensor, this makes a certain amount of care necessary in setting up the calculation. It is quite easy to write down plausible-looking approximations which give incorrect results. For example, in the low-velocity limit, the particle four-velocity u approximates a constant vector $(1, 0, 0, 0)$, but we cannot replace u in the energy-momentum tensor with $(1, 0, 0, 0)$ because it is multiplied by the possibly large factor $<da/d\tau, w>$.

Given $r > 0$, let

$$S := \{ (x, y, z) \in R^3 \mid x^2+y^2+z^2 = R^2 \}$$

denote the standard sphere of radius r in R^3. Since r will be fixed throughout, by appropriate choice of time units we may assume that $R = 1$. We are going to compute the average energy-momentum (relative to a particular energy-momentum tensor) radiated through an infinitesimal piece $C(\theta)$ of S at a given angle θ with the line of oscillation (Figure 5-6). The particle executes simple harmonic motion along the x-axis between $x = 0$ and $x = 2b$.

By "average", we mean coordinate-time average over one or more periods of oscillation.

The result for $T = T_{usual}$ is well-known to be proportional to $\sin^2\theta$, even without averaging. This follows almost immediately from 4.2(22): the rate of radiation through a retarded sphere of radius r is proportional to

$$-<a_\perp, a_\perp> = -\frac{<a, a> \sin^2\theta}{r^2} .$$

In the limit we are considering, the retarded spheres aproximate laboratory spheres, and so the same radiation rate holds for laboratory spheres. However, we cannot draw similar conclusions for a tensor which contains $<\frac{da}{d\tau}, w>$ for the reason discussed above, and the analysis for T_{L-D} is considerably more delicate.

Let $C(\theta)$ denote a small "disc" of radius δ on the sphere S whose center lies at an azimuthal angle of θ relative to the line of oscillation (see Figure 6). We shall compute the integral of $\perp T$ over a product

$$TC(\theta) := (t_1, t_2) \times C(\theta) = \{ (t, \vec{c}) \in \mathbf{M} \mid t_1 \leq t \leq t_2, \vec{c} \in C(\theta) \} .$$

The time-averaged rate of energy radiation per unit area at the angle θ is defined as:

(5) $$\lim_{\mu(C(\theta)) \to 0} \frac{1}{(t_2-t_1)\mu(C(\theta))} \int_{TC(\theta)} \perp T$$

where $\mu(C(\theta))$ denotes the area of $C(\theta)$.

The difficulty in computing the integral directly is that given $x \in TC(\theta)$, it is awkward to obtain expressions for $\tau_{ret}(x)$, and hence for u_x, w_x, etc. Instead, we use a more indirect strategy. We shall define a map

$$\phi : (\tau_1, \tau_2) \times C(\theta) \longmapsto TC(\theta) = (t_1, t_2) \times C(\theta)$$

and compute the integral (5) by pulling back $\perp T$ under ϕ_*. Here the interval (τ_1, τ_2) is thought of as an interval of proper times, and points in $C(\theta)$ are regarded as directions, or spatial unit vectors, in the laboratory frame.

The map ϕ is defined as follows. Given a proper time τ and a unit vector $\vec{c} \in C(\theta)$, let the particle send out a forward light ray in the direction of $\vec{c}$ at proper time τ. More formally, this light ray is the half-line

Appendix to 5.7   Energy flux vector

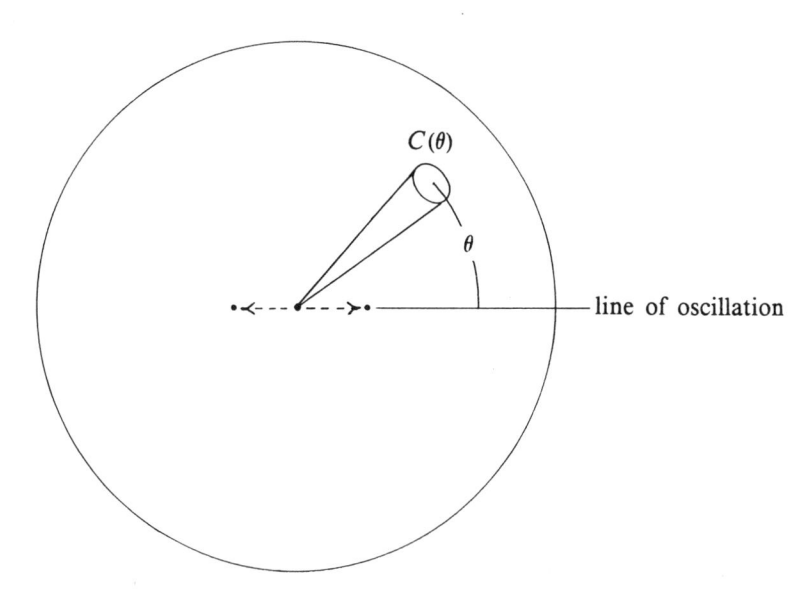

Figure 5-7

$$\{ z(\tau) + (\lambda, \lambda\vec{c}) \in \mathbf{M} \mid 0 \leq \lambda < \infty \},$$

where $\tau \longmapsto z(\tau)$ denotes as usual the particle's worldline parametrized by proper time τ. Let $\bar{t} = \bar{t}(\tau, \vec{c})$ denote the coordinate time at which the light ray crosses the laboratory sphere S, and define

(6) $\qquad \phi(\tau, \vec{c}) := (\bar{t}(\tau, \vec{c}), \vec{c}) \in \mathbf{M}$.

Unlike the maps used in Section 4.4, the restriction of ϕ to a space section $\{\tau\} \times C(\theta)$ is not necessarily an isometry, but this will cause no difficulty. The reader may be wondering what is gained by switching the integration from $TC(\theta) := (t_1, t_2) \times C(\theta)$ to the apparently similar $(\tau_1, \tau_2,) \times C(\theta)$, and the explanation is that there are simple expressions for $\tau_{ret}(y)$, u_y, etc., at points y of the form $y = \phi(\tau, \vec{c})$. For example, $\tau_{ret}(y) = \tau$!

Let $\tau \longmapsto z(\tau) = (z^0(\tau), \vec{z}(\tau))$ denote the particle's worldline, and fix proper times τ_1, τ_2 such that τ_1 and τ_2 differ by an integral number of periods and

(7) $\qquad \vec{z}(\tau_1) = 0 = \vec{z}(\tau_2)$.

Let $u(\tau) := dz/d\tau$ denote the four-velocity, and recall from 1.7(5) that

(8) $\qquad u = \gamma(v) \cdot (1, \vec{v})$,

where $v := d\vec{z}/dt$ is the coordinate velocity. The reason for the displacement by b in (1) was to assure that

(9) $\qquad \vec{v}(\tau_1) = 0 = \vec{v}(\tau_2)$.

Conditions (7) and (9) simplify the calculation by guaranteeing that the instantaneous laboratory-frame spheres $\{t_i\} \times S$, where $t_i := \bar{t}(\tau_i, \vec{c})$, $i = 1, 2$, are also retarded spheres and that the image under ϕ of $\{\tau_i\} \times C(\theta)$ is $\{t_i\} \times C(\theta)$. Thus we can now write

$$\tag{10} \int_{(t_1,t_2) \times C(\theta)} \!\bot T \;=\; \int_{(\tau_1,\tau_2) \times C(\theta)} \phi^\dagger(\bot T) \;.$$

Now we can easily outline the calculation. Let Ω_3 denote the volume 3-form on $(\tau_1, \tau_2) \times C(\theta)$, considered as a product manifold with orientation chosen so that $\phi : (\tau_1, \tau_2) \times C(\theta) \longmapsto TC(\theta)$ is orientation-preserving, where $TC(\theta)$ is given the orientation induced from **M** (i.e. prefixing an outward-pointing vector to a positively oriented basis for a tangent space to $TC(\theta)$ produces a positively oriented basis for **M**). Then

$$\tag{11} \phi^\dagger(\bot(u+w)^*) \;=\; \beta \Omega_3$$

for some function β. If an energy-momentum tensor T is defined by

$$\tag{12} T \;:=\; <\!\frac{da}{d\tau}, w\!> \frac{(u+w) \otimes (u+w)^*}{4\pi r^2}$$

(the constants in T_{L-D} are omitted for notational simplicity), then

$$\tag{13} \int_{TC(\theta)} \!\bot T \;=\; \int_{(\tau_1,\tau_2) \times C(\theta)} \phi^\dagger(\bot T)$$

$$= \int_{C(\theta)} \frac{d\sigma(\vec{c})}{4\pi} \int_{\tau_1}^{\tau_2} d\tau \left[\frac{\partial}{\partial \tau}\!\left(\frac{<\!a,w\!>(u+w)\beta}{r^2} \right) - <\!a, \frac{\partial w}{\partial \tau}\!> \frac{(u+w)\beta}{r^2} \right.$$

$$\left. - <\!a,w\!> \frac{\partial}{\partial \tau}\!\left(\frac{u+w}{r^2}\right)\beta - \frac{<\!a,w\!>(u+w)}{r^2} \frac{\partial \beta}{\partial \tau} \right] ,$$

where $d\sigma(\vec{c})$ denotes the usual area measure on the unit sphere. In (13), the quantities, u, a, w, which were originally vector fields on **M** defined via the "backward light cone construction", are viewed as pulled back to functions on $(\tau_1, \tau_2) \times C(\theta)$ via the map ϕ:

$$w(\tau, \vec{c}) \;:=\; w_{\phi(\tau,\vec{c})} \quad , \text{ etc.}$$

Note that

$$u(\tau, \vec{c}) \;=\; u(\tau) \;,$$

where we hope the reader will prefer the slight ambiguity of notation to carrying some distinguishing mark such as $\tilde{u}(\tau, \vec{c}) := u_{\phi(\tau,\vec{c})} = u(\tau)$ throughout the computations to follow, and similarly for a. Also note that for any $\tau, \vec{c} \in (\tau_1, \tau_2) \times C(\theta)$,

$$\tag{14} \phi(\tau, \vec{c}) \;=\; z(\tau) \;+\; r(\tau, \vec{c})(u(\tau) + w(\tau, \vec{c})) \;,$$

with $r(\tau, \vec{c}) > 0$ the retarded distance.

The term

$$\frac{\partial}{\partial \tau}(<\!a,w\!>(u+w)\beta)$$

vanishes when integrated over a period, so we need only evaluate the rest. For this purpose, we shall need

Appendix to 5.7 — Energy flux vector

$$\frac{\partial w}{\partial \tau}, \quad \beta, \quad \frac{\partial \beta}{\partial \tau} .$$

We shall not calculate exact expressions for all of these, but rather approximations which are valid in the low-velocity, low-amplitude limit. To help keep the formulae of decent length, we introduce the abbreviation " $\approx$ ", as, for example, in

(15) $\qquad u \approx (1, \vec{0}) ,$

to indicate that u may be replaced by $(1, \vec{0})$ in the final step of the calculation, which assumes the limit $b \to 0$, $b\omega \to 0$, $b\omega^2 = K > 0$. This replacement will occur *only* in the last step; in the intermediate steps the exact expressions must be retained. For example, from $u \approx (1, \vec{0})$, it does not follow that $a := du/d\tau \approx 0$.

If we write $a = (a^0, \vec{a})$, then from 1.7(9),

(16) $\qquad a \approx (0, \vec{a}) .$

Also, if we write $z(\tau) := (z^0(\tau), \vec{z}(\tau))$, then we have the intuitively obvious and easily verified (cf. (28) below) relations

(17) $\qquad z^0(\tau) \approx \tau , \quad \vec{z}(\tau) \approx \vec{0}$

$\qquad w(\tau, \vec{c}) \approx \vec{c} , \quad <a, w> \approx \vec{a} \cdot \vec{c} , \quad r(\tau, \vec{c}) \approx R = 1 ,$

which state that in the limit considered, the retarded quantities approach the corresponding laboratory quantities.

Now we compute $\partial w/\partial \tau$, β, $\partial \beta/\partial \tau$ along with some subsidiary quantities which will be needed. It will be convenient to define

(18) $\qquad \psi(t, \vec{c}) := \phi(\tau, \vec{c}) - z(\tau) = r(u + w) .$

Since $\phi(\tau, \vec{c})$ lies on $TC(\theta)$, we may write

(19) $\qquad \phi(\tau, \vec{c}) = (\bar{t}(\tau, \vec{c}), \vec{c}) ,$

where this defines the function

$\qquad \tau, \vec{c} \longmapsto \bar{t}(\tau, \vec{c}) ,$

and then

(20) $\qquad \psi(\tau, \vec{c}) = (\bar{t}(\tau, \vec{c}) - z^0(\tau), \vec{c} - \vec{z}(\tau))$

Since ψ is a null vector,

(21) $\qquad \bar{t} - z^0 = |\bar{t} - z^0| = |\vec{c} - \vec{z}| = [(\vec{c}-\vec{z}) \cdot (\vec{c}-\vec{z})]^{1/2} ,$

and it follows that

(22) $\qquad \dfrac{\partial(\bar{t}-z^0)}{\partial \tau} = -\dfrac{1}{|\vec{c}-\vec{z}|} \dfrac{d\vec{z}}{d\tau} \cdot (\vec{c}-\vec{z}) = -\dfrac{1}{|\vec{c}-\vec{z}|} \cdot \gamma \cdot \vec{v} \cdot (\vec{c}-\vec{z}) ,$

where we have recalled from 1.7(6) that

(23) $$\frac{d\vec{z}}{d\tau} = \frac{d\vec{z}}{dt}\cdot\frac{dt}{d\tau} = \vec{v}\cdot\gamma(\vec{v}) \ .$$

Setting $\vec{v} = 0$ in (22) gives

(24) $$\frac{\partial(\bar{t}-z^0)}{\partial\tau} \approx 0 \ ,$$

a relation which will be used repeatedly. Also from (22),

(25) $$\frac{\partial^2(\bar{t}-z^0)}{d\tau^2} \approx -\vec{a}\cdot\vec{c} \ .$$

We shall need to write out a large number of relations like (25), and the reader who wants to check the calculation will find it useful to keep in mind that

(26) $$\frac{dt}{d\tau} = \gamma \approx 1 \ , \quad \frac{d^2 t}{d^2\tau} = \frac{d\gamma}{d\tau} \approx 0 \ , \quad \frac{d\vec{v}}{d\tau} \approx \frac{d\vec{v}}{dt} \approx \vec{a} \ , \quad \bar{t} - z^0 \approx 1 \ ,$$

and

(27) $$\frac{d^2\gamma}{d\tau^2} = \frac{d}{d\tau}\left(\gamma^3\cdot<\frac{d\vec{v}}{d\tau},\vec{v}>\right) \approx \frac{d}{d\tau}<\frac{d\vec{v}}{d\tau},\vec{v}> \approx <\frac{d\vec{v}}{d\tau},\frac{d\vec{v}}{d\tau}> \approx -\vec{a}\cdot\vec{a} \ .$$

Taking the inner product with $u = \gamma\cdot(1,\vec{v})$ of the equality

$$(\bar{t},\vec{c}) - (z^0,\vec{z}(\tau)) = r\cdot(u+w)$$

(cf. (18) and (19)) gives

(28) $$r(\tau,\vec{c}) = \gamma\cdot[\bar{t}(\tau,\vec{c}) - z^0(\tau) - (\vec{c}-\vec{z}(\tau))\cdot\vec{v}(\tau)] \ ,$$

and hence $r \approx 1$,

(29) $$\frac{\partial r}{\partial\tau} = \frac{\partial\gamma}{\partial\tau}\cdot[\bar{t} - z^0 - (\vec{c}-\vec{z})\cdot\vec{v}] + \gamma\cdot[\frac{\partial(\bar{t}-z^0)}{\partial\tau} - (\vec{c}-\vec{z})\cdot\frac{d\vec{v}}{d\tau} + \frac{d\vec{z}}{d\tau}\cdot\vec{v}]$$

and

(30) $$\frac{\partial r}{\partial\tau} \approx -\vec{c}\cdot\vec{a} \ .$$

Continuing, from (20) ,

(31) $$\frac{\partial\psi}{\partial\tau} = (\frac{\partial(\bar{t}-z^0)}{\partial\tau}, \frac{\partial(\vec{c}-\vec{z})}{\partial\tau}) \ ,$$

and so from (24),

(32) $$\frac{\partial\psi}{\partial\tau} \approx 0 \ .$$

Further, from (25), (23), and (26),

(33) $$\frac{\partial^2\psi}{\partial\tau^2} = (\frac{\partial^2(t-t^0)}{\partial\tau^2}, -\frac{d^2\vec{z}}{d\tau^2}) \approx (-\vec{a}\cdot\vec{c}, -\vec{a}) \ ,$$

Appendix to 5.7 Energy flux vector 235

(34) $$\frac{\partial \phi}{\partial \tau} = \frac{dz}{d\tau} + \frac{\partial \vec{\psi}}{\partial \tau} \approx u \approx (1, \vec{0}) ,$$

and recalling that $a \approx (0, \vec{a})$,

(35) $$\frac{\partial^2 \phi}{\partial \tau^2} = a + \frac{\partial^2 \vec{\psi}}{\partial \tau^2} \approx (-\vec{a} \cdot \vec{c}, \vec{0}) .$$

Let $\vec{c}_0$ denote the center of the "disc" $C(\theta)$. Introduce a coordinate system $\vec{c} \longmapsto s_1(\vec{c}), s_2(\vec{c})$, $\vec{c} \in C(\theta)$, for $C(\theta)$ such that at $\vec{c}_0$, the corresponding tangent vectors

$$\left(\frac{\partial}{\partial s_1}\right)_{c_0} , \quad \left(\frac{\partial}{\partial s_2}\right)_{c_0}$$

are orthonormal. (Of course, we cannot require this throughout $C(\theta)$.) When we want to identify $\partial / \partial s_i$ with a concrete vector in R^3, we write $\vec{s}_i$ for $\partial / \partial s_i$, $i = 1, 2$. Thus the $\vec{s}_i$ are orthogonal unit vectors tangent to the sphere at the point $\vec{c}_0$, and $\vec{s}_i \cdot \vec{c}_0 = 0$. We shall often need to calculate the derivatives of various quantities, such as ϕ, with respect to s_i at $\vec{c}_0$. We shall denote these derivatives, e.g., as

$$\frac{\partial \phi}{\partial s_1} , \quad \frac{\partial \phi}{\partial s_2}$$

without naming the point $\vec{c}_0$ at which the derivative is taken to avoid cluttering the notation. Beginning the calculation of the derivatives with respect to s_i, we have from (21),

(36) $$\frac{\partial (\bar{t} - z^0)}{\partial s_i} = |\vec{c} - \vec{z}|^{-1} \left(\frac{\partial (\vec{c} - \vec{z})}{\partial s_i}\right) \cdot (\vec{c} - \vec{z}) = |\vec{c} - \vec{z}|^{-1} \vec{s}_i \cdot (\vec{c} - \vec{z}) ,$$

and so

(37) $$\frac{\partial (\bar{t} - z^0)}{\partial s_i} \approx \vec{s}_i \cdot \vec{c} = 0 .$$

Continuing,

(38) $$\frac{\partial \phi}{\partial s_i} = \frac{\partial \vec{\psi}}{\partial s_i} = \left(\frac{\partial (t - z^0)}{\partial s_i}, \frac{\partial (\vec{c} - \vec{z})}{\partial s_i}\right) = \left(\frac{\partial (t - z^0)}{\partial s_i}, s_i\right) ,$$

and so

(39) $$\frac{\partial \phi}{\partial s_i} \approx (0, \vec{s}_i) .$$

Differentiating (38) with respect to τ and using (36) (*not* (37)), gives

(40) $$\frac{\partial^2 \phi}{\partial \tau \partial s_i} = \left(\frac{\partial^2 (\bar{t} - z^0)}{\partial \tau \partial s_i}, \vec{0}\right) \approx 0 .$$

Now we can compute β. If s_1, s_2 is chosen so that $\frac{\partial}{\partial \tau}, \frac{\partial}{\partial s_1}, \frac{\partial}{\partial s_2}$ is positively oriented, then (11) defines β by

(41) $$\beta = \beta\Omega_3(\frac{\partial}{\partial\tau}, \frac{\partial}{\partial s_1}, \frac{\partial}{\partial s_2}) = \phi^\dagger(\bot(u+w)*)(\frac{\partial}{\partial\tau}, \frac{\partial}{\partial s_1}, \frac{\partial}{\partial s_2}) = \bot(u+w)*(\frac{\partial\phi}{\partial\tau}, \frac{\partial\phi}{\partial s_1})$$
$$= \Omega(u+w, \frac{\partial\phi}{\partial\tau}, \frac{\partial\phi}{\partial s_1}, \frac{\partial\phi}{\partial s_2}) ,$$

where Ω denotes the volume 4-form on **M**, and the last equality comes from 2.2(15). From (17), (34), (39), and (41) we immediately obtain the limiting form of β:

(42) $$\beta \approx \Omega((1, \vec{c}), (1, \vec{0}), \vec{s}_1, \vec{s}_2) \approx \Omega((0, \vec{c}), (1, \vec{0}), \vec{s}_1, \vec{s}_2) \approx 1 .$$

To get $\frac{\partial\beta}{\partial\tau}$, we first need $\partial(u+w)/\partial\tau$. Using (30) and (32) we have

(43) $$\frac{\partial(u+w)}{\partial\tau} = \frac{\partial(r^{-1}\psi)}{\partial\tau} = -r^{-2}\cdot\frac{\partial r}{\partial\tau}\cdot\psi + r^{-1}\frac{\partial\psi}{\partial\tau} \approx \vec{c}\cdot\vec{a}(1, \vec{c}) .$$

Note for future reference that it follows immediately from (43) and (30) that

(44) $$\frac{\partial}{\partial\tau}\left(\frac{u+w}{r^2}\right) \approx 3\vec{a}\cdot\vec{c}\cdot(1, \vec{c}) ,$$

and also that

(45) $$\frac{\partial w}{\partial\tau} \approx \vec{c}\cdot\vec{a}(1, \vec{c}) - a \approx \vec{c}\cdot\vec{a}(1, \vec{c}) - (0, \vec{a}) .$$

Finally,

(46) $$\frac{\partial\beta}{\partial\tau} = \Omega(\frac{\partial(u+w)}{\partial\tau}, \frac{\partial\phi}{\partial\tau}, \frac{\partial\phi}{s_1}, \frac{\partial\phi}{s_2}) + \Omega(u+w, \frac{\partial^2\phi}{\partial\tau^2}, \frac{\partial\phi}{\partial s_1}, \frac{\partial\phi}{\partial s_2})$$
$$+ \Omega(u+w, \frac{\partial\phi}{\partial\tau}, \frac{\partial^2\phi}{\partial\tau\partial s_1}, \frac{\partial\phi}{\partial s_2}) + \Omega(u+w, \frac{\partial\phi}{\partial\tau}, \frac{\partial\phi}{\partial s_1}, \frac{\partial^2\phi}{\partial\tau\partial s_2}) ,$$
$$\approx \Omega(\vec{c}\cdot\vec{a}\cdot(1, \vec{c}), (1, \vec{0}), (0, \vec{s}_1), (0, \vec{s}_2)) +$$
$$\Omega((1, \vec{c}), -\vec{a}\cdot\vec{c}\cdot(1, \vec{0}), (0, \vec{s}_1), (0, \vec{s}_2)) + 0 + 0 \approx 0 .$$

Now we return to (13) to extract the final result. We do each term in (13) separately. Recall that the first term vanishes by periodicity. The last term vanishes by (46). The second term in the inner integral is, by (45),

$$\int_{\tau_1}^{\tau_2} -\langle a, \frac{\partial w}{\partial\tau}\rangle \frac{(u+w)}{r^2}\beta\, d\tau \approx \int_{\tau_1}^{\tau_2} [(\vec{c}\cdot\vec{a})^2 - \vec{a}\cdot\vec{a}](1, \vec{c})\, d\tau .$$

For the third and final term, recall from (17) that $\langle a, w\rangle \approx -\vec{a}\cdot\vec{c}$, so from (44),

(48) $$\int_{\tau_1}^{\tau_2} -\langle a, w\rangle \frac{\partial}{\partial\tau}\left(\frac{u+w}{r^2}\right)\beta\, d\tau \approx \int_{\tau_1}^{\tau_2} 3(\vec{a}\cdot\vec{c})^2(1, \vec{c})\, d\tau .$$

Now $(\vec{c}\cdot\vec{a})^2 = \vec{a}\cdot\vec{a}\cos^2\theta$, and so (13) can be written in the limit as

(49) $$\int_{TC(\theta)} \perp T \approx \int_{C(\theta)} (4\cos^2\theta - 1)(1, \vec{c}) \frac{d\sigma(\vec{c})}{4\pi} \int_{\tau_1}^{\tau_2} \vec{a}(\tau)\cdot\vec{a}(\tau)\, d\tau \ .$$

This is the result we were seeking. Taking the energy component of (49) gives the time-averaged rate of energy radiation at the angle θ of (5) as

(50) $$\lim_{\mu(C(\theta))\to 0} \frac{1}{(t_2-t_1)\mu(C(\theta))} \int_{TC(\theta)} \perp T \approx (4\cos^2\theta - 1)\cdot \frac{1}{t_2-t_1}\int_{t_1}^{t_2} \vec{a}\cdot\vec{a}\, dt$$

$$= (4\cos^2\theta - 1)\cdot\text{constant} \ .$$

For simple harmonic motion, the constant is proportional to the square of the maximum acceleration (and inversely proportional to the square of the distance from the radiator if we put back in the radius of the sphere S).

Given (50), the analogous result for Pryce's tensor

5.7(40) $$T_{pryce} := \frac{q^2}{4\pi r^2}[-9<a,w>^2 + 2<\frac{da}{d\tau}, u+w> - <a,a>](u+w)\otimes(u+w)^* \ ,$$

can be written down almost immediately. Notice that since $<a, u> = 0$,

$$<\frac{da}{d\tau}, u> = -<a, \frac{du}{d\tau}> = -<a, a> \ ,$$

so we need only integrate

(51) $$\lim_{\mu(C(\theta))\to 0} \frac{q^2}{(t_2-t_1)\mu(C(\theta))} \int_{TC(\theta)} [-9<a,w>^2 - 3<a,a>] \frac{(u+w)\otimes\perp(u+w)^*}{r^2}$$

$$\approx (1,\vec{c})(-9\cos^2\theta + 3)\frac{q^2}{(t_2-t_1)} \int_{t_1}^{t_2} -<a,a>\, dt \ .$$

(Incidentally, this integral is also easy to do exactly by the methods of Section 4.4). Hence

(52) $$\lim_{\mu(C(\theta))\to 0} \frac{q^2}{(t_2-t_1)\mu(C(\theta))} \int_{TC(\theta)} \perp T_{pryce}$$

$$\approx (1,\vec{c})[-9\cos^2\theta + 2(4\cos^2\theta - 1) + 3]\frac{q^2}{t_2-t_1}\int_{t_1}^{t_2} -<a,a>\, dt$$

$$= \sin^2\theta\, (1,\vec{c})\frac{q^2}{t_2-t_1}\int_{t_1}^{t_2} -<a,a>\, dt \ ,$$

which does have the "correct" $\sin^2\theta$ dependence.

Appendix on units.

An excellent general discussion of units in electromagnetic theory can be found in an appendix to [Jackson], and another useful treatment which nicely complements Jackson's is an appendix to [Panofsky and Phillips]. Both are highly recommended. This appendix concentrates on how to convert from the system of units used in this book (which differ from "unrationalized natural units", only in that we do not set $\hbar := 1$) to *Gaussian units,* which are commonly used for concrete problems involving the motion of charged particles. †
Conversion from Gaussian to other commonly used systems, such as MKSA, are well explained in the above references. Our discussion will begin on a very concrete and elementary level, since the issues involved are not as familiar to mathematicians as to physicists and, though elementary, can be very confusing. The main ideas are most easily understood from a three-dimensional, non-relativistic point of view, and we shall begin in this context and later generalize to a fully relativistic treatment.

By way of introduction, suppose that we want to explore the relations between force, mass, and acceleration. We have a clock to measure time (in seconds, say) and a measuring rod to measure distance (in centimeters, say). With these we can also measure the derived quantities velocity v and acceleration A via the usual definitions. Suppose that we have a large collection of apparently identical objects such as lead weights from the same mold. We could declare these to be unit masses and measure other masses by collision experiments as explained in Section 1.8. It would then be found that two unit masses fastened together (by a fastener of negligible mass) had a mass of two units; note that within this framework this would be a genuine experimental physical law rather than a mathematical tautology. Similarly, dividing a unit mass in half is found experimentally to give a body of mass 1/2, etc.

To measure force, suppose that we have a large collection of identical springs. We could then define a unit force as that exerted by one spring when fully compressed. The force exerted by two springs connected in parallel as in the diagram would then be defined as two force units, etc.

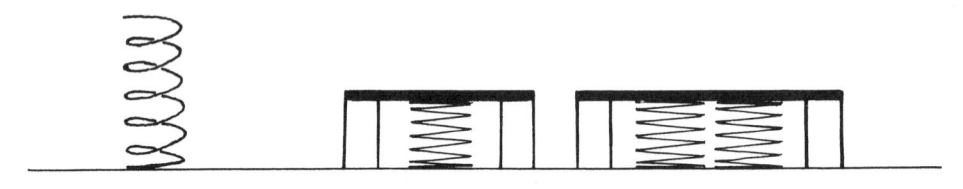

uncompressed standard spring   one unit of force   two units of force

It is not quite as easy to define fractional units of force in this framework, but the reader can no doubt think of several reasonable definitions.

Now we can try to verify experimentally Newton's Third Law

$$F = mA$$

---

† The term "Gaussian units" is used slightly differently by different authors. For example, one of the "Gaussian" Maxwell equations given by [Panofsky and Phillips, p. 466] differs from that given by [Jackson, p. 818] due to a difference of a factor of c in the definition of the charge current vector $\vec{J}$. We use the conventions of [Jackson].

with F force, m mass, and A acceleration. Of course, the first thing that we notice is that it doesn't hold, unless we have been improbably lucky in our choice of standard masses and springs. What does hold, of course, is

(1) $$F = kmA \; ,$$

where k is a constant which depends on the initial choice of standard masses and springs. At this point someone is sure to suggest changing the size of the standard mass mold or the strength of the standard springs so as to make $k = 1$ in (1). If we do this, (1) becomes a combination of a *definition* (of "unit mass" in terms of "unit force" or vice versa) and a *physical law*. For example, (1) asserts that attaching three springs in parallel will give $3/2$ units of acceleration to two unit masses fastened together. The latter is an experimental fact rather than a mathematical tautology. (For instance, it wouldn't hold if the springs were attached in series rather than in parallel.)

The same basic ideas apply to choices of units for electromagnetic theory, but the details are more complicated owing to the fact that there are more quantities to be defined and consequently more reasonable ways to insert or remove constants analogous to k. The analog of the standard springs for electromagnetism is the force of repulsion of two "standard" charged particles separated by a standard distance, and the analog of (1) is the Coulomb law

(2) $$mA = F = k\frac{q^2}{r^2}$$

for the force F between two particles of equal charge q separated by a distance r. As before, given units of time and distance, we still have the freedom to define either units of mass or charge so that $k := 1$. As the Coulomb law 5.2(13) shows, this was the choice implicitly made in the system of "natural" units used in this book. (That this choice was *derived* rather than made explicitly is due to the fact that certain relations between the units are implicit in the particular form chosen for the Maxwell-Lorentz equations, as will be seen below.) The choice $k := 1$ is also made in the Gaussian system, but other systems define mass and charge independently and retain k as an experimentally determined constant.

Even given our choice of Coulomb law,

(3) (Coulomb law in our units and in Gaussian units) $\quad mA = \dfrac{q^2}{r^2} \; ,$

there is further freedom in the choice of units for mass and charge which we shall now examine in detail.

Throughout the book we have chosen units of distance and time so that the velocity of light is unity. However, the Gaussian system of units uses seconds for time units and centimeters, rather than light-seconds, for distance. It also uses grams for mass units and so includes the c.g.s. (centimeter-gram-second) system. To proceed in our comparison we must explicitly introduce the speed of light, c. In our units, $c = 1$, and in c.g.s. units

(4) $$c = 2.998 \times 10^{10} \text{ cm/sec} \; ,$$

which is conveniently close to 3×10^{10} cm/sec. We shall think of c as a parameter which characterizes the length unit in terms of a given time unit (seconds, in this case) and correspondingly view changing the value of c as equivalent to changing the length unit. That the centimeter, *per se,* happens to be the length unit used in the Gaussian system is of course unimportant; what is important is that the length unit has been chosen independently of the time unit so that the velocity of light is not necessarily unity.

If r is a distance in light-seconds (the unit of distance in our system), let

(5) $$\bar{r} := cr$$

denote the corresponding distance in centimeters. Similarly, given any quantity such as mass m, acceleration A, and charge q as measured in our system, denote the corresponding quantity as measured in the Gaussian system by the same symbol barred; e.g. $\bar{m}$, $\bar{A}$, $\bar{q}$. From first principles,

(6) $$\bar{A} = cA \ .$$

Substituting in (3), we have

$$\bar{m}cA = \bar{m}\bar{A} = \frac{\bar{q}^2}{\bar{r}^2} = \left(\frac{\bar{q}}{q}\right)^2 \cdot \frac{1}{c^2} \cdot \frac{q^2}{r^2} = \left(\frac{\bar{q}}{q}\right)^2 \cdot \frac{1}{c^2} \cdot mA \ ,$$

from which we see that

(7) $$\frac{\bar{m}}{m} = \frac{1}{c^3}\left(\frac{\bar{q}}{q}\right)^2 \ .$$

This is the basic relation between units of mass, charge, and distance, assuming that time is measured in the same units in both systems. In practice, we are given c and the Gaussian values of $\bar{m}$ and $\bar{q}$ for a particular particle, say an electron, and can then choose the charge q and mass m in our system as we please subject to (7). Of course, for consistency we then use the same ratios $\frac{q}{\bar{q}}$, $\frac{m}{\bar{m}}$ for all other particles as well. In other words, given c, $\bar{m}$, and $\bar{q}$, we still have the freedom to choose arbitrarily a unit of charge or of mass, but not both. †

The magnitude E of the electric field due to a static point charge q is defined in all commonly used systems of units as

(8) $$E := k\frac{q}{r^2} \ ,$$

where k is the same constant that appears in (2). This guarantees that the Coulomb force (2) satisfies the special case

$$mA = qE$$

of the Lorentz equation. Since $k = 1$ in the present context (natural and Gaussian units), definition (8) determines the transformation law for E:

(9) $$\bar{E} = c^{-2}\frac{\bar{q}}{q}E \ .$$

Our use of bars for Gaussian quantities makes it typographically difficult to use the vector symbol for vector Gaussian quantities, and so for the rest of this appendix **we shall denote three-dimensional vector quantities by boldface.** Now we examine the transformation properties of the Maxwell equation (cf. section 3.6)

---

† If the unit of time is allowed to be changed as well, there is further freedom which is explored in Exercise 3.14. In that case, any two of the three units of mass, charge, and time may be arbitrarily specified.

(10) $$\nabla \cdot \mathbf{E} = 4\pi \rho_{lab} .$$

From first principles, the corresponding objects in the Gaussian system satisfy

(11) $$\overline{\nabla} = c^{-1}\nabla , \qquad \overline{\rho}_{lab} = c^{-3}\frac{\overline{q}}{q}\rho_{lab} ,$$

the last relation arising from the fact that a unit volume in the "barred" system contains c^{-3} as many particles as our unit volume, but the charge of each particle is increased by $\frac{\overline{q}}{q}$.

Given (10) and (11), the corresponding equation

(12) $$\overline{\nabla} \cdot \overline{\mathbf{E}} = 4\pi \overline{\rho}_{lab}$$

for the Gaussian barred quantities will hold if and only if $\mathbf{E}$ transforms according to (9). This observation could be viewed as a second derivation of (9), but it is better to view it as a derivation of the Gaussian Maxwell equation (12). From this point of view, it is simply an accident that (12) is obtained by barring all quantities in (10). If this procedure had produced an equation not consistent with (9), we would have simply modified (12) by inserting the appropriate factor of c ; obviously there is no divine law that says that the Maxwell equations in the Gaussian system must be obtained by barring all quantities in our old Maxwell equations. Indeed, some of the other Maxwell equations and also the Lorentz equation do end up being modified by factors of c in the Gaussian system.

The three-dimensional (but relativistically exact) Lorentz equation in the Gaussian system is taken to be

(13) $$\frac{d\overline{\mathbf{p}}}{dt} = \overline{q}(\overline{\mathbf{E}} + \frac{\overline{\mathbf{v}}}{c}\times \overline{\mathbf{B}}) ,$$

where of course $\overline{\mathbf{v}}$ denotes the velocity in cm/sec and

$$\overline{\mathbf{p}} := \overline{m}\gamma_c(\overline{\mathbf{v}})\overline{\mathbf{v}}$$

is the space component of the relativistic momentum, with

$$\gamma_c(\overline{\mathbf{v}}) := \left(1 - \frac{\overline{\mathbf{v}}\cdot\overline{\mathbf{v}}}{c^2}\right)^{-1/2} .$$

Notice that when $\overline{\mathbf{v}} = 0$, (13) is consistent with the Coulomb law (2) and the transformation law (9), and that the vector version of (9),

(14) $$\overline{\mathbf{E}} = c^{-2}\frac{\overline{q}}{q}\mathbf{E} ,$$

follows from (7) and (13). Similarly, (7) and (13) imply the transformation law

(15) $$\overline{\mathbf{B}} = c^{-2}\frac{\overline{q}}{q}\mathbf{B}$$

for $\overline{\mathbf{B}}$. In fact, the reason for modifying the original Lorentz equation by replacing $\mathbf{v}$ by $\mathbf{v}/c$ was to make $\mathbf{B}$ transform according to the same law as $\mathbf{E}$. Since Lorentz transformations mix up the components of $\mathbf{E}$ and $\mathbf{B}$, the latter requirement is natural in anticipation of a convenient relativistic generalization.

The transformation laws (14) and (15) together with the *a priori* relations (11) and the original Maxwell equations then uniquely determine the Maxwell equations in the Gaussian system; they are

(16) (Gaussian Maxwell equations)

(i) $\quad \overline{\nabla} \cdot \overline{\mathbf{E}} = 4\pi \overline{\rho}_{lab}$

(ii) $\quad \overline{\nabla} \times \overline{\mathbf{B}} = \dfrac{1}{c}[4\pi \overline{\rho}_{lab} \overline{\mathbf{v}} + \dfrac{\partial \overline{\mathbf{E}}}{\partial t}]$

(iii) $\quad \overline{\nabla} \cdot \overline{\mathbf{B}} = 0$

(iv) $\quad \overline{\nabla} \times \overline{\mathbf{E}} = -\dfrac{1}{c} \cdot \dfrac{\partial \overline{\mathbf{B}}}{\partial t}$.

Though expressed in three-dimensional form, the above transformation laws are relativistically exact and completely determine the relations between the two systems of units. In making numerical conversions, there is still a free parameter, $q/\overline{q}$ or $m/\overline{m}$, which may be chosen arbitrarily. A simple choice is $q := \overline{q}$, in which case our system uses the same units of charge as the Gaussian system but different units of mass. The charge q_e of the electron is then

(17) $\quad q_e = \overline{q}_e = 4.80 \times 10^{-10}$ statcoulombs.

Since the mass $\overline{m}_e$ of the electron in grams is

(18) $\quad \overline{m}_e = 9.11 \times 10^{-28}$ grams,

its mass m_e in the unnamed mass units of our system corresponding to this choice of charge unit is

(19) $\quad m_e = c^3 \overline{m}_e = 2.45 \times 10^4$.

Notice that given c, the ratio

(20) $\quad \dfrac{q_e^2}{m_e} = c^{-3} \dfrac{\overline{q}_e^2}{\overline{m}_e} = 9.38 \times 10^{-24}$

is independent of the choice of mass and charge units. This ratio appears prominently in many conversions and "dimensionless" constants.

The above three-dimensional Gaussian equations may be translated into four-dimensional notation as follows. We continue to work in **M** with a given Lorentz coordinatization and the usual metric. Let $\tau \longmapsto z(\tau) = (z^0(\tau), \mathbf{z}(\tau))$ be the worldline of a particle with charge q, mass m, and four-velocity $u := dz/d\tau$. Define a new "Gaussian four-velocity" $\overline{u}$ by

(22) $\quad \overline{u} := cu$.

Of course, this is not really a four-velocity in the sense in which the term was used in the main text because it does not have unit norm. We take as four-dimensional Gaussian Lorentz equation

(23) $\quad \overline{m} \dfrac{d\overline{u}}{d\tau} = \dfrac{\overline{q}}{c} \hat{F}(\overline{u})$,

where $\bar{F} = (\bar{F}_{ij})$ is to be a new field tensor related to the old F by

(24) $$\bar{F} := c^{-2}\frac{\bar{q}}{q}F \ .$$

This choice of transformation law makes (23) the same as both the previous relativistic Lorentz equation

$$\frac{du}{d\tau} = q\hat{F}(u)$$

and the three-dimensional Gaussian Lorentz equation (13).

Let $x = (x^0, x^1, x^2, x^3) = (t, x^1, x^2, x^3)$ denote the given coordinates in **M**, and write

$$\partial_i := \frac{\partial}{\partial x_i} \ ,$$

and

(25) $$\bar{\partial}_i := c^{-1}\partial_i \qquad \text{for } i = 1, 2, 3 \ .$$

Clearly, the $\bar{\partial}_i$ for $i = 1, 2, 3$ are the operations of differentiating with respect to a new length which is c times the old length. The four-dimensional Gaussian Maxwell equations are taken to be the following.

(26) (Gaussian Maxwell equations)

(i) $$d\bar{F} = 0 \ ,$$

which may be written in components as

$$\partial_i \bar{F}_{jk} + \partial_j \bar{F}_{ki} + \partial_k \bar{F}_{ij} = 0 \ .$$

Note that the differentiations in (i) are with respect to the old coordinates. Obviously, (i) is equivalent to the usual Maxwell equation $dF = 0$.

(ii) $$\delta\bar{F} = 4\pi\bar{\rho}\,\bar{u}^* \ ,$$

in components,

$$\partial_i \bar{F}^i{}_j = 4\pi\bar{\rho}\,\bar{u}_j \ ,$$

with $\bar{\rho}$ the proper charge density, defined analogously to (11) by $\bar{\rho} := c^{-2}\rho\bar{q}/q$. Again, the differentiations are with respect to the old coordinates.

If (i) and (ii) are written out in components in terms of the $\bar{\partial}_i$ for $i = 1, 2, 3$, the three-dimensional Gaussian Maxwell equations (16) result. A more usual way to write equations (26) is to define

(27) $$\bar{\partial}_0 := c^{-1}\frac{\partial}{\partial t} \ ,$$

in which case they become

(28) (Gaussian Maxwell equations as usually written)

(i) $\quad \overline{dF} = 0$, $\quad$ in components, $\quad \bar{\partial}_i \bar{F}_{jk} + \bar{\partial}_j \bar{F}_{ki} + \bar{\partial}_k \bar{F}_{ij} = 0$.

(ii) $\quad \overline{\delta F} = \dfrac{4\pi \bar{\rho} \bar{u}^*}{c}$, $\quad$ in components, $\quad \bar{\partial}_i \bar{F}^i{}_j = \dfrac{4\pi \bar{\rho} \bar{u}_j}{c}$.

Note that in (28), $\bar{\partial}_0$ is *not* time differentiation in either system, a point which is easy to forget and, once forgotten, can cause considerable confusion.

Solutions to Exercises

Solution 1.1.

A null vector obviously cannot be a member of an orthonormal basis.

Solution 1.2.

The proof by induction on the dimension of the space V is almost identical to the usual one for positive definite spaces. Any orthonormal set $e_1, \cdots, e_n$ is linearly independent and hence can be embedded in a (not necessarily orthonormal) basis $e_1, \cdots, e_m, f_{m+1}, \cdots, f_n$. Replacing f_{m+j} by

$$f_{m+j} - \sum_{i=1}^{m} \frac{<e_i, f_{m+j}>}{<e_i, e_i>} e_i, \quad j = 1, \cdots, n-m,$$

we may assume that f_{m+j} is orthogonal to e_i for $1 \leq i \leq m$, $1 \leq j \leq n-m$. The restriction of the inner product to the subspace F spanned by the f_{m+j} is nondegenerate: if $g \in F$ were orthogonal to all $f \in F$, then by construction of F, g would be orthogonal to all $v \in V$. Hence by induction (the case of dimension 1 being trivial), F has an orthonormal basis $e_{m+j}, \cdots, e_n$ and then $e_1, \cdots, e_m, e_{m+1}, \cdots, e_n$ is an orthonormal basis for V.

Solution 1.3.

In any given Lorentz frame, n_1 is of the form $n_1 = \alpha(1, \vec{v})$ with α a nonzero scalar and $|\vec{v}| = 1$, and after passing to a new Lorentz frame related to the original by a spatial rotation (rotate $\vec{v}$ into a positive multiple of $(0, 1, 0, 0)$), we may assume that $n_1 = \alpha(1, 1, 0, 0)$. In this new frame, $n_2 = \beta(1, \vec{w}) = \beta(1, w^1, w^2, w^3)$ with $|\vec{w}| = 1$, and $<n_1, n_2> = \alpha\beta(1-w^1)$, which can vanish only if $w^1 = 1$ and $n_2 = (\beta/\alpha)n_1$.

Since any non-null vector can be divided by its norm to obtain a unit vector, for the last statement, we need only show that a subspace of dimension at least two cannot consist entirely of null vectors. Let e_1 and e_2 be in the subspace and linearly independent. If either is not a null vector, we are done. If both are null vectors, then $e_1 + e_2$ is not a null vector because then $<e_1+e_2, e_1+e_2> = 2<e_1, e_2>$, which is nonzero because two linearly independent null vectors cannot be orthogonal.

Solution 1.4.

By Exercise 3, such a subspace S contains a unit vector u, which we may assume timelike (otherwise we're done). Let v be any nonzero vector in S independent of u, and define $w := v - <v, u>u$, so that w is nonzero and orthogonal to u. By Exercise 2, we may embed u in an orthonormal basis u, e_1, e_2, e_3, and since w is orthogonal to u, $w = -\sum_{i=1}^{3} <w, e_i>e_i$. Then $<w, w> = \sum_{i=1}^{3} <w, e_i>^2<e_i, e_i> < 0$, since $<e_i, e_i> < 0$ for $i = 1, 2, 3$ (because any orthonormal basis for $\mathbf{M}$ has precisely three negative norm elements).

Solution 1.5.

(i) The subspace spanned by any null vector.

(ii) The subspace spanned by $(1, 1, 0, 0)$ and $(0, 0, 1, 0)$. The first vector is nonzero and orthogonal to all vectors in the subspace, an impossibility in a subspace which has an orthonormal basis.

(iii) Add to (ii) the vector $(0, 0, 0, 1)$.

Solution 1.6.

(a) Suppose the coordinates of u are $u = (s, \vec{w}) = s(1, \vec{w}/s)$. Since u is timelike, $|\vec{w}/s| < 1$, and since u is a unit vector, $s = \pm\gamma(\vec{w}/s)$.

(b) If $u = \gamma(\vec{v})(1, \vec{v})$, then use the boost 1.4(8) associated with v.

(c) By (b), we may work in a coordinatization in which $u = (1, \vec{0})$. If S is any subspace containing u, then for any $v = (v^0, \vec{v}) \in S$, also $(0, \vec{v}) \in S$. Hence if we write $\mathbf{M} = R^1 \oplus R^3$ with respect to this coordinatization, we have $S = R^1 \oplus (S \cap R^3)$. The restriction of the metric to $S \cap R^3$ is negative definite and so $S \cap R^3$ has an orthonormal basis; adjoining u to this orthonormal basis produces an orthonormal basis for S.

Solution 1.7.

If S has no null vector, then we can apply the Gram-Schmidt process to any basis to obtain an orthonormal basis e_1, e_2, e_3. If the norms of two of these, say e_1 and e_2 had opposite sign, then $e_1 + e_2$ would be a null vector, so the inner product is either positive definite or negative definite on S. By exercise 2, we can adjoin a unit vector e_0 to this orthonormal basis for S to obtain an orthonormal basis for $\mathbf{M}$. Since any two orthonormal bases have the same number of positive norm and negative norm elements, e_1, e_2, and e_3 have negative norm while e_0 has positive norm, and S is "space" relative to the Lorentz coordinatization defined by this orthonormal basis. The Lorentz transformation L is, of course, the linear transformation which takes the basis $\{e_i\}_{i=0}^{3}$ to the "standard" basis $\{s_i\}$ for $\mathbf{M}$ (cf. the end of Section 1.5).

Solution 1.8

Let S be a three-dimensional subspace of $\mathbf{M}$. We shall show the following.

(i) If S has a timelike vector u, then it is isomorphic to the subspace spanned by

$$(1, 0, 0, 0), (0, 1, 0, 0), (0, 0, 1, 0),$$

which is "three-dimensional Minkowski space".

(ii) If S contains no nonzero null vector, then S is isomorphic to the subspace spanned by

$$(0, 1, 0, 0), (0, 0, 1, 0), (0, 0, 0, 1).$$

An alternate characterization of such subspaces is that the restriction of the metric to them is negative definite; such subspaces are called *spacelike*.

(iii) If S is neither of type (i) nor (ii), then S is isomorphic to the subspace spanned by

$$(1, 1, 0, 0), \ (0, 0, 1, 0), \ (0, 0, 0, 1).$$

Note that the metric is degenerate on such an S, since the first vector is orthogonal to all vectors in the subspace.

It is clear that these three classes are disjoint, since the metric is not degenerate in (i) and (ii), and (ii) is distinguished from (i) by the property that in case (ii), S contains no null vector.

First we show (i). Supose S has a timelike vector u. By Exercise 6(a), after a preliminary change of coordinatization we may assume that

$$u = (1, 0, 0, 0).$$

Write $\mathbf{M} = R^1 \oplus R^3$ relative to this coordinatization, and consider R^1 and R^3 as subspaces of $\mathbf{M}$ in the obvious way. Since $u \in S$, $S = R^1 \oplus (S \cap R^3)$. Now $S \cap R^3$ has an orthonormal basis e_1, e_2 because the metric on R^3 is negative definite, and passing to a coordinatization defined by an orthonormal basis u, e_1, e_2, e_3 for $\mathbf{M}$ (cf. Exercise 2) gives S as described in (i) with respect to this coordinatization and hence isomorphic to (i) with respect to the original coordinatization.

If S has no null vector, we are in case (ii) by Exercise 7, so assume S has a null vector n, which, after a preliminary spatial rotation, may be taken to be

$$n = (1, 1, 0, 0).$$

We shall show that S is as described in (iii) in this coordinatization. Let $w = (w^0, w^1, w^2, w^3)$ be any vector in S independent of n. Try to contruct a timelike vector (which would put us in case (i)) of the form $n + \lambda w$:

$$<n+\lambda w, n+\lambda w> \ = \ 2\lambda <n, w> + \lambda^2 <w, w> \ = \ 2\lambda(w^0 - w^1) + \lambda^2 <w, w>.$$

If $w^0 \neq w^1$, this can be made positive by taking λ sufficiently small and of the same sign as $(w^0 - w^1)$. Hence we may assume that $w^0 = w^1$, in which case $w = w^0 n + (0, 0, w^2, w^3)$. This implies that if $w \in S$, then also $(0, 0, w^2, w^3) \in S$, and from this it follows easily that $(0, 0, 1, 0)$ and $(0, 0, 0, 1)$ are in S. (If n, w, v is any basis for S, then so is $n, w - w^0 n, v - v^0 n$, and suitable linear combinations of the last two vectors will give $(0, 0, 1, 0)$ and $(0, 0, 0, 1)$).

Solution 1.9

Replacing u and v by scalar multiples of themselves, we may assume both are unit vectors. By Exercise 6(b), we may assume that $u = (1, 0, 0, 0)$. By 6(a), v is necessarily of the form $v = \pm \gamma(\vec{v})(1, \vec{v})$, and

$$|<u, v>| \ = \ \gamma(\vec{v}) \ \geq \ 1 \ = \ <u, u>^{1/2} <v, v>^{1/2}.$$

Solution 1.10

By a preliminary translation, we may take $q = 0$. Then $(p^0)^2 < (p^1)^2 + (p^2)^2 + (p^3)^2$. Define $\vec{v} := p^0\vec{p}/|\vec{p}|^2$, and use the coordinatization defined by a boost B with velocity $\vec{v}$: $(p'^0, \vec{p}') := B(p^0, \vec{p})$. By 1.4(8), $p'^0 = \gamma(\vec{v})(p^0 - \vec{p}\cdot\vec{v}) = 0$.

Solution 1.11

Recall that 1.7(1') was shown to be independent of the parametrization, and use it with the worldline parametrized by proper time τ:

$$\tau = \int_a^\tau <\frac{dz}{d\tau}, \frac{dz}{d\tau}>^{1/2} d\tau$$

Differentiating with respect to τ gives $1 = <dz/d\tau, dz/d\tau>^{1/2}$.

Solution 1.12.

Recall from 1.4(8) that with $\vec{u} := \vec{v}/|\vec{v}|$,

$$B(\vec{v})(t, \vec{w}) = ((t - \vec{w}\cdot\vec{v})\gamma(\vec{v}), ((\vec{w}\cdot\vec{u})\vec{u} - \vec{v}t)\gamma(\vec{v}) + \vec{w} - (\vec{w}\cdot\vec{u})\vec{u}) .$$

Hence, using $(U^{-1}\vec{w})\cdot\vec{v} = \vec{w}\cdot U\vec{v}$, and $\gamma(U\vec{v}) = \gamma(\vec{v})$ (both because U is orthogonal), we have

$$RB(\vec{v})R^{-1}(t, \vec{w}) = R((t - U^{-1}\vec{w}\cdot\vec{v})\gamma(\vec{v}), ((U^{-1}w\cdot\vec{u})\vec{u} - \vec{v}t)\gamma(\vec{v})$$
$$+ U^{-1}\vec{w} - (U^{-1}\vec{w}\cdot\vec{u})\vec{u})$$
$$= ((t - \vec{w}\cdot U\vec{v})\gamma(U\vec{v}), ((w\cdot U\vec{u})U\vec{u} - U\vec{v}t)\gamma(U\vec{v}) + \vec{w} - (\vec{w}\cdot U\vec{u})U\vec{u})$$
$$= B(U\vec{v})(t, \vec{w}) .$$

Solution 1.13

First suppose that L is a pure boost; we want to show that f is strictly monotonically increasing. We may choose the spatial coordinate axes e_1, e_2, e_3 so that L is a boost B with velocity v in the e_1-direction. From 1.4(8),

$$f(t) = t' = \gamma(v)\cdot(t - z^1(t)v) , \text{ and hence}$$

$$\frac{df}{dt} = \gamma(v)(1 - \frac{dz^1}{dt}\cdot v) > 0 ,$$

since all velocities are smaller than 1. The corresponding formula for $L = BR$ with B the same as before is

$$\frac{df}{dt} = \gamma(v) = \gamma(v)\cdot(1 - R^1{}_j \frac{dz^j}{dt} v) > 0 .$$

Solution 1.14

Recall from 1.4(3) that

(i) $$t = \gamma(v)(t' + vx')$$
$$x = \gamma(v)(x' + vt') \ .$$

Parametrize the worldline by coordinate time t. The particle's velocity in the original, unprimed frame is

$$\frac{dx}{dt} = \frac{\partial x}{\partial x'} \cdot \frac{dx'}{dt} + \frac{\partial x}{\partial t'} \cdot \frac{dt'}{dt} \ .$$

Here the total derivatives, dx/dt, dx'/dt, and dt'/dt refer to differentiation on the wordline parametrized by coordinate time. (For example, to each coordinate time t corresponds a unique point $(t, x(t))$ on the worldline, which has an x'-coordinate given by (i), and this defines x' as a function of t.) So,

(ii) $$\frac{dx}{dt} = \gamma(v)\frac{dx'}{dt} + v\gamma(v)\frac{dt'}{dt} = (\gamma(v)\frac{dx'}{dt'} + v\gamma(v))\frac{dt'}{dt}$$
$$= (w + v)\gamma(v)\frac{dt'}{dt} \ ,$$

and

(iii) $$\frac{dt'}{dt} = \frac{\partial t'}{\partial x} \cdot \frac{dx}{dt} + \frac{\partial t'}{\partial t} = -v\gamma(v)\frac{dx}{dt} + \gamma(v) \ .$$

Substituting (iii) in (ii) and solving for $\frac{dx}{dt}$ yields the solution:

$$\frac{dx}{dt} = \frac{w + v}{1 + wv} \ .$$

Solution 1.15.

By Exercise 13, the relation between the coordinate velocities in the original and the barred frames is

$$\bar{v}_1 = \frac{v_1 - v}{1 - v_1 v} \ , \text{ etc. We have}$$

$$\bar{v}_1 + \bar{v}_2 = \frac{v_1 - v}{1 - v_1 v} + \frac{v_2 - v}{1 - v_2 v} = \frac{v_1 - v}{1 - v_1 v} + \frac{-v_1 - v}{1 + v_1 v}$$

$$= -2v\frac{1 - v_1^2}{1 - v_1^2 v^2} \ .$$

Equating this with the similar expression obtained by replacing the initial with the final velocities gives

$$2v\frac{1 - v_1^2}{1 - v_1^2 v^2} = 2v\frac{1 - (v_1')^2}{1 - (v_1')^2 v^2} \ ,$$

250   Solutions 1

and cross-multiplying and simplifying yields (3).

Solution 1.16.

We present two solutions. The first, based on a systematic analysis of a spacetime diagram, contains geometrical ideas which occur in many situations. The second is a quick way of seeing the answer to this particular problem essentially without computation.

<u>First Solution:</u> Consider the spacetime diagram shown in which the worldline of the rocket passes through the spacetime origin.

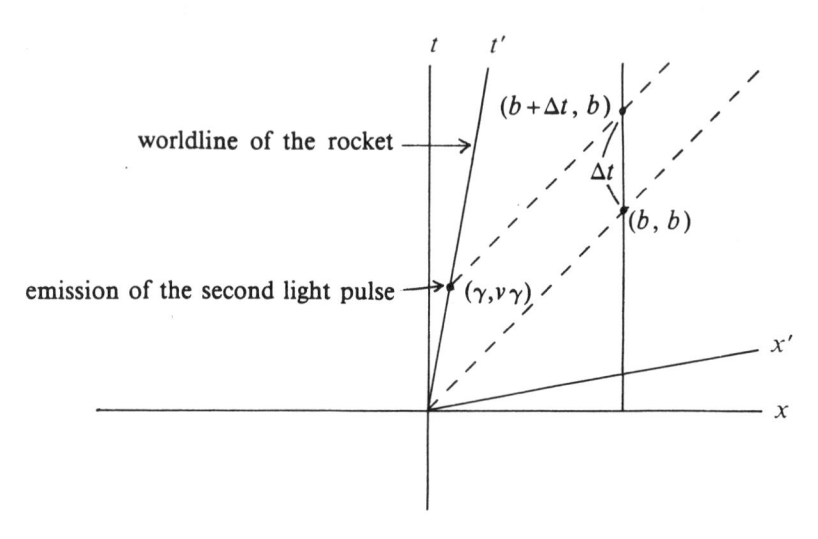

Diagram 1.16

We consider two light pulses, one emitted by the rocket as it passes the origin and the other emitted one proper-time unit later. The dashed 45° lines are the worldlines of these light pulses. If the stationary observer is located at x-coordinate b, then they reach his worldline at points (b, b) and $(b+\Delta t, b)$. The worldline of the rocket is the t'-axis in a Lorentz frame in which it is at rest. This rest frame, with axes labelled $t'-x'$, is related to the original $t-x$ Earth frame by the boost 1.4(2):

$$t' = \gamma(v)(t - vx)$$

$$x' = \gamma(v)(x - vt) \ .$$

The event $t' = 1$, $x' = 0$ at the emission of the second light pulse corresponds to the point $(\gamma, v\gamma)$ in the Earth frame. The vector from the event of second emission to its reception at $(b + \Delta t, b)$ is of the form $\lambda(1, 1)$ for some scalar λ. Hence by vector addition,

$$(b + \Delta t, b) = (\gamma, v\gamma) + \lambda(1, 1) \ .$$

Replacing λ by $\mu := \lambda - b$, this becomes

$$(\Delta t, 0) = (\gamma, v\gamma) + \mu(1, 1) \quad , \text{and so } \mu = -v\gamma \quad \text{and}$$

$$\Delta t = \gamma + \mu = (1 - v)\gamma = \frac{1-v}{(1-v^2)^{1/2}} = \left(\frac{1-v}{1+v}\right)^{1/2}.$$

Second Solution. A quich way to see the answer is to think first about the corresponding Newtonian problem in which a universal time is available. This is also the relativistic situation if we stipulate that all measurements are to be made within the Earth frame. If the flashes were emitted at intervals of, say, ΔT seconds in the Earth frame, then between the emission of any two flashes the distance of the rocket from the observer diminishes by a distance $v \Delta T$, so that the flash arrives $v \Delta T$ seconds earlier than it would have had the rocket been stationary. Hence the observed time interval Δt between flashes would be

$$\Delta t = \Delta T - v \Delta T = (1-v)\Delta T,$$

and this is the formula for the classical Doppler shift when the observer is stationary in the medium of transmission (applicable, for example, to sound waves assuming that the air is stationary in the Earth frame). The only feature of this analysis which does not apply directly in the relativistic setting is that the emission interval ΔT is measured in the Earth frame, and so does not correspond to the emission interval in the rocket frame. However, the relation between the two is just the classical time dilation 1.4(6): if the emission interval is one time unit in the rocket's rest frame, then $\Delta T = \gamma$. Hence

$$\Delta t = (1 - v)\Delta T = \frac{(1-v)}{(1-v^2)^{1/2}} = \left(\frac{1-v}{1+v}\right)^{1/2}.$$

With complete understanding of the underlying physical relationships and some practice, one can often quickly intuit in a similar way the correct answers to many elementary problems. However, one is much less likely to make mistakes using a diagram analysis. The beginner may find it profitable to first solve such problems using a rigorous diagram analysis so as to be confident of the answer and then build intuition by trying to think of a fast way to do it. (Some may prefer the opposite order!) It is always unwise to rely on intuition which is not absolutely crystal clear; there are some phenomena in this area which are very hard to see intuitively, and even experts sometimes make mistakes.

Solution 1.17.

This follows immediately from 1.7(1), since the integrand is always less than 1. Of course, that 1.7(1) actually does give time as measured by the moving particle is itself nothing more than a physical *assumption* (sometimes called the "clock hypothesis") albeit a very reasonable one. On the other hand, it is certainly not hard to imagine that the rate of a clock in its rest frame might depend on its proper acceleration or higher derivatives of its worldline. The reader who wondered why such a seemingly trivial problem followed a fairly hard one like the Doppler shift and puzzled over it is quite likely thinking in the right way. In a sense (*) is "obvious" (no one is likely to dispute that it follows from 1.7(1)!), but in a deeper sense it is not.

Solution 1.18

Replacing T by the affine transformation $\tilde{T}$ defined by $\tilde{T}(x) := T(x) - T(0)$, we may assume that $b = 0$, and the problem reduces to showing that any linear transformation C which preserves the light cone (i.e. maps null vectors to null vectors) is a scalar multiple of a Lorentz transformation. Let $e_0 := (1, 0, 0, 0)$ and let $s := (0, \vec{w})$ be any unit spacelike vector orthogonal to e_0. By considering the equation

(1) $\qquad 0 = \langle C(e_0 \pm s), C(e_0 \pm s) \rangle = \langle Ce_0, Ce_0 \rangle \pm 2 \langle Ce_0, Cs \rangle + \langle Cs, Cs \rangle$,

we conclude that

$$\langle Ce_0, Cs \rangle = 0 \quad \text{and} \quad \langle Cs, Cs \rangle = -\langle Ce_0, Ce_0 \rangle.$$

In particular, the restriction of C to "space" $S := \{ (0, \vec{w}) \mid \vec{w} \in R^3 \}$ relative to this coordinatization is a scalar multiple by $|\langle Ce_0, Ce_0 \rangle|^{1/2}$ of an isometry (if $\langle Ce_0, Ce_0 \rangle > 0$) or an anti-isometry (if $\langle Ce_0, Ce_0 \rangle < 0$). In a moment we shall show that $\langle Ce_0, Ce_0 \rangle$ is positive, and suppose momentarily that this is the case, so that $Ce_0 = \lambda \cdot (1, \vec{v})$ for some $\vec{v}$ with $|\vec{v}| < 1$ and some scalar λ. Apply the same analysis with C replaced by $R := (\gamma(\vec{v})/\lambda) BC$, where B is the boost 1.4(8) with velocity $\vec{v}$, so that $Re_0 = e_0$, and conclude that R maps space S (the orthogonal complement of e_0) onto itself and is an isometry there; i.e. R is a spatial rotation and consequently C is a scalar multiple of a Lorentz transformation.

To complete the proof we need only establish that for our original C, indeed $\langle Ce_0, Ce_0 \rangle > 0$. Suppose this is false. (Incidentally, it *can* be false in two-dimensional Minkowski space, as shown by the transformation $C(t, x) := (x, t)$ which interchanges "time" and "space".) It is clear from (1) that $\langle Ce_0, Ce_0 \rangle \neq 0$, and, replacing C by a scalar multiple of itself, we may assume that $\langle Ce_0, Ce_0 \rangle = -1$, in which case the restriction of C to S is an anti-isometry; i.e. $\langle Cs_1, Cs_2 \rangle = -\langle s_1, s_2 \rangle$ for all $s_1, s_2 \in S$. In particular, $\langle Cs, Cs \rangle$ is timelike for all $s \in S$. Taking s_1, s_2 orthogonal vectors in S gives two *orthogonal* timelike vectors Cs_1, Cs_2, which contradicts the "backward Schwartz" inequality of Exercise 9.

Solution 1.19

Any Lorentz frame related to the Earth frame by a boost

$$t' = \gamma(v)(t - vx)$$

$$x' = \gamma(v)(x - vt)$$

with $v > 1/2$, so that the x'-axis is between the plane's worldline and the line $t = x$. (See diagram.)

Diagram 1.19

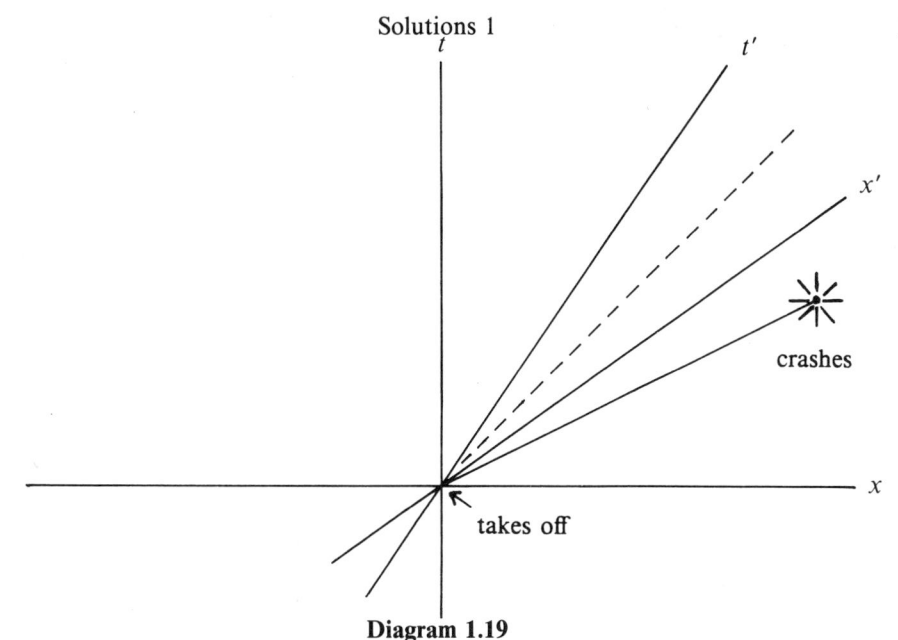

Solution 1.20.

$$\frac{d\vec{p}}{dt} = m\frac{d\gamma}{dt}\vec{v} + m\gamma\frac{d\vec{v}}{dt} = m\gamma^3(\vec{v}\cdot\frac{d\vec{v}}{dt})\vec{v} + m\gamma\vec{A}$$
$$= m\gamma^3(\vec{v}\cdot\vec{A})\vec{v} + m\gamma\vec{A} \quad ,$$

which can be parallel to $\vec{A}$ only if either $\vec{A}$ is parallel to $\vec{v}$ or $\vec{A}$ is orthogonal to $\vec{v}$. Thus Newton's law (3) can only hold in these two very special cases. If $\vec{A}$ is in the direction of $\vec{v}$, then $(\vec{v}\cdot\vec{A})\vec{v} = (\vec{v}\cdot\vec{v})\vec{A}$, and

$$\frac{d\vec{p}}{dt} = m\gamma(\gamma^2\vec{v}\cdot\vec{v} + 1)\vec{A} = m\gamma^3\vec{A} \quad .$$

Solution 1.21.

We can write $Le_0 = \pm\gamma(\vec{v})(1,\vec{v})$ (cf. Exercise 6). Replacing L by $-L$, if necessary, we may assume that "+" holds and shall show that $L = BR$. Let B be the boost associated with velocity $\vec{v}$. Then $BLe_0 = e_0$. Since BLe_i must be orthogonal to BLe_0 for $i \neq 0$, it follows that BL preserves "space" relative to this coordinatization and hence BL is a spatial rotation R and $L = B^{-1}R$, B^{-1} being the boost with velocity $-v$.

Solution 1.22.

(a) Write $f := \frac{dt}{d\tau}$, $g := \frac{dx}{d\tau}$, so that $u = (f, g)$. Separating into components the equation

$$\frac{du}{d\tau} = Aw \quad ,$$

and recalling that $\gamma = \frac{dt}{d\tau}$ gives the system:

(i) $\qquad \frac{df}{d\tau} = Av\gamma = A\frac{dx}{dt}\frac{dt}{d\tau} = Ag \qquad$, and

(ii) $$\frac{dg}{d\tau} = A\gamma = A\frac{dt}{d\tau} = Af .$$

The general solution is easily written down with f and g as linear combinations of $e^{A\tau}$ and $e^{-A\tau}$, but we do not want the general solution; since u has norm 1, we are only interested in solutions which satisfy

$$f^2 - g^2 = \langle u, u \rangle = 1 .$$

By simple inspection, for any constant k,

$$f := k \cosh A\tau$$

$$g := k \sinh A\tau$$

will work, and going back to the original equation and adjusting the arbitrary constants so as to give the desired initial conditions gives

$$t(\tau) = A^{-1} \sinh A\tau$$

$$x(\tau) = A^{-1}(\cosh A\tau - 1 + bA) .$$

Since the worldlines are hyperbolas in two-dimensional Minkowski space, as shown in the diagram, this is known as *hyperbolic motion*.

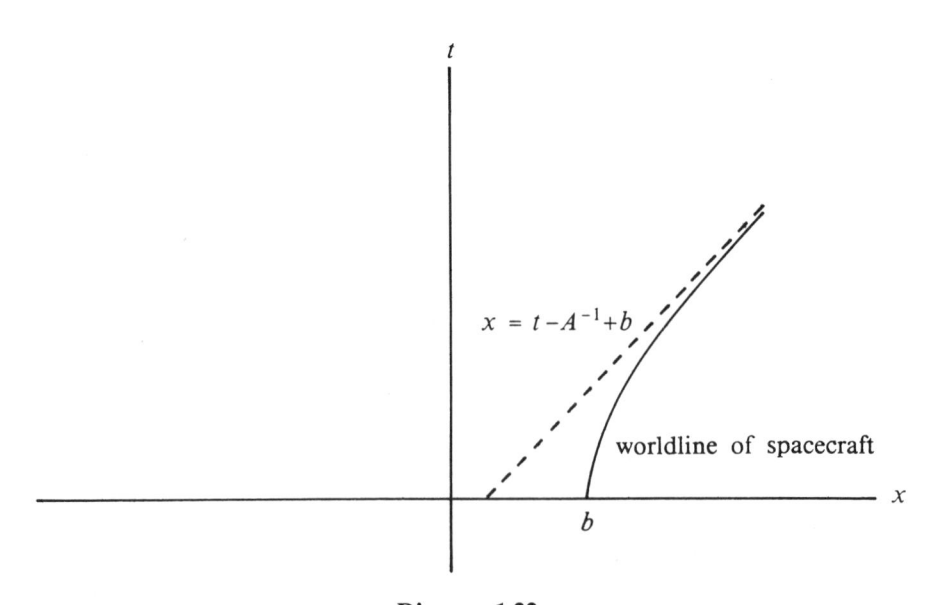

Diagram 1.22

(b) Note that the worldline becomes asymptotic to the forward light cone $x = t - A^{-1} + b$ as $\tau \longmapsto \pm\infty$. Since forward light cones are the worldlines of photons, this means that the photon whose worldline is the asymptote can never reach the rocket, and the same is of course true of any photon which begins earlier than that one. Thus the friend's signal will never reach the rocket if $b \geq A^{-1}$. After the blastoff, the rocket can never receive a message from the friend. However, the friend can receive from the rocket, as is apparent from the diagram.

Solution 1.23.

Let $\vec{v} := (v \cos \phi, -v \sin \phi)$ be the particle's final velocity. The basic equation is

$$(h\nu, h\nu, 0, 0) + (m, 0, 0, 0)$$
$$= (h\nu', h\nu' \cos \theta, h\nu' \sin \theta, 0) + m\gamma(v)(1, v \cos \phi, -v \sin \phi, 0).$$

Given m and ν, separating into components gives three equations for the three unknowns ν', θ, ϕ. The solution is most neatly obtained as follows. Define $\vec{n} := (1, 0, 0)$ as the unit vector in the direction of the incident photon and $\vec{n}' := (\cos \theta, \sin \theta, 0)$ the direction of the scattered photon. Then the above becomes the two equations

(3) $\qquad m + h(\nu-\nu') = m\gamma \qquad$ and

(4) $\qquad h\nu\vec{n} - h\nu'\vec{n}' = m\vec{v}\gamma$.

Now $(1 - \vec{v}\cdot\vec{v})\gamma^2 = 1$, so if we square both equations and subtract, the terms involving m^2 cancel, and we obtain

$$2mh(\nu-\nu') + h^2(\nu-\nu')^2 - h^2(\nu^2+\nu'^2) + 2h^2\nu\nu'\vec{n}\cdot\vec{n}' = 0,$$

which simplifies to

$$mh(\nu-\nu') - h^2\nu\nu' + h^2\nu\nu'\vec{n}\cdot\vec{n}' = 0.$$

Since $\vec{n}\cdot\vec{n}' = \cos \theta$, we have

$$\frac{1}{\nu'} - \frac{1}{\nu} = \frac{h}{m}(1 - \cos\theta) .$$

Solution 2.1

For any $v \in V$,

$$u^*(v) := \langle u, v \rangle = u^\lambda g_{\lambda\sigma} v^\sigma = g^{\lambda\alpha} f_\alpha g_{\lambda\sigma} v^\sigma$$
$$= (g^{\lambda\alpha} g_{\lambda\sigma}) f_\alpha v^\sigma = \delta^\alpha_\sigma f_\alpha v^\sigma = f_\alpha v^\alpha = f(v) .$$

Note that the symmetry of g^{ij} was used.

A slightly simpler solution is

$$u^*(v) = u^\beta v_\beta = g^{\beta\alpha} f_\alpha v_\beta = f_\alpha v^\alpha = f(v),$$

but the latter may smack of black magic to those unaccustomed to the way the inner product components are implicit in the covariant components v_i of a contravariant vector v^i. The first solution only uses the "natural" components v^i and f_i.

Solution 2.2

(a) We have

(*) $$v^\alpha e_\alpha = \bar{v}^\alpha \bar{e}_\alpha = A^\alpha{}_\beta v^\beta B^\gamma{}_\alpha e_\gamma .$$

For any i, the e_i-component of (*) is

$$v^i = A^\alpha{}_\beta v^\beta B^i{}_\alpha = (B^i{}_\alpha A^\alpha{}_\beta) v^\beta .$$

Since this is true for all vectors v,

$$B^i{}_\alpha A^\alpha{}_j = \delta^i{}_j ; \quad \text{i.e. } B = A^{-1} .$$

(b) (i) $$\delta^i{}_j = \bar{e}^i(\bar{e}_j) = \bar{e}^i(B^\alpha{}_j e_\alpha) = B^\alpha{}_j \bar{e}^i(e_\alpha) .$$

Hence the matrix $(\bar{e}^i(e_j))$ is the inverse of the matrix B; i.e. $(\bar{e}^i(e_j)) = A$. Since from first principles, $\bar{e}^i = \bar{e}^i(e_\alpha)e^\alpha$ (true for any form!),

$$\bar{e}^i = \bar{e}^i(e_\alpha)e^\alpha = A^i{}_\alpha e^\alpha .$$

(ii) Given (i), the calculation of the relation between the $\bar{f}_i$ and the f_j should be the same as (a) with upper and lower index positions reversed (because V may be viewed as the dual of V^*). Hence we expect that

$$\bar{f}_i = f_\alpha B^\alpha{}_i .$$

As a check, we define $\bar{f}_i$ in this way and compute that it has the desired property:

$$\bar{f}_i \bar{e}^i = f_\beta B^\beta{}_i \bar{e}^i = f_\beta B^\beta{}_i A^i{}_\alpha e^\alpha = f_\beta \delta^\beta{}_\alpha e^\alpha = f_\alpha e^\alpha .$$

Solution 2.3

Let $e_1, e_2, \cdots, e_n$ be a basis for V, σ a q-form, and let $v_1, v_2, \cdots, v_q \in V$ be any q vectors. For each i, write v_i as a linear combination

$$v_i = \sum_{\alpha=1}^n c_i^\alpha e_\alpha ,$$

and substitute in σ:

$$\sigma(v_1, \cdots, v_q) = \sigma(\sum_{\alpha_1} c_1^{\alpha_1} e_{\alpha_1}, \cdots, \sum_{\alpha_q} c_q^{\alpha_q} e_{\alpha_q})$$

$$= \sum_{\alpha_1} \cdots \sum_{\alpha_q} c_1^{\alpha_1} \cdots c_q^{\alpha_q} \sigma(e_{\alpha_1}, \cdots, e_{\alpha_q}) .$$

When $q > n$, at least two of the integers $\alpha_1, \cdots, \alpha_q$ must be equal, and since σ is alternating, each of the $\sigma(e_{\alpha_1}, \cdots, e_{\alpha_q})$ must vanish.

Solution 2.4

(a) If ω is any n-form, then for any n vectors $v_1, \cdots, v_n$ written as in (b), we have by multilinearity:

$$(1) \qquad \omega(v_1, v_2, \cdots, v_n) = \sum_{j_1=1}^{n} \sum_{j_2=1}^{n} \cdots \sum_{j_n=n}^{n} a_1^{j_1} a_2^{j_2} \cdots a_n^{j_n} \omega(e_{j_1}, e_{j_2}, \cdots, e_{j_n}) .$$

Since ω is alternating, $\omega(e_{j_1}, e_{j_2}, \cdots, j_n)$ vanishes whenever two or more of the j_k's are equal and otherwise is

$$\pm \omega(e_1, e_2, \cdots, e_n) ,$$

where "+" holds if $j_1, \cdots, j_n$ is an even permutation of $1, \cdots, n$ and "-" otherwise. Hence ω is completely determined by (1) and the value of $\omega(e_1, \cdots, e_n)$. It follows that if α is any nonzero n-form, we have

$$\omega = \frac{\omega(e_1, \cdots, e_n)}{\alpha(e_1, \cdots, e_n)} \cdot \alpha .$$

Thus $\dim(V_n^0)$ must be 0 or 1; that it is not zero will follow from (b), which shows in particular that $e^1 \wedge \cdots \wedge e^n \neq 0$.

(b) Using (1) with $\omega = e^1 \wedge \cdots \wedge e^n$, we see that

$$(e^1 \wedge \cdots \wedge e^n)(v_1, \cdots, v_n) = \sum_{\sigma} \text{sgn}(\sigma) a_{1\sigma(1)} a_{2\sigma(2)} \cdots a_{n\sigma(n)} ,$$

which is a classical formula for (or definition of) the determinant of the matrix (a_{ij}).

Solution 2.5

Let $q \geq 2$, let $\alpha_1, \alpha_2, \cdots, \alpha_q$ be arbitrary 1-forms on a vector space V, and suppose

$$(*) \qquad \alpha_1 \wedge \alpha_2 \wedge \cdots \wedge \alpha_q = \beta_1 \otimes \beta_2 \otimes \cdots \otimes \beta_q$$

for some 1-forms $\beta_1, \cdots, \beta_q$, with both sides nonzero. At least two of the β's, say β_i and β_j, must be linearly independent; otherwise the right side would be symmetric and the left side antisymmetric.

Since β_i and β_j are independent, there exists a vector v_i in the nullspace of β_i and a vector v_j in the nullspace of β_j such that $\beta_i(v_j) \neq 0 \neq \beta_j(v_i)$. (Proof: Any form is determined up to a scalar multiple by its nullspace, so if two forms are linearly independent, then their nullspaces are different, and, since both nullspaces are of the same dimension, neither can be contained in the other.) For each $k \neq i, j$, choose a vector u_k such that $\beta_k(u_k) \neq 0$, and define $u_j := v_i$ and $u_i := v_j$. Applying the right side of (*) to the q-tuple $u_1, u_2, \cdots, u_i, \cdots, u_j, \cdots, u_q$, gives a nonzero scalar, but interchanging u_i and u_j gives zero, contradicting the antisymmetry of the left side.

Solution 2.6

This is obvious if $\{e_i\}$ happens to be orthonormal. For the general case, we shall show that if $\{\bar{e}_i\}$ is any other positively oriented basis, $\{\bar{e}^i\}$ the dual basis, and $\bar{G}$ the inner product matrix with respect to the new basis, then

$$|\det(\bar{G})|^{1/2}\, \bar{e}^1 \wedge \cdots \wedge \bar{e}^n \;=\; |\det(G)|^{1/2}\, e^1 \wedge \cdots \wedge e^n \;.$$

To this end, let A be the matrix of the linear transformation which sends e_i to $\bar{e}_i$, $1 \leq i \leq n$. Then $\det A\, \bar{e}^1 \wedge \cdots \wedge \bar{e}^n = e^1 \wedge \cdots \wedge e^n$ (cf. after 2.2(6)). On the other hand, $\bar{G} = A^T G A$, where T denotes the transpose, so $\det(\bar{G}) = (\det A)^2 \det(G)$. Note that (*) really has little to do with the metric g *per se*, but rather exploits the transformation properties of its matrix under a change of basis.

Solution 2.7

Let $\beta_1, \beta_2, \cdots, \beta_q \in V^*$ be given 1-forms, and consider the real-valued, alternating, q-linear map $\phi : V^* \times V^* \times \cdots \times V^*$ (q copies) $\longmapsto \mathbf{R}$ defined by

$$\phi(\alpha_1, \alpha_2, \cdots, \alpha_q) := \det(\langle \alpha_i, \beta_j \rangle)$$

for all $\alpha_1, \cdots, \alpha_q \in V^*$. By the Universal Mapping Property of $\Lambda_q(V)$ [Warner, p. 57], there exists a unique linear map $\tilde{\phi} : \Lambda_q(V) \longmapsto \mathbf{R}$ such that

$$\tilde{\phi}(\alpha_1 \wedge \alpha_2 \wedge \cdots \wedge \alpha_q) \;=\; \phi(\alpha_1, \cdots, \alpha_q) \;.$$

Note that $\tilde{\phi}$ is an element of the dual space $\Lambda_q(V)^*$ of $\Lambda_q(V)$.

Of course, $\tilde{\phi}$ depends on the initial choice of $\beta_1, \cdots, \beta_q$. Let us denote this by writing

$$\tilde{\phi}(\,\cdot\,, \cdots, \,\cdot\,) \;=\; \Phi(\beta_1, \cdots, \beta_q)(\,\cdot\,, \cdots, \,\cdot\,) \;.$$

This defines Φ as an alternating q-linear map

$$\Phi : V^* \times V^* \times \cdots \times V^* \longmapsto \Lambda_q(V)^* \;.$$

Again using the universal property, we obtain a linear map

$$\tilde{\Phi} : \Lambda_q(V) \longmapsto \Lambda_q(V)^*$$

such that for all $\beta_1, \cdots, \beta_q \in V^*$,

$$\tilde{\Phi}(\beta_1 \wedge \cdots \wedge \beta_q) := \Phi(\beta_1, \cdots, \beta_q) \;.$$

Observe that for all $\beta_1, \cdots, \beta_q \in V^*$,

(1) $$\tilde{\Phi}(\beta_1 \wedge \cdots \wedge \beta_q)(\alpha_1 \wedge \cdots \wedge \alpha_q) \;=\; \det(\langle \alpha_i, \beta_j \rangle) \;.$$

Consider the bilinear map $B(\,\cdot\,,\,\cdot\,) : \Lambda_q(V) \times \Lambda_q(V) \longmapsto \mathbf{R}$ naturally associated with $\tilde{\Phi}$; this map sends a pair λ, μ of q-forms to

$$B(\lambda, \mu) := \tilde{\Phi}(\mu)(\lambda) \;.$$

By (1), we want to show that $B(\,\cdot\,,\,\cdot\,)$ is the inner product map, and by bilinearity it is enough to show this when λ and μ are drawn from the basis 2.2(2) consisting of all wedge products of the e^i, say

$$\lambda := e^{s_1} \wedge \cdots \wedge e^{s_q}, \qquad 1 \leq s_1 < s_2 < \cdots < s_q \leq n,$$

$$\mu := e^{t_1} \wedge \cdots \wedge e^{t_q}, \qquad 1 \leq t_1 < t_2 < \cdots < t_q \leq n.$$

Of course, $\langle \lambda, \mu \rangle = 0$ unless $\lambda = \mu$, in which case $\langle \lambda, \mu \rangle = 1$. On the other hand,

$$B(\lambda, \mu) := \det(\langle e^{s_i}, e^{t_j} \rangle)$$

(i.e. the matrix whose i, j entry is $\langle e^{s_i}, e^{t_j} \rangle$). If $\lambda \neq \mu$, then this matrix has a zero row corresponding to the s_i which is not a t_j for any j, and so has determinant zero. If $\lambda = \mu$, the matrix is the identity and so has determinant 1.

Of course, this problem could also be done by direct indicial tensor analysis, but the solution is likely to be much more complicated than the above. The Universal Mapping Property is an efficient tool.

Solution 2.8

(a) Choose a dual basis $e^1, e^2, \cdots, e^n$, and write

$$\sigma = \sum_{i_1 < i_2 < \cdots < i_n} c_{i_1 i_2 \cdots i_{n-p}} e^{i_1} \wedge \cdots \wedge e^{i_{n-p}}.$$

Suppose that one of the c's, say $c_{j_1 j_2 \cdots j_{n-p}}$ is nonzero. Let $k_1 < k_2 < \cdots < k_p$ be the complement of $j_1, j_2, \cdots, j_{n-p}$ in $1, 2, \cdots, n$, and define $\beta := e^{k_1} \wedge \cdots \wedge e^{k_p}$. Since *any* set of indices $i_1 < i_2 < \cdots < i_{n-p}$ *except* $j_1, \cdots, j_{n-p}$ contains one of the k_a's, we have for any such set $e^{k_1} \wedge \cdots \wedge e^{k_p} \wedge e^{i_1} \wedge \cdots \wedge e^{i_{n-p}} = 0$, and so

$$\beta \wedge \sigma = \pm c_{j_1 \cdots j_{n-p}} \Omega \neq 0.$$

(b) If γ_1, γ_2, are two $(n-p)$-forms satisfying (*), then we have

(**) $\qquad \beta \wedge (\gamma_1 - \gamma_2) = 0 \qquad$ for all p-forms β,

and we may apply (a) to conclude that $\gamma_1 - \gamma_2 = 0$.

Solution 2.9

(a) <u>Solution 1</u>. A very quick solution can be obtained by defining *any* positive definite inner product $\langle \cdot, \cdot \rangle$ on V and any orientation and then defining $\perp$ as the Hodge duality operation relative to this inner product and orientation. If α is any nonzero $(n-1)$-form, then $\perp \alpha$ is a 1-form which may be considered as a multiple of the first element e^1 of an orthonormal dual basis $e^1 := \perp \alpha \langle \perp \alpha, \perp \alpha \rangle^{-1/2}, e^2, \cdots, e^n$ for V^*. Then α is obviously decomposable because

$$\alpha = \pm \perp \perp \alpha = \pm \langle \perp \alpha, \perp \alpha \rangle^{1/2} e^2 \wedge \cdots \wedge e^n.$$

Though very slick, this solution has the unsatisfying feature of dragging in an arbitrary inner product which has nothing to do with anything in the original problem. The following solution avoids this.

Solution 2. Let $e^1, e^2, \cdots, e^n$ be a basis for V^*, and set $\Omega := e^1 \wedge \cdots \wedge e^n$. Let α be a given $(n-1)$-form, and write

$$\alpha = \sum_{k=1}^n a_k e^1 \wedge \cdots \wedge \hat{e}^k \wedge \cdots \wedge e^n ,$$

where the hat indicates that e^k is omitted. The coefficient a_k can be picked out by noting that

$$e^k \wedge \alpha = (-1)^{k-1} a_k \Omega .$$

We would like to choose the basis $\{e^i\}$ so that all but one of the a_k vanish.

For any 1-form $\beta \in V^*$, $\beta \wedge \alpha$ is a multiple of Ω; denoting this multiple as $g(\beta)$, so that

$$\beta \wedge \alpha = g(\beta) \Omega ,$$

defines a 1-form g on V^*. By general principles, the nullspace of g is of dimension at least $n-1$, and so we can choose a basis $f^1, f^2, \cdots, f^{n-1}, f^n$ for V^* such that $f^1, f^2, \cdots, f^{n-1}$ are in the nullspace of g. Then if we write

$$\alpha = \sum_{k=1}^n b_k f^1 \wedge \cdots \wedge \hat{f}^k \wedge \cdots \wedge f^n$$

with respect to this new basis, we have for $k = 1, 2, \cdots, n-1$,

$$(-1)^{k-1} b_k \Omega = f^k \wedge \alpha = g(f^k) \Omega = 0 ,$$

so α is decomposed as $\alpha = b_n f^1 \wedge \cdots \wedge f^{n-1}$.

(b) This is trivial for spaces of dimension ≤ 2 and for q-forms on a space of dimension 3 when $q \neq 2$; the remaining case of $q = 2$ on a space of dimension 3 is covered by (a).

(c) Let e^1, e^2, e^3, e^4 be a basis for the space of 1-forms, and let α be an arbitrary 2-form. Then

$$\alpha = \sum_{1 \leq i < j \leq 4} \alpha_{ij} e^i \wedge e^j$$

$$= e^1 \wedge \sum_{j=2}^4 \alpha_{1j} e^j + \sum_{2 \leq i < j \leq 4} \alpha_{ij} e^i \wedge e^j .$$

The first term on the right is obviously decomposable, and the second term (which has the same algebraic properties as a 2-form on a vector space of dimension 3) is decomposable by the argument of part (a).

(d) Continuing from (c), define $\omega := e^1 \wedge e^2 + e^3 \wedge e^4$. Then ω cannot be decomposable because $\omega \wedge \omega = 2 e^1 \wedge e^2 \wedge e^3 \wedge e^4 \neq 0$, but any decomposable form α satisfies $\alpha \wedge \alpha = 0$.

Solution 2.10

The answers are:

(a) $\perp e^0 = e^1 \wedge e^2 \wedge e^3$; (b) $\perp e^1 = e^0 \wedge e^2 \wedge e^3$; (c) $\perp e^2 = -e^0 \wedge e^1 \wedge e^3$;

(d) $\perp(e^0 \wedge e^3) = -e^1 \wedge e^2$; (e) $\perp(e^1 \wedge e^2) = e^0 \wedge e^3$; (f) $\perp(e^1 \wedge e^2 \wedge e^3) = e^0$;

(g) $\perp(e^0 \wedge e^2 \wedge e^3) = e^1$; (h) $\perp(e^0 \wedge e^1 \wedge e^2 \wedge e^3) = -1$.

An efficient way to arrive at the answers is as follows. We know that the dual of a wedge product of orthonormal basis vectors is $\pm$ the wedge product of the remaining basis vectors, so the only problem is to determine the sign. The latter can be read off from the special case

$$\alpha \wedge \perp \alpha = \langle \alpha, \alpha \rangle \Omega$$

of the defining relation 2.2(7). For instance, $(\perp e^0 \wedge e^3) = \pm e^1 \wedge e^2$, so one considers

$$(e^0 \wedge e^3) \wedge (\pm e^1 \wedge e^2) = \langle e^0 \wedge e^3, e^0 \wedge e^3 \rangle \Omega = -\Omega \quad,$$

and chooses the sign appropriately.

Solution 2.11

Choose a positively oriented orthonormal dual basis $e^1, e^2, \cdots, e^n$ such that β is a multiple of e^1 and γ is in the subspace spanned by e^1 and e^2. Clearly, there is no loss of generality in assuming that $\beta = e^1$. Write $\gamma = \gamma_\perp + a\beta$ with $\gamma_\perp$ orthogonal to β. Since $\beta \wedge \gamma = \beta \wedge \gamma_\perp$, the left side of (*) is independent of a, and a short calculation shows that the right side is also independent of a. Hence we may take $a = 0$ and $\gamma = e^2$. Now $\perp(e^1 \wedge e^2) = e^3 \wedge e^4 \wedge \cdots \wedge e^n$, and

$$\alpha \wedge \perp(e^1 \wedge e^2) = \alpha_1 e^1 \wedge e^3 \wedge e^4 \wedge \cdots \wedge e^n + \alpha_2 e^2 \wedge e^3 \wedge \cdots \wedge e^n \quad,$$

which vanishes, along with the right side of (*), when α is orthogonal to e^1 and e^2. Hence, by linearity, we may assume that $\alpha = \alpha_1 e^1 + \alpha_2 e^2$. Then

$$\perp(\alpha \wedge \perp(e^1 \wedge e^2)) = (-1)^n (\alpha_1 e^2 - \alpha_2 e^1) = (-1)^n (\langle \alpha, e^1 \rangle e^2 - \langle \alpha, e^2 \rangle e^1) \quad.$$

Solution 2.12

By linearity, it is sufficient to prove this for α a wedge product of elements of the dual basis $\{e^i\}$. Reordering the basis and replacing α by $-\alpha$ if necessary, we may assume that the basis is positively oriented and that

$$\alpha = e^1 \wedge e^2 \wedge \cdots \wedge e^p \quad.$$

Then $\perp \alpha = s_1 e^{p+1} \wedge e^{p+2} \wedge \cdots \wedge e^n$, where s_1 is a sign. To determine s_1, consider the relation

$$\alpha \wedge \perp \alpha = \langle \alpha, \alpha \rangle \Omega = \prod_{i=1}^{p} \langle e^i, e^i \rangle e^1 \wedge e^2 \wedge \cdots \wedge e^p \wedge e^{p+1} \wedge \cdots \wedge e^n \quad,$$

which shows that $s_1 = \prod_{i=1}^{p} \langle e^i, e^i \rangle$. Similarly,

$$\perp(e^{p+1} \wedge \cdots \wedge e^n) = s_2 e^1 \wedge \cdots \wedge e^p \quad , \text{with}$$

$$s_2 = \prod_{i=p+1}^{n} <e^i, e^i> \frac{e^1 \wedge \cdots \wedge e^p \wedge e^{p+1} \wedge \cdots \wedge e^n}{e^{p+1} \wedge \cdots \wedge e^n \wedge e^1 \wedge \cdots \wedge e^p} \quad ,$$

where the symbolic "quotient" of the two n-forms on the right (which differ by at most a sign) denotes $+1$ if they are equal and -1 otherwise. Inspection shows that $e^{p+1} \wedge \cdots \wedge e^n \wedge e^1 \wedge \cdots \wedge e^p$ can be changed into $e^1 \wedge \cdots \wedge e^p \wedge e^{p+1} \wedge \cdots \wedge e^n$ by $p(n-p)$ interchanges of the e^i. (Think of bringing e^1 to the front by $n-p$ interchanges, then $n-p$ more to bring e^2 to its proper place to the left of e^1, etc.) Hence
$$\perp(e^{p+1} \wedge \cdots \wedge e^n) = \prod_{i=p+1}^{n} <e^i, e^i> (-1)^{p(n-p)} e^1 \wedge \cdots \wedge e^p \quad. \text{ Finally, note that since } \{e_i\} \text{ is}$$
orthonormal, $e^i = <e_i, e_i> e_i{}^*$, so $<e^i, e^i> = <e_i, e_i>$.

Solution 2.13

Let $\{e_i\}_{i=1}^n$ be any basis, and let $G = (g_{ij})$ denote the matrix of the inner product with respect to this basis. It is routine to check that the non-degeneracy of the inner product is equivalent to $\det(G) \neq 0$.

Let L be an isometry with matrix $\tilde{L} = (L^i{}_j)$ with respect to the given basis. Then for all vectors $v = (v^i) \in V$,

$$v^i g_{ij} v^j \;=\; <v, v> \;=\; <Lv, Lv> \;=\; <e_i L^i{}_\alpha v^\alpha, e_j L^j{}_\beta v^\beta> \;=\; v^\alpha L^i{}_\alpha g_{ij} L^j{}_\beta v^\beta \quad .$$

We may view this as an ordinary matrix equation, $v^T G v = v^T \tilde{L}^T G \tilde{L} v$ with $v = (v^i)$ a column vector, where T denotes the transpose. Since this matrix equation holds for all v,

$$G = \tilde{L}^T G \tilde{L} \quad .$$

Thus

$$\det(G) \;=\; \det(\tilde{L}^T) \det(G) \det(\tilde{L}) \;=\; (\det(\tilde{L}))^2 \det(G) \quad ,$$

whence $\det(L) = \det(\tilde{L}) = \pm 1$.

Solution 2.14

(a) Let $\{x^i\}$ be any coordinate system around p. If $w_p = \sum w^i (\partial/\partial x_i)_p$ is any vector at p (so the w^i are simply scalars), then

$$q \longmapsto \sum w^\alpha \left(\frac{\partial}{\partial x_\alpha}\right)_q$$

extends w_p to the coordinate neighborhood.

(b)(i)

$$v_p(\alpha(fw)) - \alpha(D_{v_p}(fw)) \;=\; v_p(f\alpha(w)) - \alpha(v_p(f)w) - f(p)\alpha(D_{v_p} w)$$

$$=\; v_p(f)\alpha(w) + f(p)v_p(\alpha(w)) - v_p(f)\alpha(w) - f(p)\alpha(D_{v_p} w)$$

$$=\; f(p) \cdot [v_p(\alpha(w)) - \alpha(D_{v_p} w)] \quad .$$

(ii) Given a vector *field* w, write $w = \sum w^i(\partial/\partial x_i)$ in the coordinate neighborhood; here the $w^i(\cdot)$ are functions. Then

(**)
$$v(\alpha(w)) - \alpha(D_v w) = \sum_i [v(\alpha(w^i \frac{\partial}{\partial x_i})) - \alpha(D_v(w^i \frac{\partial}{\partial x_i}))]$$

$$= \sum_i w^i(p)[v(\alpha(\frac{\partial}{\partial x_i})) - \alpha(D_v(\frac{\partial}{\partial x_i}))]$$

where the second equality is just (i) with f replaced by w^i. The right side of (**) depends only on $w^i(p)$, and so depends only on w_p.

Solution 2.15

Using 2.5(5) with $\alpha := dx^j$, $v := v^i \frac{\partial}{\partial x_i}$, $w := \sum_i w^i \frac{\partial}{\partial x_i}$, we have

$$(D_{\frac{\partial}{\partial x_i}}(dx^j))(\frac{\partial}{\partial x_m}) = \frac{\partial}{\partial x_i}(dx^j(\frac{\partial}{\partial x_m})) - dx^j(D_{\frac{\partial}{\partial x_i}}\frac{\partial}{\partial x_m})$$

$$= 0 - dx^j(\sum_k \Gamma^k_{im} \frac{\partial}{\partial x_k}) = -\Gamma^j_{im}.$$

Solution 2.16

Let $\{e^i\}$ be a dual basis. Since both sides are linear in λ and β, it is sufficient to prove this for $\lambda = e^i$, $\beta = e^{j_1} \wedge \cdots \wedge e^{j_q}$. We may also assume that the indices $i, j_1, \cdots, j_q$ are distinct (otherwise both sides vanish), and reordering the basis, if necessary, we may assume that $\lambda = e^1$ and $\beta = e^2 \wedge \cdots \wedge e^{q+1}$, in which case

$$\lambda \wedge \beta := e^1 \wedge e^2 \wedge \cdots \wedge e^{q+1}.$$

By linearity, it is sufficient to verify 2.5(15) with $v_k = e_{i_k}$ selected from the basis vectors. If the set $e_{i_1}, e_{i_2}, \cdots, e_{i_{q+1}}$ is not a permutation of $1, 2, \cdots, q+1$, both sides vanish. Since both sides are antisymmetric, we need only check the formula for $v_1 := e_1, v_2 := e_2, \cdots, v_{q+1} := e_{q+1}$, in which case the left side gives 1 and all terms on the right vanish except the first term

$$(-1)^{1+1} e^1(e_1) \cdot (e^2 \wedge \cdots \wedge e^{q+1})(e_2, \cdots, e_{q+1}) = 1.$$

Solution 2.17

$$\delta v := \sum_{k=1}^n (D_{\frac{\partial}{\partial x_i}} v)(dx^i) = \sum_{k=1}^n (D_{\frac{\partial}{\partial x_i}} v)^i.$$

Now
$$D_{\frac{\partial}{\partial x_i}} v = D_{\frac{\partial}{\partial x_i}}(\sum_\alpha v^\alpha \frac{\partial}{\partial x_\alpha})$$

$$= \sum_\alpha \frac{\partial v^\alpha}{\partial x_i} \cdot \frac{\partial}{\partial x_\alpha} + \sum_\alpha v^\alpha D_{\frac{\partial}{\partial x_i}}(\frac{\partial}{\partial x_\alpha}) = \sum_\alpha \frac{\partial v^\alpha}{\partial x_i} \cdot \frac{\partial}{\partial x_\alpha} + \sum_\alpha \sum_\beta v^\alpha \Gamma^\beta_{i\alpha} \frac{\partial}{\partial x_\beta}$$

so

$$\delta v = \sum_i \frac{\partial v^i}{\partial x_i} + \sum_{\alpha,i} v^\alpha \Gamma^i_{i\alpha} \ .$$

Solution 2.18

By linearity, it is enough to establish this for decomposable $T = w_1 \otimes \cdots \otimes w_p \otimes \beta_1 \otimes \cdots \otimes \beta_q$. Since for such a T,

$$T(\alpha_1, \cdots, \alpha_p, v_1, \cdots, v_q) = \alpha_1(w_1) \cdots \alpha_p(w_p) \beta_1(v_1) \cdots \beta_q(v_q) \ ,$$

the result follows immediately from the definition 2.5(5) which ensures that $(D_v \alpha)(w) = v(\alpha(w)) - \alpha(D_v w)$.

Solution 2.19

(a) Let $p \in M$. Since g is a tensor field of type 0,2, Dg is the tensor field of type 0,3 defined for all $u_p, v_p, w_p \in M_p$ by 2.5(10):

$$Dg(u_p, v_p, w_p) := (D_{w_p} g)(u_p, v_p) \ .$$

By Exercise 18, if we extend u and v to vector fields defined in a neighborhood of p, then

$$(D_{w_p} g)(u_p, v_p) = w_p(g(u, v)) - g(D_{w_p} u, v_p) - g(u_p, D_{w_p} v) = 0 \ ,$$

where the last equality is the defining relation 2.7(1) for the Riemannian connection, and this shows that $Dg = 0$.

(b) By Exercise 18 and part (a), at p we have,

$$\frac{\partial g_{ij}}{\partial x_k}(p) = \frac{\partial}{\partial x_k} g(\frac{\partial}{\partial x_i}, \frac{\partial}{\partial x_j})$$

$$= (D_{\frac{\partial}{\partial x_k}} g)(\frac{\partial}{\partial x_i}, \frac{\partial}{\partial x_j}) + g(D_{\frac{\partial}{\partial x_k}} \frac{\partial}{\partial x_i}, \frac{\partial}{\partial x_j}) + g(\frac{\partial}{\partial x_i}, \frac{D_{\partial}}{\partial x_k} \frac{\partial}{\partial x_j})$$

$$= 0 + \Gamma^\alpha_{ki} g_{\alpha j} + g_{i\alpha} \Gamma^\alpha_{kj} = 0 \ .$$

Solution 2.20

(a) Let $\{e_i\}$ be a positively oriented orthonormal basis. Use the linear coordinates $\{x^i\}$ defined by expansion of an arbitrary element of M as $\sum x^i e_i$, so that $\{dx^i\}$ is an orthonormal dual basis for the tangent space at any point. By linearity and after possible reordering of the basis, it is sufficient to establish (*) for α of the form

$$\alpha = f \, dx^1 \wedge dx^2 \cdots \wedge dx^p \quad \text{with } f \in C^\infty(M) \ .$$

From general principles we know that $\perp \alpha = \pm f dx^{p+1} \wedge \cdots \wedge dx^n$, and considering the relation

$$\alpha \wedge \perp \alpha = \langle \alpha, \alpha \rangle dx^1 \wedge \cdots \wedge dx^n$$

and setting $\epsilon_i := \langle dx^i, dx^i \rangle$, we see that

$$\lrcorner \alpha = (\prod_{i=1}^{p} \epsilon_i) f \, dx^{p+1} \wedge \cdots \wedge dx^n \ .$$

Then

$$d \lrcorner \alpha = \prod_{i=1}^{p} \epsilon_i \sum_{j=1}^{p} \frac{\partial f}{\partial x_j} dx^j \wedge dx^{p+1} \wedge \cdots \wedge dx^n \ ,$$

and

$$\lrcorner d \lrcorner \alpha = \prod_{i=1}^{p} \epsilon_i \sum_{j=1}^{p} \frac{\partial f}{\partial x_j} \lrcorner (dx^j \wedge dx^{p+1} \wedge \cdots \wedge dx^n) \ .$$

Again, $\lrcorner (dx^j \wedge dx^{p+1} \wedge \cdots \wedge dx^n) = s \, dx^1 \wedge \cdots \wedge \hat{dx}^j \wedge \cdots \wedge dx^p$, where the hat means that dx^j is omitted, and s is a sign obtained as follows. First compute

$$dx^j \wedge dx^{p+1} \wedge \cdots \wedge dx^n \wedge dx^1 \wedge \cdots \wedge \hat{dx}^j \wedge \cdots \wedge dx^p$$

$$= (-1)^{(j-1)(n-p+1)+(p-j)(n-p)} dx^1 \wedge \cdots \wedge dx^n \ .$$

To see this, think of obtaining the wedge product on the right from that on the left by successively moving each of $dx^1 \cdots dx^{j-1}$ to its proper place to the left of $dx^j \wedge dx^{p+1} \cdots \wedge dx^n$, which can be done with $n-p+1$ interchanges for each, or a total of $(j-1)(n-p+1)$. Similarly, moving each of the $dx^{j+1} \wedge \cdots \wedge dx^p$ through the $dx^{p+1} \wedge \cdots \wedge dx^n$ can be done with a total of $(p-j)(n-p)$ interchanges. The sign can be simplified by noting that since $p(p+1) \equiv 0 \pmod{2}$,

$$(j-1)(n-p+1) + (p-j)(n-p) \equiv j(n-p+1) + (n-p+1) + (p-j)(n-p)$$

$$\equiv j(n-p) + j + (n-p) + 1 + (p-j)(n-p) \equiv (p+1)(n-p) + j + 1$$

$$\equiv (p+1)n + j + 1 \pmod{2}.$$

There is also a contribution

$$\epsilon_j \prod_{i=p+1}^{n} \epsilon_i \quad \text{to} \quad s \quad \text{from the norm of} \quad dx^j \wedge dx^{p+1} \wedge \cdots \wedge dx^n \ . \quad \text{Hence, since}$$

$\prod_{i=1}^{n} \epsilon_i = (-1)^b$,

$$\lrcorner d \lrcorner \alpha = (-1)^b (-1)^{(p+1)n} \sum_{j=1}^{p} \epsilon_j (-1)^{j+1} \frac{\partial f}{\partial x_j} dx^1 \wedge \cdots \wedge \hat{dx}^j \wedge \cdots \wedge dx^p.$$

On the other hand, since $dx^1 \wedge \cdots dx^p$ is the antisymmetrization 2.2(1) of $dx^1 \otimes \cdots \otimes dx^p$, we have

$$\alpha = \sum_{j=1}^{p} (-1)^{j+1} f \, dx^j \otimes dx^1 \wedge \cdots \hat{dx}^j \wedge \cdots \wedge dx^p \ .$$

(To see this, when summing 2.2(1), first sum over all permutations π such that $\pi(1) = j$. Raising the first index on $dx^j \otimes dx^1 \wedge \cdots \wedge \hat{dx}^j \wedge \cdots \wedge dx^p$ gives

$$\epsilon_j f \frac{\partial}{\partial x_j} \otimes dx^1 \wedge \cdots \wedge \hat{dx}^j \wedge \cdots \wedge dx^p \quad ,$$

and differentiating and contracting gives

$$\epsilon_j \frac{\partial f}{\partial x_j} dx^1 \wedge \cdots \wedge \hat{dx}^j \wedge \cdots \wedge dx^p \quad .$$

Thus

$$\delta(\alpha) = \sum_{j=1}^{n} (-1)^{j+1} \epsilon_j \frac{\partial f}{\partial x_j} dx^1 \wedge \cdots \wedge \hat{dx}^j \wedge \cdots \wedge dx^p \quad ,$$

and comparing with the expression above for $\lrcorner d \lrcorner \alpha$ shows that

$$\delta(\alpha) = (-1)^b (-1)^{n(p+1)} \lrcorner d \lrcorner \alpha \quad .$$

(b) If M is a manifold the above proof does not apply directly because it is not in general possible to choose a coordinate system $\{x^i\}$ such that $\{dx^i\}$ is orthogonal at *every* point. It is possible to obtain this at any given point p, but the relation $\lrcorner (dx^1)_q \wedge \cdots \wedge (dx^p)_q = \pm (dx^{p+1})_q \wedge \cdots \wedge (dx^n)_q$ will hold only at $q = p$, so we cannot differentiate it directly. As mentioned in the text, a way around this difficulty is to use for the computation a positively oriented coordinate system $\{x^i\}_{i=1}^n$ such that the connection coefficients Γ_{ij}^k vanish at the point $p \in M$ at which we want to verify (*) and with the additional property that the forms $\{(dx_i)_p\}_{i=1}^n$ are orthonormal. It is a standard fact that such coordinate systems exist [Hicks, p. 131]. (However, the vanishing of the Γ_{ij}^k cannot in general be obtained in a neighborhood of p, but only at p itself.) It follows from 2.5(6) that the vanishing of the $\Gamma_{ij}^k(p)$ implies that

(1) $\qquad (D(dx^i))_p = 0 \quad ,$

and also

(2) $\qquad \dfrac{\partial g_{ij}}{\partial x_i}(p) = 0 \qquad$ for all i, j, k

from Exercise 19.

The problem which arises in trying to generalize the inner product space proof and its resolution by the use of the above special coordinates can be seen most clearly by working out a special case. We shall work out the special case $\alpha := f dx^2$ in a two-dimensional M. Once the problem and solution for this case are seen, it will be clear how the general case follows in a similar way. If the general proof were written down immediately, it would appear formidable due to the proliferation of indices, and it would be hard to extract the main idea.

In our special case, since $(g_{ij}(p))$ is a diagonal matrix, raising an index on $\alpha := f(dx^2)_p$ gives the vector $fg^{22}(\partial/\partial x_2)_p$, and so at p,

$$\delta(\alpha) = \delta(fg^{22} \frac{\partial}{\partial x_2}) = \left[\frac{D_\partial}{\partial x_1}(fg^{22} \frac{\partial}{\partial x_2})\right]^1 + \left[\frac{D_\partial}{\partial x_2}(fg^{22} \frac{\partial}{\partial x_2})\right]^2$$

$$= \left[\frac{\partial}{\partial x_1}(fg^{22}) \frac{\partial}{\partial x_2} + fg^{22} D_{\frac{\partial}{\partial x_1}} \frac{\partial}{\partial x_2}\right]^1 + \left[\frac{\partial}{\partial x_2}(fg^{22}) \frac{\partial}{\partial x_2} + fg^{22} D_{\frac{\partial}{\partial x_2}} \frac{\partial}{\partial x_2}\right]^2$$

$$= fg^{22} \Gamma_{12}^1 + \frac{\partial f}{\partial x_2} g^{22} + f \frac{\partial g^{22}}{\partial x_2} + fg^{22} \Gamma_{22}^2$$

$$= \frac{\partial f}{\partial x_2} g^{22} \ .$$

From this it should be clear that for a general p-form α, the calculation of $(\delta(\alpha))$ in terms of the dx^i will algebraically parallel the corresponding computation in which $\{x^i\}$ are orthonormal linear coordinates in an inner product space; all the additonal derivatives which occur in the manifold case obligingly vanish at p.

Now let us look at the calculation of $\lrcorner d\lrcorner \alpha$ in the above special case. Let Ω denote the volume n-form. From Exercise 6,

(3) $$\Omega = |\det(G)|^{1/2} dx^1 \wedge dx^2 \ ,$$

where G denotes the matrix $G := (g_{ij})$. Now from 2.2(15),

$$(\lrcorner \alpha)_j = \Omega^i{}_j \alpha_i = g^{ik}\Omega_{kj}\alpha_i = g^{2k}\Omega_{kj}f \qquad ; \text{i.e.}$$

(4) $$\lrcorner \alpha = g^{2k}\Omega_{k1}dx^1 + g^{2k}\Omega_{k2}dx^2 = g^{22}\Omega_{21}dx^1 + g^{21}\Omega_{12}dx^2 \ .$$

At p itself, $g^{21} = 0$ and $\lrcorner \alpha = g^{22}\Omega_{21}dx^1 = -g^{22}dx^1$, which is what held globally when M was an inner product space, but when we apply d to (4) in the present case, derivatives of $g^{21}\Omega_{12}$ will appear in addition. However, (2) and (3) show that in this special coordinate system, these additional derivatives vanish at p, so the computation again parallels that of the inner product case.

Solution 2.21

(a) Let $\{x^i\}$ be a local coordinate system. Then

$$\delta(fv) := \sum_\alpha (D_{\frac{\partial}{\partial x_\alpha}} v)^\alpha = \sum_\alpha \frac{\partial f}{\partial x_\alpha} v^\alpha + f \sum_\alpha (D_{\frac{\partial}{\partial x_\alpha}} v)^\alpha$$

$$= v(f) + f\delta(v) \ .$$

(b) $$(\delta(v \otimes w))^j = \sum_\alpha (D_{\frac{\partial}{\partial x_\alpha}} (v \otimes w))^{\alpha j}$$

$$= \sum_\alpha ((D_{\frac{\partial}{\partial x_\alpha}} v) \otimes w)^{\alpha j} + \sum_\alpha (v \otimes D_{\frac{\partial}{\partial x_\alpha}} w)^{\alpha j}$$

$$= \sum_\alpha (D_{\frac{\partial}{\partial x_\alpha}} v)^\alpha w^j + \sum_\alpha v^\alpha (D_{\frac{\partial}{\partial x_\alpha}} w)^j$$

$$= \delta(v)w^j + (D_v w)^j \ .$$

(c) Recall from 2.8(11) that $\delta(v^* \otimes \text{anything}) := \delta(v \otimes \text{anything})$, so

$$\delta(v^* \wedge w^*) = \delta(v^* \otimes w^* - w^* \otimes v^*) := \delta(v \otimes w^*) - \delta(w \otimes v^*) \ .$$

Now by essentially the same computation as in (b) but with the second index lowered and recalling from 2.7(2) that covariant differentiation commutes with raising and lowering indices,

$$\delta(v \otimes w^*) = \delta(v)w^* + D_v(w^*) = \delta(v)w^* + (D_v w)^* \ .$$

Hence

$$\delta(v^* \wedge w^*) = \delta(v)w^* - \delta(w)v^* + (D_v w - D_w v)^*$$
$$= \delta(v)w^* - \delta(w)v^* + [v, w]^* .$$

Solution 2.22

(a) The orientations on ∂V are as drawn; the arrows point in the direction of a positively oriented basis vector for the one-dimensional tangent space.

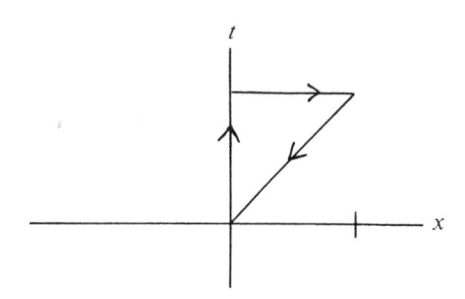

(b) We compute the integral over the light cone in detail, the other integrals being similar. We may coordinatize L by the map

$$s \longmapsto (1-s, 1-s) \qquad 0 \le s \le 1 .$$

The tangent vector $\dfrac{\partial}{\partial s}$ to this curve has components $(-1, -1)$ with respect to the standard basis for $\mathbf{M}^2$ and constitutes a positively oriented basis for the tangent space. (The simpler coordinatization $s \longmapsto (s, s)$ is a "negatively oriented" coordinatization, and if it were used below, the sign would come out wrong.) In order to integrate $\perp w^*$ over L using the definition of integration of a 1-form over a 1-manifold, we need to express $\perp w^*$ in the form $\perp w^* = f ds$ for some function $f = f(s)$, and obviously, $f = \perp w^*(\partial/\partial s)$. By 2.2(15), we have

$$\perp w^*(\partial/\partial s) = \Omega(w, \partial/\partial s) = \Omega((!w^0, w^1), (-1, -1))$$
$$= w^1 - w^0 ;$$

i.e. when $\perp w^*$ is considered as a 1-form on L, we have

$$\perp w^* = (w^1 - w^0) ds .$$ Thus, writing for typographical convenience $w^i(t, x)$ in place of $w^i_{(t,x)}$, we have

$$\int_L \perp w^* = \int_0^1 [w^1(1-s, 1-s) - w^0(1-s, 1-s)] ds .$$

In a similar way,

$$\int_T \perp w^* = \int_0^1 -w^1(t, 0) \, dt$$

and

$$\int_S \bot w^* = \int_0^1 w^0(1, x)\, dx \ .$$

(c) We have

$$\delta(w) = \frac{\partial w^0}{\partial t} + \frac{\partial w^1}{\partial x} \ .$$

Thus

$$\int_V \delta(w)\, \Omega = \int_V \frac{\partial w^0}{\partial t}\, dtdx + \int_V \frac{\partial w^1}{\partial x}\, dtdx$$

$$= \int_0^1 dx \int_x^1 \frac{\partial w^0}{\partial t}\, dt + \int_0^1 dt \int_0^t \frac{\partial w^1}{\partial x}\, dx$$

$$\int_0^1 [w^0(1, x)\, dx - w^0(x, x)]\, dx + \int_0^1 [w^1(t, t) - w^1(t, 0)]\, dt \ .$$

Solution 2.23

By the definition of Hodge dual, we also have for any p-forms λ, μ,

$$(\lambda, \mu) = \int_M \lambda \wedge \bot \mu \ .$$

Hence for any p-form α and $(p+1)$-form β,

$$(d\alpha, \beta) = \int_M (d\alpha) \wedge \bot \beta = \int_M d(\alpha \wedge \bot \beta) - (-1)^p \int_M \alpha \wedge d(\bot \beta) \ .$$

By Stokes' Theorem,

$$\int_M d(\alpha \wedge \bot \beta) = \int_{\partial M} \alpha \wedge \bot \beta = 0$$

since ∂M is empty. (A manifold M is also a manifold with boundary whose boundary is empty.) Hence, recalling from Exercise 20 and 2.2(11) that

$$\delta(p\text{-form}) = (-1)^{b+(p+1)n} \bot d \bot (p\text{-form}) \ ,$$

where b is the number of negative norm directions in M,

and

$$\bot^2(p\text{-form}) = (-1)^{p(n-p)+b}(p\text{-form}) \ , \text{ we have}$$

$$(d\alpha, \beta) = (-1)^{p+1} \int_M \alpha \wedge d(\bot \beta)$$

$$= (-1)^{p+1}(-1)^{(n-p)p+b} \int_M \alpha \wedge \bot \bot d \bot \beta$$

$$= (-1)^{p+1}(-1)^{(n-p)p+b}(-1)^{b+((p+1)+1)n}\int_M \alpha \wedge \bot \delta\beta$$

$$= -(\alpha, \delta\beta) \ .$$

Solution 2.24

(a) Denoting the typical point of R^3 by (x^1, x^2, x^3), we have

$$v := v^1 \frac{\partial}{\partial x_1} + v^2 \frac{\partial}{\partial x_2} + v^3 \frac{\partial}{\partial x_i}$$

and, since the since the $\partial/\partial x_i$ are orthonormal,

$$v^* := v^1(\frac{\partial}{\partial x_1})^* + v^2(\frac{\partial}{\partial x_2})^* + v^3(\frac{\partial}{\partial x_i})^* = v^1 dx^1 + v^2 dx^2 + v^3 dx^3 \ .$$

Then

$$dv^* = (\frac{\partial v^2}{\partial x_1} - \frac{\partial v^1}{\partial x_2}) dx^1 \wedge dx^2 + (\frac{\partial v^3}{\partial x_1} - \frac{\partial v^1}{\partial x_3}) dx^1 \wedge dx^3 + (\frac{\partial v^3}{\partial x_2} - \frac{\partial v^2}{\partial x_3}) dx^2 \wedge dx^3.$$

On the other hand,

$$(\nabla \times \vec{v})^* = (\frac{\partial v^3}{\partial x_2} - \frac{\partial v^2}{\partial x_3}) dx^1 + (\frac{\partial v^1}{\partial x_3} - \frac{\partial v^3}{\partial x_1}) dx^2 + (\frac{\partial v^2}{\partial x_1} - \frac{\partial v^1}{\partial x_2}) dx^3) \ ,$$

so we need only check $\bot(dx^1) = dx^2 \wedge dx^3$, etc.

(b) Recalling from Exercise 2.20 that $\bot d \bot(\vec{w}^*) = \delta(\vec{w}^*) = \delta(\vec{w}) = 0$, we have $d\bot(\vec{w}^*) = 0$. By the Scholium, there exists locally a 1-form β such that

(i) $\qquad d\beta = \bot(\vec{w}^*) \quad$ and

(ii) $\qquad \delta\beta = \alpha \ .$

Let $\vec{v}$ denote the corresponding vector field with $\beta = \vec{v}^*$. Then (ii) is the same as $\nabla \cdot \vec{v} = \alpha$, and part (a) and (i) show that

$$(\nabla \times \vec{v})^* = \bot d(\vec{v}^*) = \bot d\beta$$

$$= \bot\bot(\vec{w}^*) = \vec{w}^* \ .$$

Solution 3.2.

Let e_0, e_1, e_2, e_3 be the orthonormal basis with repect to which

3.4(2) $$(F^i{}_j) = \begin{bmatrix} 0 & E^1 & E^2 & E^3 \\ E^1 & 0 & B^3 & -B^2 \\ E^2 & -B^3 & 0 & B^1 \\ E^3 & B^2 & -B^1 & 0 \end{bmatrix}.$$

is written. First look at the 3×3 submatrix in the lower right corresponding to the magnetic field. This is an antisymmetric linear transformation operating on the Euclidean space R^3, and it is easy to show that for any such there exists an orthonormal basis with respect to which its matrix is of the form

$$\begin{bmatrix} 0 & 0 & 0 \\ 0 & 0 & -c \\ 0 & c & 0 \end{bmatrix}.$$

(Proof: Since the characteristic polynomial has degree 3, it has a real eigenvalue which must be zero because the matrix is antisymmetric. Splitting the corresponding eigenvector off in an orthogonal direct sum, which must be invariant because the transformation is antisymmetric, reduces the problem to the 2×2 case, which is trivial.) Hence there exists an orthonormal set f_1, f_2, f_3 spanning the same subspace as e_1, e_2, e_3 such that the matrix of $\hat{F}$ with respect to the basis e_0, f_1, f_2, f_3 is of the form

(**) $$\begin{bmatrix} 0 & \bar{E}^1 & \bar{E}^2 & \bar{E}^3 \\ \bar{E}^1 & 0 & 0 & 0 \\ \bar{E}^2 & 0 & 0 & \bar{B}^1 \\ \bar{E}^3 & 0 & -\bar{B}^1 & 0 \end{bmatrix}.$$

Changing f_2 and f_3 while keeping them orthonormal and orthogonal to f_1 will not change the form of (**). Hence we may choose f_2 so that it is a multiple of $\bar{E}^2 e_2 + \bar{E}^3 e_3$, which brings the matrix of $\hat{F}$ with respect to e_0, f_1, f_2, f_3 to the desired form (*).

Solution 3.3.

We start with the canonical form

$$\hat{F} := \begin{bmatrix} 0 & E^1 & E^2 & 0 \\ E^1 & 0 & 0 & 0 \\ E^2 & 0 & 0 & B^1 \\ 0 & 0 & -B^1 & 0 \end{bmatrix},$$

of Exercise 2, written with respect to a basis e_0, e_1, e_2, e_3. Note that $E^1 \neq 0$ because $0 \neq \vec{E}\cdot\vec{B} = E^1 B^1$. The matrix

$$L := \begin{bmatrix} \gamma(v) & 0 & 0 & v\gamma(v) \\ 0 & 1 & 0 & 0 \\ 0 & 0 & 1 & 0 \\ v\gamma(v) & 0 & 0 & \gamma(v) \end{bmatrix}$$

is the matrix of a boost with velocity v in the e_3-direction, and

$$L^{-1}\hat{F}L = \begin{bmatrix} 0 & \gamma E^1 & \gamma E^2 + v\gamma B^1 & 0 \\ \gamma E^1 & 0 & 0 & v\gamma E^1 \\ \gamma E^2 + v\gamma B^1 & 0 & 0 & v\gamma E^2 + \gamma B^1 \\ 0 & -v\gamma E^1 & -v\gamma E^2 - \gamma B^1 & 0 \end{bmatrix}.$$

We shall first try to make the new electric and magnetic fields $\vec{E}' := \gamma E^1 e_1 + (\gamma E^2 + v\gamma B^1)e_2$ and $\vec{B}' := (v\gamma E^2 + \gamma B^1)e_1 - v\gamma E^1 e_2$ parallel; having accomplished this, a spatial rotation will produce the desired form (†). They will be parallel if and only if

$$0 = \det \begin{bmatrix} \gamma E^1 & v\gamma E^2 + \gamma B^1 \\ \gamma E^2 + v\gamma B^1 & -v\gamma E^1 \end{bmatrix} = -v\gamma^2(E^1)^2 - \gamma^2(vE^2 + B^1)(E^2 + vB^1) .$$

Solving, we obtain

$$\frac{v}{1+v^2} = \frac{-B^1 E^2}{(E^1)^2 + (E^2)^2 + (B^1)^2} .$$

We need a solution with $|v| < 1$. This is possible if and only if the right side has absolute value less than $1/2$, which is true when $E^1 \neq 0$ (because $0 \le (B^1 \pm E^2)^2 = (B^1)^2 \pm (2B^1 E^2) + (E^2)^2$, so $|B^1 E^2| \le ((B^1)^2 + (E^2)^2)/2$).

Solution 3.4.

(a) Squaring (*) of Exercise 2 yields:

(i) $$\hat{F}^2 = \begin{bmatrix} (E^1)^2 + (E^2)^2 & 0 & 0 & E^2 B^1 \\ 0 & (E^1)^2 & E^1 E^2 & 0 \\ 0 & E^1 E^2 & (E^2)^2 - (B^1)^2 & 0 \\ -B^1 E^2 & 0 & 0 & -(B^1)^2 \end{bmatrix} ,$$

Since $\vec{E} \cdot \vec{B} = E^1 B^1$ with respect to the basis of Exercise 2, we also have

(ii) $$\hat{F}(\hat{\perp}F) = \begin{bmatrix} -\vec{E}\cdot\vec{B} & 0 & 0 & 0 \\ 0 & -\vec{E}\cdot\vec{B} & 0 & 0 \\ 0 & 0 & -\vec{E}\cdot\vec{B} & 0 \\ 0 & 0 & 0 & -\vec{E}\cdot\vec{B} \end{bmatrix} ,$$

It is apparent from (i) that

(iii) $$\vec{E}\cdot\vec{E} - \vec{B}\cdot\vec{B} = (E^1)^2 + (E^2)^2 - (B^1)^2 = \frac{1}{2}\mathrm{tr}(\hat{F}^2) .$$

Since the matrix (*) of Exercise 2 was obtained from the general matrix 3.4(2) for $\vec{F}$, by a purely spatial rotation (which will only rotate $\vec{E}$ and $\vec{B}$ and thus will not change $\vec{E}\cdot\vec{E}$ and $\vec{B}\cdot\vec{B}$), (iii) holds not only for the basis with respect to which (*) was written, but also for any orthonormal basis. Similarly, we see from (ii) that $\vec{E}\cdot\vec{B}$ does not depend on the orthonormal basis with respect to which the matrix is written.

(b) If $\vec{E}\cdot\vec{B} \neq 0$, then (ii) shows that $\hat{F}^{-1} = -(\vec{E}\cdot\vec{B})^{-1}(\underline{\ast}\hat{F})$. Conversely, if $\vec{E}\cdot\vec{B} = 0$, the second or fourth column of the matrix (*) of Exercise 2 vanishes, so $\hat{F}$ can't be invertible.

Solution 3.5.

(a) The potential 1-form $A^* = (A_i)$ is

$$A^* = \phi\, dt + \sum_{i=1}^{3} A_i dx^i = \phi\, dt - \sum_{i=1}^{3} A^i dx^i .$$

Hence $F = dA^* = \sum_{i=1}^{3}\left[-\frac{\partial\phi}{\partial x_i} - \sum_{i=1}^{3}\frac{\partial A^i}{\partial t}\right] dt \wedge dx^i$

$+ (\frac{\partial A^1}{\partial x_2} - \frac{\partial A^2}{\partial x_1}) dx^1 \wedge dx^2 + (\frac{\partial A^1}{\partial x_3} - \frac{\partial A^3}{\partial x_1}) dx^1 \wedge dx^3 + (\frac{\partial A^2}{\partial x_3} - \frac{\partial A^3}{\partial x_2}) dx^2 \wedge dx^3 .$

Comparing with 3.4(1) gives $E^i = -\frac{\partial\phi}{\partial x_i} - \frac{\partial A^i}{\partial t}$ and $B^i = (\nabla\times\vec{A})^i$.

(b) By exercise 2.24, if $\vec{B}$ satisfies $\nabla\cdot\vec{B} = 0$ and is time-independent, there exists a time-independent vector field $t, \vec{x} \longmapsto \vec{A}(\vec{x})$ such that

$$\nabla\times\vec{A} = \vec{B} .$$

The Maxwell equation

$$\nabla\times\vec{E} = -\frac{\partial\vec{B}}{\partial t} = 0$$

together with Exercise 2.24(a) and the result of Section 2.9 (or classical vector analysis) shows that there exists a time-independent function ϕ such that

$$\vec{E} = -\nabla\phi \; ;$$

equations (i) and (ii) of part (a) now imply that $A := (\phi, \vec{A})$ is a time-independent potential for the given field F.

Solution 3.6.

(a) In the present 3-dimensional language, the equations mentioned above (taking the time-zero hyperplane as S) read:

2.9(25)(i) $\qquad \nabla \times \vec{A} = \vec{B} \qquad$ at time $t = 0$,

2.9(25)(ii) $\qquad \dfrac{\partial \vec{A}}{\partial t} = -\nabla \phi - \vec{E}$, and

2.9(27)(ii) $\qquad \dfrac{\partial \phi}{\partial t} = -\nabla \cdot \vec{A} + \delta(A)$.

The equation 2.9(27)(i) is automatically true. Note that 2.9(27)(ii) is also tautologous if the specification of δA is not required.

(b) If we want $\nabla \cdot \vec{A} = 0$, then we see from 2.9(25)(ii) that ϕ must satisfy

(1) $\qquad \Delta \phi := \nabla \cdot (\nabla \phi) = -\dfrac{\partial}{\partial t}(\nabla \cdot \vec{A}) - \nabla \cdot \vec{E}$

$\qquad \qquad \qquad = -\nabla \cdot \vec{E} = -4\pi \rho_{lab}$,

which is Poisson's equation. Choose ϕ to satisfy (1); for example, we may take [John, p.100]

$$\phi(t, \vec{x}) = \int_{R^3} \dfrac{\rho_{lab}(t, \vec{x}-\vec{y})}{|\vec{y}|} d^3\vec{y} ,$$

and then integrate 2.9(25)(ii) for $\vec{A}$ (this is trivial!) under the initial condition

(2) $\qquad \nabla \cdot \vec{A} = 0 \qquad$ at time $t = 0$.

Then it follows that $\nabla \cdot \vec{A} = 0$ for all times, since

$$\dfrac{\partial}{\partial t}(\nabla \cdot \vec{A}) = \nabla \cdot \left(\dfrac{\partial \vec{A}}{\partial t}\right) = \nabla \cdot (-\nabla \phi - \vec{E}) = 0 .$$

(c) If F is a free field, then this ϕ vanishes identically.

Solution 3.7.

As noted in Section 1.5, any Lorentz transformation L which preserves the time orientation can be written as a product $L = BR$ of a boost B and a spatial rotation R, so it is sufficient to establish invariance just for boosts and spatial rotations, and the latter is immediate.

To see invariance under boosts, let B be a boost, which by rotation invariance may be taken in the x-direction: $B(t, x, y, z) = (t', x', y', z')$ with

$\qquad t' := \gamma(v)(t - vx)$

$\qquad x' := \gamma(v)(x - vt)$

$\qquad y' := y \quad ; \quad z' := z$.

If we transfer B from the cone to R^3 via the above identification J, it becomes the transformation $\tilde{B} := J^{-1}BJ$ defined as follows, where we set $\vec{x} = (x, y, z)$ and $r := |\vec{x}|$:

$$\tilde{B}(x, y, z) = J^{-1}((r-vx)\gamma, (x-vr)\gamma, y, z) = ((x-vr)\gamma, y, z) \ .$$

Let $\tilde{\mu}$ denote the measure $d^3\vec{x}/|\vec{x}| = d^3\vec{x}/r$ on R^3. The invariance of μ under B is the same as the invariance of $\tilde{\mu}$ under $\tilde{B}$. By standard measure-theoretic arguments, $\tilde{\mu}(\tilde{B}(E)) = \tilde{\mu}(E)$ follows for all measurable subsets E of R^3 if it is true for all open subsets. For any such open E, since the Jacobian determinant of $\tilde{B}$ is $\partial/\partial x((x-vr)\gamma) = (1-vx/r)\gamma$, we have

$$\tilde{\mu}(\tilde{B}(E)) := \int_{\tilde{B}(E)} \frac{d^3\vec{x}}{r} = \int_E \frac{1}{r \circ \tilde{B}} \cdot (1 - \frac{vx}{r})\gamma \, d^3\vec{x} \ .$$

Now

$$(r \circ \tilde{B})(x, y, z) = ((x-vr)^2\gamma^2 + y^2 + z^2)^{1/2}$$

$$= (\gamma^2 x^2 - 2\gamma^2 xvr + \gamma^2 v^2(x^2 + y^2 + z^2) + y^2 + z^2)^{1/2}$$

$$= \gamma \cdot (x^2(1 + v^2) - 2xvr + y^2 + z^2)^{1/2}$$

$$= \gamma \cdot (x^2 v^2 - 2xvr + r^2)^{1/2} = \gamma \cdot (r - xv) \ .$$

Hence

$$\tilde{\mu}(\tilde{B}(E)) = \int_E \frac{1}{(r-xv)\gamma} \cdot \gamma \cdot \frac{1}{r} \cdot (r-vx) \, d^3\vec{x}$$

$$= \int_E \frac{d^3\vec{x}}{r} = \tilde{\mu}(E) \ .$$

That is, $\tilde{\mu}$ is invariant under $\tilde{B}$, and so μ is invariant under B.

Solution 3.8.

Since $\delta f = 0$ for 0-forms f,

$\Box f = \delta df$, and using Exercise 2.23,

$$\int_M \Box f = \int_M (\delta df, 1) = -\int_M (df, d(1)) = 0 \ ,$$

since obviously $d(1) = 0$, where 1 denotes the function constantly equal to one.

Solution 3.9.

In standard spherical polar coordinates, for any function g,

$$\Box g = \frac{\partial^2 g}{\partial t^2} - \frac{1}{r^2}\frac{\partial}{\partial r}(r^2 \frac{\partial g}{\partial r}) - \frac{1}{r^2\sin\theta}\frac{\partial}{\partial \theta}(\sin\theta \frac{\partial g}{\partial \theta}) - \frac{1}{r^2\sin^2\theta}\frac{\partial^2 g}{\partial \phi^2} \ .$$

For fixed r, the part involving the angle variables is the Laplacian for the 2-sphere S_r of radius r, and Exercise 8 (or direct integration) shows that its integral over the sphere S_r vanishes. Hence if we integrate the integrand on the right of (5) first over a sphere and define

$$h(t, r) := \frac{1}{4\pi} \int g(t, r, \theta, \phi) \sin\theta \, d\theta d\phi \ ,$$

we have

$$\int_{R^3} \frac{1}{4\pi |\vec{x}|} (\Box g)(|\vec{x}|, \vec{x}) \, d^3\vec{x} = \int_0^\infty [\frac{\partial^2 h}{\partial t^2} - \frac{\partial^2 h}{\partial r^2} - \frac{2}{r}\frac{\partial h}{\partial r}](r, r) \frac{4\pi r^2 dr}{4\pi} \ .$$

Define functions f, p, q of one real variable by

$$f(r) := h(r, r) \qquad p(r) := \frac{\partial h}{\partial t}(r, r) \qquad q(r) := \frac{\partial h}{\partial r}(r, r) \ ,$$

so that

$$\frac{dp}{dr} = \frac{\partial^2 h}{\partial t^2}(r, r) + \frac{\partial^2 h}{\partial t \partial r}(r, r) \ ,$$

$$\frac{dq}{dr} = \frac{\partial^2 h}{\partial t \partial r}(r, r) + \frac{\partial^2 h}{\partial r^2}(r, r) \quad , \text{ and}$$

$$\frac{df}{dr} = \frac{\partial h}{\partial t}(r, r) + \frac{\partial h}{\partial r}(r, r) = p(r) + q(r) \ .$$

Then

$$r[\frac{\partial^2 h}{\partial t^2} - \frac{\partial^2 h}{\partial r^2} - \frac{2}{r}\frac{\partial h}{\partial r}](r, r) = r[\frac{dp}{dr} - \frac{dq}{dr} - \frac{2}{r}q(r)]$$

$$= \frac{d(rp)}{dr} - \frac{d(rq)}{dr} - [p(r) + q(r)] = \frac{d}{dr}[rp(r) - rq(r) - f(r)] \ .$$

Hence

$$\int_0^\infty [\frac{\partial^2 h}{\partial t^2} - \frac{\partial^2 h}{\partial r^2} - \frac{2}{r}\frac{\partial h}{\partial r}] \frac{4\pi r^2 dr}{4\pi}$$

$$= \int_0^\infty \frac{d}{dr}[rp(r) - rq(r) - f(r)] \, dr = f(0) = h(0, 0) = g(0, \vec{0}) \ .$$

"Solution" 3.10.

First note that for a one-dimensional delta-function and any constant $c > 0$,

$$\delta(cx) = \frac{1}{c}\delta(x) \ ,$$

since the substitution $x \longmapsto cx$ may be viewed as shrinking the "graph" of δ in the x-direction by a factor of c. Generalizing, if f is a function with $f(0) = 0$ and $f'(0) \neq 0$, then "near x",

$$\delta(f(x)) = \frac{1}{|f'(0)|}\delta(x) \ ;$$

which can be seen from the substitution $u = f(x)$ in the integral

$$\int_{-\epsilon}^{\epsilon} \delta(f(x))g(x)\,dx = \int_{f(-\epsilon)}^{f(\epsilon)} \delta(u)g(x)\frac{du}{f'(x)} = \frac{g(0)}{|f'(0)|} \ .$$

Generalizing still further, if f is a function with a simple zero at $x = a$ (i.e. $f(a) = 0$ and $f'(a) \neq 0$) and g a function with a simple zero at $b \neq a$, then

$$\delta(f(x)g(x)) = \frac{1}{|f'(a)g(a)|}\delta(x-a) + \frac{1}{|f(b)g'(b)|}\delta(x-b) \ ,$$

assuming that $g(a) \neq 0 \neq f(b)$. Applying this to

$$<x,x> = t^2 - |\vec{x}|^2 = (t-|\vec{x}|)(t+|\vec{x}|)$$

considered as a distribution in t for fixed $\vec{x} \neq 0$, we get

$$\delta(<x,x>) = \frac{\delta(t-|\vec{x}|)}{2|\vec{x}|} + \frac{\delta(t+|\vec{x}|)}{2|\vec{x}|} \ .$$

What about the terrible singularity at $(0,\vec{0})$, where $(t-|\vec{x}|)$ and $(t+|\vec{x}|)$ have a common zero? Don't ask! I have never found a simple way to make direct translations from manipulations using $\delta(<x,x>)$ to rigorous distribution theory proofs. Usually, I end up going back to 3.7(9) and establishing whatever is necessary directly from that. Moreover, the resulting argument invariably seems easier to follow than corresponding delta-function manipulations.

Solution 3.11.

$$\frac{\partial L^{\alpha jk}}{\partial x_\alpha} = \frac{\partial T^{\alpha k}}{\partial x_\alpha}x^j + T^{\alpha k}\frac{\partial x^j}{\partial x_\alpha} - \frac{\partial T^{\alpha j}}{\partial x_\alpha}x^k - T^{\alpha j}\frac{\partial x^k}{\partial x_\alpha}$$

$$= 0 + T^{\alpha k}\delta^j{}_\alpha - 0 - T^{\alpha j}\delta^k{}_\alpha = T^{jk} - T^{kj} \ .$$

Solution 3.12.

Let $\vec{v}$ be the coordinate velocity and $u = (u^0, \vec{u}) = (\gamma(\vec{v}), \vec{v}\gamma(\vec{v}))$ the four-velocity. From inspection of the first component of 3.4(3), we see that

$$0 = \frac{du^0}{d\tau} = \frac{d\gamma}{d\tau} = \frac{d}{d\tau}(1 - |\vec{v}|^2)^{-1/2} ,$$

so the speed $|\vec{v}|$ is constant.

Solution 3.13.

(a) Take the magnetic field to have strength B in the z-direction of 3-space, and write $v = |\vec{v}|$. From the three-dimensional Lorentz equation

3.4(6)　　　$$\frac{d(M\vec{v})}{dt} = q(\vec{E} + \vec{v} \times \vec{B}) = q\vec{v} \times \vec{B}$$

and the fact that $|\vec{v}|$, hence M, is constant, it follows that the path satisfies the differential equation

(1)　　　$$\frac{d\vec{v}}{dt} = \frac{q}{M} \vec{v} \times \vec{B} .$$

A curve $t \longmapsto (x(t), y(t), z(t))$ in three dimensions whose velocity satisfies the differential equation (1) is uniquely determined by its initial position and velocity. The uniform circular motion

(2)　　　$$x(t) = r \cos \frac{vt}{r}$$

$$y(t) = r \sin \frac{vt}{r}$$

$$z(t) = 0$$

is seen by direct substitution to satisfy the equation if r and v are related by

$$r = \frac{Mv}{qB} ,$$

so this uniform circular motion must be the general solution up to a spatial translation and rotation.

(b) If motion in the z-direction is allowed, a solution to (1) with arbitrary initial z-coordinate z_0 and initial z-velocity α is obtained by replacing $z = 0$ in (2) with

$$z(t) = z_0 + \alpha t .$$

(c) Yes, this is a contradiction. If the Lorentz equation holds, a charged particle in a pure magnetic field cannot lose (kinetic) energy. Thus, assuming conservation of energy, it cannot radiate energy. That radiation from particles moving in a pure magnetic field is observed in particle accelerators (Section 5.6) indicates a need to modify the Lorentz equation.

Solution 3.14

(a) That $\bar{u} := dz/d\bar{\tau} = (dz/d\tau)\cdot d\tau/d\bar{\tau} = k^{-1}u$ is immediate. To see that $\bar{u}^* = ku^*$, compute that for any vector v,

$$\bar{u}^*(v) := \bar{g}(\bar{u}, v) = k^2 g(\bar{u}, v) = k^2 k^{-1} g(u, v) = ku^*(v).$$

(b) First we observe that for any 2-form G,

(5) $$\bar{\delta}G = k^{-2}\delta G.$$

This is seen most quickly from the component expression for δ :

$$(\bar{\delta}G)_j := \frac{\partial}{\partial x_\lambda}(\bar{g}^{\lambda\nu}G_{\nu j}) = k^{-2}\frac{\partial}{\partial x_\lambda}(g^{\lambda\nu}G_{\nu j}) = k^{-2}(\delta G)_j \ .$$

(When verifying this in a general spacetime for the next problem, don't forget to use covariant derivatives in place of the partial derivatives above. When considering the case of nonconstant k, the difference is crucial.)

Given this, we obtain from the inhomogeneous Maxwell equation (3),

$$k^{-2}\phi\delta F = \bar{\delta}\bar{F} = 4\pi\bar{\rho}\bar{u}^* = 4\pi\frac{\bar{q}}{q}k^{-3}k\rho u^* \ ,$$

so

$$\phi(k, \frac{\bar{q}}{q}, \frac{\bar{m}}{m}) = \frac{\bar{q}}{q} \ .$$

That is,

(6) $$\bar{F} = \frac{\bar{q}}{q}F \ .$$

Now from the Lorentz equation (2) we obtain

$$\frac{\bar{m}}{m}k^{-2}\cdot m\frac{du^i}{d\tau} = \bar{m}\frac{d\bar{u}^i}{d\bar{\tau}} = \bar{q}\cdot\bar{g}^{i\lambda}\bar{F}_{\lambda\nu}\bar{u}^\nu = \frac{\bar{q}^2}{q}k^{-2}k^{-1}q\, g^{i\lambda}F_{\lambda\nu}u^\nu \ ,$$

so we obtain (4):

$$\frac{\bar{m}}{m} = k^{-1}\left(\frac{\bar{q}}{q}\right)^2 \ .$$

(c) Let q_e denote the charge of the electron in the original system of units and m_e its mass. It is clear from (4) that the choice

$$k := \frac{m_e}{q_e^2}$$

will permit the choices $\bar{m}_e := 1$ and $\bar{q}_e := 1$.

The values of mass and charge for the electron given in the problem are in the Gaussian (and c.g.s.) system in which the velocity of light is 3.0×10^{10} cm/sec and must first be converted to our units in which the velocity of light is unity before the above transformation laws can be applied. This is carried out in detail in the Appendix, and the result (equation (20)) is that in our units with time measured in seconds,

$$k^{-1} = 9.4 \times 10^{-24} \text{ sec} .$$

Thus the new time unit which makes $\bar{q}_e = \bar{m}_e = 1$ is 9.4×10^{-24} sec.

Solution 3.15

(a) We first show that for any 2-form G,

$$\bar{\delta} G = k^{-2} \delta .$$

We cannot use the argument of the solution given for the previous exercise because multiplying the metric by the conformal factor $k(x)^2$ alters the Riemannian covariant derivative. Instead of computing the new covariant derivative, we use the relation

(1) $$\delta = -\underline{\perp} d \underline{\perp}$$

of Exercise 2.20.

Let e_0, e_1, e_2, e_3 be a positively oriented tetrad of vector fields defined in a neighborhood of the point at which we want to evaluate (1) and orthonormal with respect to g. Define a new tetrad

$$(\bar{e}_i)_x := k(x)^{-1}(e_i)_x , \quad x \in M ,$$

which is orthonormal with respect to $\bar{g}$. Note that

$$\bar{e}_i{}^* = k e_i{}^* .$$

Let Ω and $\bar{\Omega}$ denote the volume 4-forms relative to g and $\bar{g}$, respectively. Then

$$\Omega = -e_0{}^* \wedge e_1{}^* \wedge e_2{}^* \wedge e_3{}^*$$

(the minus appears because three of the e_i have negative norm) and

$$\bar{\Omega} = -\bar{e}_0{}^* \wedge \bar{e}_1{}^* \wedge \bar{e}_2{}^* \wedge \bar{e}_3{}^* = k^4 \Omega .$$

Let $\underline{\perp}$ and $\bar{\underline{\perp}}$ denote the Hodge duality operations with respect to Ω and $\bar{\Omega}$, respectively. We first show that for any 2-form $G = (G_{ij})$, $\bar{\underline{\perp}} G = \underline{\perp} G$. (This is *only* true for 2-forms!) The general 2-form is a linear combination of 2-forms of the type $e_i{}^* \wedge e_j{}^*$, so it is enough to establish this for $G = e_i{}^* \wedge e_j{}^*$. For notational simplicity we do the special case $G = e_0{}^* \wedge e_1{}^*$, from which the general case will be apparent. From general principles (cf. Section 2.2), we have

$$\bar{\underline{\perp}}(e_0{}^* \wedge e_1{}^*) = C e_2{}^* \wedge e_3{}^*$$

for some constant C (varying from point to point), and we can read off C by considering the defining relation

$$e_0{}^* \wedge e_1{}^* \wedge \bar{\underline{\perp}}(e_0{}^* \wedge e_1{}^*) = \bar{g}(e_0{}^* \wedge e_1{}^*, e_0{}^* \wedge e_1{}^*) \bar{\Omega}$$

$$= 2 \bar{g}(e_0{}^*, e_0{}^*) \bar{g}(e_1{}^*, e_1{}^*) \bar{\Omega} .$$

Now

$$\bar{g}(e_i{}^*, e_i{}^*) = k^{-2}\bar{g}(\bar{e}_i{}^*, \bar{e}_i{}^*) = k^{-2}g(e_i{}^*, e_i{}^*)$$

by orthonormality, and so

$$e_0{}^* \wedge e_1{}^* \wedge \overline{\perp}(e_0{}^* \wedge e_1{}^*) = 2k^{-2}g(e_0{}^*, e_0{}^*)k^{-2}g(e_1{}^*, e_1{}^*)k^4\Omega$$

$$= 2g(e_0{}^*, e_0{}^*)g(e_1{}^*, e_1{}^*)\Omega = e_0{}^* \wedge e_1{}^* \wedge \perp(e_0{}^* \wedge e_1{}^*) \ .$$

Hence $\overline{\perp}(e_0{}^* \wedge e_1{}^*) = \perp(e_0{}^* \wedge e_1{}^*)$.

From this we see immediately that for any 2-form F,

$$\bar{\delta}F = -\overline{\perp}d\overline{\perp}F = -\overline{\perp}d\perp F \ .$$

Now $d\perp F$ is a 3-form, and a calculation similar to the above shows that for any 3-form G,

$$\overline{\perp}G = k^{-2}\perp G \ ,$$

so for any 2-form F,

(2) $$\bar{\delta}F = k^{-2}\delta F \ .$$

Given (2), the check of the inhomogenous Maxwell equation is routine:

$$\bar{\delta}\bar{F} = k^{-2}\delta\bar{F} = k^{-2}\rho u^* = k^{-3}\rho\bar{u}^*$$

$$= \bar{\rho}\bar{u}^* \ .$$

Equation (2) is noteworthy. It is obvious, *a priori*, that $\overline{\perp}F$ will be a power of k times $\perp F$. In fact, calculation similar to the above shows that for any p-form α,

$$\overline{\perp}\alpha = k^{4-2p}\perp\alpha \ .$$

What is remarkable about the special case of 2-forms is that the power $4-2p$ of k vanishes so that the differentiation in $\perp d\perp$ is unaffected by the replacement of $\overline{\perp}$ by $\perp$. Note also that (2) would not follow if we defined $\bar{F} = k^\beta F$, etc., as in the previous exercise because the differentiation in δ would hit the k^β; only for $\beta = 0$ does this work.

Solution 4.1

After a preliminary spatial translation, we may assume that the backward light cone has vertex the origin and that $\vec{x}_0 := \vec{x}(0) \neq 0$. First we show that the worldline must intersect the cone. Let us say that the particle is *outside* the cone if at any given coordinate time $t < 0$ we have $|\vec{x}(t)| > |t|$, and that it is *inside* if $|\vec{x}(t)| < t$. If there were no intersection of the worldline with the cone, the particle would always be outside (since it is outside at $t = 0$ and if it were ever inside, then the set of times for which the particle is outside and the set for which it is inside would disconnect the time interval $(-\infty, 0]$, an impossibility.) Hence we may assume that

$$|\vec{x}(t)| > |t| \quad \text{for all } t < 0 \ .$$

From this we see that between any given time $t < 0$ and time 0 the particle has travelled a distance of at least

$$|\vec{x}(t) - \vec{x}_0| \geq |\vec{x}(t)| - |\vec{x}_0| > |t| - |\vec{x}_0| \ .$$

Its average speed over the time interval $[t, 0]$ is then at least

$$(1) \qquad \frac{|t| - |\vec{x}_0|}{|t|} = 1 - \frac{|\vec{x}_0|}{|t|} \to 1 \quad \text{as } t \to -\infty .$$

The elementary estimate

$$(2) \qquad \frac{|\vec{x}_0 - \vec{x}(t)|}{|t|} = \frac{1}{|t|} \left| \int_t^0 \vec{v}(s) \, ds \right| \leq \frac{1}{|t|} \int_t^0 |\vec{v}(s)| \, ds$$

shows that its instantaneous speed cannot always be strictly less than its average speed, and thus (1) contradicts the assumption that $|\vec{v}(t)| \leq \alpha < 1$. Hence an intersection must occur. Exercise 1.22, in which it was noted that a particle with nonzero constant proper acceleration has a worldline asymptotic to a light cone, shows that the assumption that $|\vec{v}|$ be bounded away from 1 cannot be omitted.

Now we show that only one intersection can occur. Suppose intersections occurred at two coordinate times $t_1 < t_2 < 0$. After a spatial rotation, we may assume that

$$\vec{x}(t_2) = (|t_2|, 0, 0) \quad \text{and} \quad \vec{x}(t_1) = (q, r, s) \quad \text{with } t_1^2 = q^2 + r^2 + s^2.$$

Then
$$|\vec{x}(t_2) - \vec{x}(t_1)| = \left| \int_{t_1}^{t_2} \vec{v}(t) \, dt \right| \leq (t_2 - t_1)\alpha$$

by (2). On the other hand,

$$(t_2 - t_1)^2 \alpha^2 \geq |\vec{x}(t_2) - \vec{x}(t_1)|^2$$
$$= (|t_2| - q)^2 + r^2 + s^2 = |t_2|^2 - 2|t_2|q + t_1^2$$
$$\geq |t_2|^2 - 2|t_2||t_1| + t_1^2 = (t_2 - t_1)^2 \ ,$$

a contradiction when $\alpha < 1$.

Solution 4.2

The essence of part (c) is that the integral of the terms involving F_{ext} is $O(r_0)$. For a hyperplane cap (which is a three-dimensional ball), the fact that the integral of the terms involving F_{ext} is $O(r_0)$ would be expected from the fact that the volume of such a cap is $O(r_0^3)$, while the singularity of the terms involving F_{ext} is only $O(r_0^{-2})$. Even though a light cone cap is not assigned a nonzero "volume" by the Minkowski metric, there is still an intuitive sense in which it may be considered to have a "volume" which is $O(r_0^3)$. To get the heuristic picture, think of a one-dimensional light cone in two-dimensional Minkowski space. The cutoff cone consists of two finite line segments. The "length" of these segments is not defined in a coordinate-free way (or if it is, it is zero), but relative to any given parametrization $r \mapsto r(u + w)$, $0 < r < r_0$, it may be considered to have "length" r_0. Similarly, in the four-dimensional situation, the distinguished timelike vector $u(\tau)$ at the vertex of the cone in effect provides a parametrization of the cone relative to which it has a nonzero "volume" which is $O(r_0^3)$.

Now we compute (a). We use the notation of Section 4. Recall that S_1^{std} is the "standard" sphere of radius one in Euclidean R^3. Let r_0 and τ be fixed, let $P(\epsilon) := (\epsilon, r_0) \times S_1^{std}$, and define $\psi: P(\epsilon) \times S_1^{std} \longrightarrow \mathbf{M}$ by

$$\psi(r, s) := z(\tau) + r \cdot (u(\tau) + w(\tau, s)) \quad \text{for all } r \in (\epsilon, r_0), \ s \in S_1^{std}.$$

We think of $P(\epsilon)$ as a product Riemannian manifold in the obvious way and denote its volume 3-form by Ω_3. (The choice of orientation will cancel out in the end.) Then $\psi'(\partial/\partial r) = u + w$ and if e_1, e_2 is a basis for the tangent space of S_1^{std} at some point, then by reasoning similar to that around 4.4(13), $\psi'(e_i)$ is orthogonal to u and w and $r_0^{-1}\psi'$ is an isometry when restricted to any tangent space of S_1^{std}. By 2.2(15),

$$(\psi^\dagger)_{r,s}(\perp u^*)(\frac{\partial}{\partial r}, e_1, e_2) = \perp u^*(\psi'(\frac{\partial}{\partial r}), \psi'(e_1), \psi'(e_2))$$

$$= \Omega(u, u+w, \psi'(e_1), \psi'(e_2)) = \Omega(u, w, \psi'(e_1), \psi'(e_2)) = r^2;$$

(since $\psi'(e_i)$ has length r). In other words, with proper choice of orientation on $P(\epsilon)$,

$$(\psi^\dagger)_{r,s}(\perp u^*) = r^2 \Omega_P.$$

Similarly,

$$\psi^\dagger(\perp w^*) = -r^2 \Omega_P,$$

and if f_1, f_2 are orthogonal to u, w, then

$$\psi^\dagger(\perp f_i^*) = \Omega(f_i, u+w, \psi'(e_1), \psi'(e_2)) = 0$$

because f_i is a linear combination of $\psi'(e_1)$ and $\psi'(e_2)$, which span the space orthogonal to u and w. Hence

$$\int_{C_r(\tau, \epsilon)} \frac{1}{r^4} \perp (u \otimes u^*) = \int_{P(\epsilon)} \frac{1}{r^4} u \, \psi^\dagger(\perp u^*)$$

$$= \int_{P(\epsilon)} \frac{1}{r^4} u \, r^2 \Omega_3 = u(\tau) \int_\epsilon^r dr \int_{S_1^{std}} \frac{1}{r^2} d\sigma_1(s)$$

$$= 4\pi u(\tau)(\frac{1}{\epsilon} - \frac{1}{r}).$$

For part (b), just note that the $\perp(w \otimes w^*)$ term in $\perp T$ pulls back to something odd in s and so integrates to zero, and all other terms not involving F_{ext} or $\perp(u \otimes u^*)$ pull back to zero. There are just two terms involving $\perp(u \otimes u^*)$ (one appears directly and other after writing the identity as in (27)), and applying (a) to them gives (b).

For (c), the terms involving F_{ext} may be analyzed as follows:

(i) $\perp(\hat{F}_{ext}(w) \otimes u^*)$ pulls back to a term odd in s and so contributes nothing, and the same is true for

$$\perp(u \otimes \hat{F}_{ext}(w)^*) = \perp[u \otimes (<F_{ext}(w), u>u^* + \text{vector orthogonal to } u \text{ and } w)];$$

(ii) the term $q/(4\pi r^2)\perp(-\hat{F}_{ext}(u) \otimes w^*)$ pulls back to $q/4\pi \hat{F}_{ext}(u)\Omega_P$, so

$$\oint_{C(\tau;\epsilon)} \frac{q}{4\pi r^2} \bot (-\hat{F}_{ext}(u) \otimes w^*) = \int_\epsilon^{r_0} dr \int q\hat{F}_{ext}(u(\tau)) \frac{d\sigma_1(s)}{4\pi} ,$$

$$= (r_0-\epsilon)q\hat{F}_{ext}(u(\tau)) .$$

(iii) Finally, the term

$$-\frac{q}{4\pi r^2} \bot (w \otimes \hat{F}_{ext}(u)^*) = -\frac{1}{4\pi r^2} \bot w \otimes (-<\hat{F}_{ext}(u), w>w^* + \text{vector orthogonal to } u \text{ and } w)^*$$

integrates to $(r_0-\epsilon)q\hat{F}_{ext}(u(\tau))/3$, but this is cancelled by the contribution of the term $q<\hat{F}_{ext}(u), w>I/4\pi$, as is seen by similar reasoning after writing I as in (27).

Solution 4.3

(a) Taking the differential of the relation

$$<x-z_{adv}, x-z_{adv}> = 0$$

yields as in the text

$$d\tau_{adv} = \frac{x^* - z_{adv}^*}{<x-z_{adv}, u_{adv}>} ,$$

but now the analog of 4.2(4) is

(i) $$r_{adv} = <z_{adv}-x, u_{adv}> ,$$

so

$$d\tau_{adv} = \frac{r_{adv}(w_{adv}^* - u_{adv}^*)}{-r_{adv}} = u_{adv}^* - w_{adv}^* .$$

(b) In the following, we drop the subscripts "adv" for brevity. Taking the differential of (i) gives

$$dr_{adv} = <u,u>d\tau - u^* + <z-x, a>d\tau$$
$$= <u,u>d\tau - u^* - r<a,w>d\tau$$
$$= -u^* + (1 - r<a,w>)(u^* - w^*)$$
$$= -r<a,w>u + (r<a,w> - 1)w^* .$$

Solution 4.4

Suppose we reverse the direction of time (i.e. reverse the time orientation on **M**). This makes all worldlines "run backwards", corresponding to the substitutions

$$\tau \longmapsto -\tau \quad , \quad u \longmapsto -u \;.$$

Recall that the retarded potential produced by a worldline with four-velocity u is $A^{ret} := qu_{ret}*/r_{ret}$. Assuming that reversing the direction of time does not change q it is apparent that time reversal changes A^{ret} into $-A^{adv}$ and hence $F^{ret} := dA^{ret}$ into $-F^{adv}$. Thus we expect to get F^{adv} by replacing u by $-u$ (actually, u_{ret} by $-u_{adv}$) throughout the formula 4.2(16) for F^{ret} and taking the negative of the result. The terms containing u are double-negated and do not change, but the term $w^* \wedge a_\perp^*$ changes sign, yielding 4.2(17). A similar analysis holds even if time reversal is considered to change the sign of the charge.

Solution 4.5

(a) The antisymmetry of G is a routine check, from which it follows that for any vector functions v, w,

$$\frac{d}{d\tau}\langle v(\tau), w(\tau)\rangle = \langle D_u v, w\rangle + \langle v, D_u w\rangle$$

$$= \langle G(v), w\rangle + \langle v, G(w)\rangle = 0 \;,$$

so $\langle v(\tau), w(\tau)\rangle = \langle v(0), w(0)\rangle$.

(b) Choose an orthonormal triple e_1, e_2, e_3 of spacelike vectors in $M_{z(0)}$ with each e_i orthogonal to $u(0)$, and Fermi-Walker transport it along the worldline $z(\cdot)$, obtaining for each τ an orthonormal triple $e_1(\tau), e_2(\tau), e_3(\tau)$ which is orthogonal to $u(\tau)$. Given $s = (s_1, s_2, s_3) \in S_r^{std}$, define

$$w(\tau, s) := \frac{1}{r}\sum_{i=1}^{3} s_i e_i(\tau)$$

and

$$\phi(\tau, s) := z(\tau) + r(u(\tau) + w(\tau, s)) \;.$$

Since the $e_i(\tau)$ are orthonormal for fixed τ, the map $w(\tau, \cdot)$ and hence $\phi(\tau, \cdot)$ are isometries.

Solution 4.6

If the field goes from an initial value which produces a four-acceleration a to zero in a time interval Δt, then

$$\frac{da}{d\tau} \approx \frac{-a}{\Delta t} \;,$$

so to make (4) of the same magnitude as a, we expect to have to make

$$\Delta t < (\frac{2}{3}\frac{q^2}{m}) \approx 6\times 10^{-24} \text{ sec.}$$

Solution 4.7

We use the same general procedure and notation (except as noted in the statement of the problem) as in the text. Recall that $v = v_{\tau,s}$ denotes the tangent vector $\partial/\partial\tau$ at the point $\tau, s \in P_\epsilon(\tau_1, \tau_2)$. For any orthonormal basis q_1, q_2 for $(S_\epsilon^{std})_s$ and any $h \in \mathbf{M}$,

(1) $$\psi^\dagger(\perp h^*)(v_{\tau,s}, q_1, q_2) = \Omega(h, \psi'(v), \psi'(q_1), \psi'(q_2)) \ ,$$

so we must compute

$$\psi'(v) = \frac{\partial \psi}{\partial \tau} = \frac{dz}{d\tau} + \epsilon\frac{\partial y}{\partial \tau} = u(\tau) + \epsilon\frac{\partial y}{\partial \tau} \ .$$

To compute $\partial y/\partial \tau$, differentiate the relations

$$<u(\tau), y(\tau, s)> = 0 \quad \text{and} \quad <y(\tau, s), y(\tau, s)> = -1$$

to conclude that

$$<u, \frac{\partial y}{\partial \tau}> = -<\frac{du}{d\tau}, y> = -<a, y> \quad \text{and} \quad <\frac{\partial y}{\partial \tau}, y> = 0 \ ,$$

and hence

(2) $$\psi'(v) = (1 - \epsilon<a, y>)u + \text{vector orthogonal to } u \text{ and } y \ .$$

We may also conclude by reasoning similar to that of the text that $\psi'(q_1)$ and $\psi'(q_2)$ are orthogonal to u and y, which implies that the part of $\psi'(v)$ orthogonal to u and y contributes nothing to (1). Extracting the component of y in h and substituting (2) in (1) shows that

(3) $$\psi^\dagger(\perp h^*)(v, q_1, q_2) = \pm<h, y>(1 - \epsilon<a, y>) \ ,$$

the sign depending on the orientation conventions. Using the convention of the text, in which v, q_1, q_2 is declared a positively oriented basis for $P_\epsilon(\tau_1, \tau_2)$ if and only if $y(\tau, s), u(\tau), \psi'(q_1), \psi'(q_2)$ is a positively oriented basis for $\mathbf{M}$, it follows both that ψ is orientation-preserving and that

$$\psi^\dagger(\perp h^*) = -<h, y>(1 - \epsilon<a, y>)\Omega_P \ .$$

Solution 4.8

(a) Let a proper time τ be fixed. All quantities such as $z(\cdot)$, $u(\cdot)$, $a(\cdot)$, will be assumed evaluated at τ unless otherwise specified. Their values at the retarded time $\tau_{ret} = \tau_{ret}(x) = \tau - \sigma$ relative to a given field point $x \in \mathbf{M}$ will be denoted by the subscript "ret"; e.g. $z_{ret} := z(\tau_{ret}(x))$. We have

$$z(\tau-\sigma) = z(\tau) - \sigma u(\tau) + \frac{\sigma^2}{2}a(\tau) + O(\sigma^3) \ ,$$

and so for $x = z(\tau) + \epsilon y(\tau, s)$,

(1) $$x - z_{ret} = x - z(\tau-\sigma) = \epsilon y(\tau, s) + \sigma u(\tau) - \frac{\sigma^2}{2}a(\tau) + O(\sigma^3) \ .$$

Substituting the right side in $\langle x-z_{ret}, x-z_{ret}\rangle = 0$ gives

(2) $$0 = -\epsilon^2 + \sigma^2(1 - \epsilon\langle y, a\rangle) + (\epsilon + \sigma)O(\sigma^3) \ .$$

It is physically obvious and routine to prove that the proper-time retardation σ goes to zero as the radius ϵ of the tube goes to 0. From this and (2) it follows that

$$\lim_{\epsilon \to 0} \frac{\sigma}{\epsilon} = 1 \ ,$$

which implies that

$$\sigma^2 = \frac{\epsilon^2}{(1-\epsilon\langle y, a\rangle)} + O(\epsilon^4)$$

$$= \epsilon^2 + \epsilon^3\langle y, a\rangle + O(\epsilon^4) \ ,$$

so

(3) $$\sigma = \epsilon(1 + \epsilon\langle y, a\rangle + O(\epsilon^2))^{1/2}$$

$$= \epsilon(1 + \frac{1}{2}\langle y, a\rangle\epsilon) + O(\epsilon^3) \ .$$

Recall from 4.2(3) that

$$r_{ret} := \langle u_{ret}, x-z_{ret}\rangle$$

$$= \langle u-\sigma a+O(\epsilon^2), \epsilon y+\sigma u - \frac{1}{2}\sigma^2 a+O(\sigma^3)\rangle$$

$$= \sigma - \sigma\epsilon\langle y, a\rangle + O(\sigma^3)$$

$$= \epsilon - \frac{1}{2}\epsilon^2\langle a, y\rangle + O(\epsilon^3) \ ,$$

where (3) was used in passing to the last line.

(b) The procedure will be to expand $(u+w)_{ret}$ and $(a_\perp)_{ret}$ as power series in σ. Since integration over S_ϵ^{std} will contribute a factor $4\pi\epsilon^2$, we may drop terms which are $O(\epsilon^3)$. Using (1), expand $u+w$ by writing

$$u_{ret} + w_{ret} = \frac{x-z_{ret}}{r_{ret}}$$

$$= \frac{\epsilon y + \sigma u - \frac{\sigma^2}{2}a + O(\epsilon^3)}{\epsilon(1-\frac{1}{2}\epsilon\langle a, y\rangle+O(\epsilon^2))}$$

$$= \left[y(\tau, s) + u(\tau)(1+\frac{1}{2}\epsilon\langle y, a\rangle) - \frac{1}{2}\epsilon a(\tau) + O(\epsilon^2)\right]\left[1 + \frac{1}{2}\epsilon\langle a, y\rangle + O(\epsilon^2)\right]$$

$$= (1+\frac{1}{2}\epsilon\langle y, a\rangle)y + (1+\epsilon\langle y, a\rangle)u - \frac{1}{2}\epsilon a + O(\epsilon^2) \ .$$

Since $u_{ret} = u(\tau) - \epsilon a(\tau) + O(\epsilon^2)$, we also have

(4) $\qquad w_{ret} = (1 + \frac{1}{2}\epsilon<a,y>)y + \epsilon<a,y>u + \frac{1}{2}\epsilon a + O(\epsilon^2)$,

and of course

$$a_{ret} = a(\tau) - \sigma \frac{da}{d\tau} + O(\sigma^2) = a(\tau) - \epsilon \frac{da}{d\tau} + O(\epsilon^2) .$$

Recalling that $a_\perp = a + <a,w>w$, we may expand

$$\frac{(a_\perp \otimes (u+w)^*)_{ret}}{4\pi r_{ret}^3} = \frac{1}{4\pi\epsilon^3(1-\frac{1}{2}\epsilon<a,y>+O(\epsilon^2))^3}(a + <a,y>y)\otimes(u+y)^*$$

$$+ \frac{1}{4\pi\epsilon^2}\left[-\frac{da}{d\tau} - <\frac{da}{d\tau},y>y\right](u(\tau) + y(\tau,s))^*$$

$$+ \text{ terms involving only } y, a, u + O(\epsilon^{-1}) .$$

The $O(\epsilon^{-1})$ terms get multiplied by $4\pi\epsilon^2$ when integrated over a sphere of radius ϵ and so can be ignored. Any term of the form $a \otimes anything$ or $<a,y>y \otimes anything$ obviously integrates to a multiple of a and so contributes a mass renormalization. This includes all the $O(\epsilon^{-3})$ terms, so the only terms infinite in the limit $\epsilon \to 0$ are mass renormalizations. The terms involving y, a, u are numerous, but they are all of the form considered in Section 4.4 and their integrals are therefore absorbed in the constants K_i. The two terms containing $\frac{da}{d\tau}$ integrate over the sphere to

$$-\frac{2}{3}\int_{\tau_1}^{\tau_2} \frac{da}{d\tau} d\tau + O(\epsilon) .$$

Solution 4.9

(a) This is very easy once one notices that under conditions of constant speed $|\vec{v}|$, both a and $da/d\tau$ must be purely spatial in the lab frame. To see this, recall from 1.7(5) that the four-velocity u is given in lab-frame components by $u = (\gamma(\vec{v}), \vec{v}\gamma(\vec{v}))$, and if $|\vec{v}|$ is constant, then so is $\gamma(\vec{v})$, and hence $a = (0, \gamma d\vec{v}/d\tau)$. The lab-frame energy radiation between proper times τ_1 and τ_2 predicted by (1) is the inner product of

$$-\int_{\tau_1}^{\tau_2} \frac{2}{3}q^2(\frac{da}{d\tau} + <a,a>u) d\tau$$

with $(1, \vec{0})$, and since $<da/d\tau, (1, \vec{0})> = 0$, that is obviously the same as the inner product of 4.3(16) with $(1, \vec{0})$.

(b) Suppress the two extraneous space dimensions from the notation, and write v for the coordinate velocity, so that the four-velocity is $u = (\gamma(v), v\gamma(v))$. Define

$$w := (v\gamma(v), \gamma(v)) ,$$

so that w is a spacelike unit vector orthogonal to u. Now $a := \frac{du}{d\tau}$ is also orthogonal to

u, so a must be a scalar multiple of w, say $a = Aw$. We then have

(2) $$\frac{da}{d\tau} = \frac{dA}{d\tau}w + A\frac{dw}{d\tau} .$$

Since w is a unit vector, $\frac{dw}{d\tau}$ is orthogonal to w and hence is a multiple of u. The component of $\frac{dw}{d\tau}$ on u is

$$<\frac{dw}{d\tau}, u> = -<w, \frac{du}{d\tau}> = -<w, Aw> = A ,$$

which implies that $\frac{dw}{d\tau} = Au$ and

(3) $$\frac{da}{d\tau} = \frac{dA}{d\tau}w + A^2 u .$$

We already know from 4.3(11) that (1) is orthogonal to u, and comparing with (3) shows that $A^2 = -<a, a>$ and

$$\frac{da}{d\tau} + <a, a>u = \frac{dA}{d\tau}w = \frac{dA}{d\tau} \cdot (v\gamma, v) .$$

The lab-frame energy radiation between proper times τ_1 and τ_2 predicted by (1) is then

(4) $$-\frac{2}{3}q^2 \int_{\tau_1}^{\tau_2} \frac{dA}{d\tau} v\gamma(v)\, d\tau ,$$

whereas that predicted by 4.3(16) is

(5) $$-\frac{2}{3}q^2 \int_{\tau_1}^{\tau_2} <a, a>\gamma(v)\, d\tau = \int_{\tau_1}^{\tau_2} A^2 \gamma(v)\, d\tau .$$

It is easy to see that, (4) and (5) are not always the same; for instance, if the proper acceleration is constant then (4) vanishes but (5) does not.

Solution 4.10

(a) Consider, in the notation of Section 4.4, the isometry

$$\psi : S_r^{std} \longrightarrow S_r(0)$$

defined by

$$\psi(s) = z(0) + r(u(0) + w(0, s)) .$$

Let Ω_2 denote the usual area 2-form on S_r^{std} (whose orientation will be specified below) and Ω_4 the volume 4-form on $\mathbf{M}$. Given a vector field v on $\mathbf{M}$, consider the 2-form $\alpha := \perp v^*(u, \cdot, \cdot)$ obtained by contracting the constant vector $u = u(0)$ with $\perp v^*$. From 2.2(15),

$$\alpha(y_1, y_2) := \perp v^*(u(0), y_1, y_2) = \Omega_4(v, u(0), y_1, y_2) .$$

Let us compute the pullback $\psi^\dagger(\alpha)$. We have for any tangent vectors v_1, v_2 at a point $s \in S_r^{std}$,

(7) $$\psi_s^\dagger(\alpha)(v_1, v_2) := \alpha(\psi_s'(v_1), \psi_s'(v_2))$$
$$= \Omega(v, u(0), \psi_s'(v_1), \psi_s'(v_2)) \ .$$

Looking at the right side of (7) as a linear function of v for fixed v_1, v_2, it is apparent that the nullspace of this function is the subspace of **M** spanned by $u(0)$ and tangent vectors to the sphere $S_r(0)$ at the point $\psi(s)$. The vector $w(0, s)$ is orthogonal to this subspace, and the value of (7) with $v := w(0, s)$ is ± 1, depending on the orientation chosen for S_r^{std}. Choosing the orientation so that ψ is orientation-preserving when $S_r(0)$ is given the "natural" orientation induced from the orientation on space sections of **M** (see Section 1.5) and the requirement that $w(0, s)$ be an outward normal to $S_r(0)$ (see Section 2.6), it follows that

(8) $$\psi^\dagger(\bot v^*(u, \cdot, \cdot)) := \psi^\dagger(\alpha) = -\langle v, w \rangle \Omega_2 \ .$$

Of course,

(9) $$\int_{S_r(0)} \bot T(u, \cdot, \cdot) = \int_{S_r^{std}} \psi^\dagger(\bot T(u, \cdot, \cdot)) \ ,$$

and from this and (8), one can immediately read off the integrals of the various terms of the decomposition 4.4(20) of $\bot T(u, \cdot, \cdot)$. The only possibly nonzero terms are those of the form $v \otimes \bot w^*$, which pull back to $v \Omega_2$. The integrals of some of these terms, such as $w \otimes w^*/4\pi r^4$ are immediately seen to vanish by virtue of being odd in the sphere variable. The only remaining terms with nonvanishing integrals are

$$-\frac{e^2}{4\pi r^2} \langle a_\bot, a_\bot \rangle u \otimes w^* \ , \quad \text{which integrates to} \quad -\frac{2}{3} e^2 \langle a, a \rangle u \ ,$$

and

$$\frac{e^2}{4\pi r^3} a_\bot \otimes w^* \ , \quad \text{which integrates to} \quad \frac{2e^2}{3r} a \ ,$$

yielding (1).

(b) Let $\vec{v}$ be the electron's velocity relative to the rest frame, so that the its four-velocity is $u = \gamma(1, \vec{v})$ in this frame. All components below will be with respect to the lab frame. The *proper*-time rate $dE/d\tau$ of radiation of pure energy in the lab frame is the first component of the assumed energy-momentum radiation rate $2e^2/3(-\langle a, a \rangle u + ka)$, which is

$$\frac{dE}{d\tau} = -\frac{2}{3} e^2 \langle a, a \rangle \gamma + \frac{2}{3} e^2 k a^0 \ .$$

The *coordinate*-time energy radiation rate dE/dt, obtained by multiplying this by the conversion factor $d\tau/dt = \gamma^{-1}$ between proper and coordinate time, is

(10) $$\frac{dE}{dt} = -\frac{2}{3} e^2 \langle a, a \rangle + \frac{2}{3} e^2 k a^0 \gamma^{-1} \ .$$

Note that (10) is in general frame-dependent, but is frame-independent when $k = 0$.

(c) Let $\vec{A} := d\vec{v}/dt$ denote the coordinate acceleration relative to the lab frame. From 1.7(9),

$$a = \gamma^4 \cdot (\vec{A} \cdot \vec{v}) \cdot (1, \vec{v}) + \gamma^2 \cdot (0, \vec{A}) \ ,$$

so that

(11) $$<a, a> = -\gamma^6 \cdot (\vec{A} \cdot \vec{v})^2 - \gamma^4 \cdot \vec{A} \cdot \vec{A} \ .$$

For uniform circular motion, $\vec{A}$ is directed toward the center of the circle and is orthogonal to $\vec{v}$, so that

$$a = \gamma^2 \cdot (0, \vec{A})$$

is purely spatial in the *laboratory* frame, as well as the rest frame, and

(12) $$\frac{dE}{dt} = \frac{2}{3} e^2 \gamma^4 A^2 \ ,$$

with $A := |\vec{A}|$. The centripetal acceleration A is given by the usual formula $A = v^2/R$, and (6) results.

Appendix 2

This appendix presents a very simple proof of the result of Section 2.9 for the special case of Minkowski space. I thank R. Seeley for bringing to my attention the basic idea of the proof, which later turned out to be well-known (cf., for instance, [Hestenes and Sobczyk, Section 7-4]). As a bonus, the proof gives a C^∞ version of the result for the case of Euclidean spaces, which were not covered in this generality in Section 2.9. Though the result of Section 2.9 is considerably more general for spacetimes than that to be given below (it applies to curved spacetimes and also shows how to satisfy certain initial conditions), the additional generality is not important for most of the electrodynamical issues treated in the rest of the book.

The manifold on which the differential forms are defined is assumed to be R^n with a constant metric g, which by a change of basis may be taken to be diagonal:

$$g_{ij} = \begin{cases} 0 & i \neq j \\ \pm 1 & i = j \end{cases}.$$

We impose a blanket C^∞ hypothesis on functions, form fields, etc. throughout this section.

Recall from Section 3.7 that the differential operator $\Box$ is defined on p-form fields α by

$$(2) \qquad \Box \alpha := (d\delta + \delta d)\alpha,$$

that it is given in standard coordinates on R^n on functions f by

$$(3) \qquad \Box f := \sum_{i,j=1}^n g^{ij} \frac{\partial^2 f}{\partial x_i \partial x_j} = \sum_{i=1}^n g^{ii} \frac{\partial^2 f}{\partial x_i^2},$$

and that for an arbitrary p-form

$$(4) \qquad \alpha = \sum \alpha_{i_1 \ldots i_p} dx^{i_1} \wedge \cdots \wedge dx^{i_p},$$

with the $\alpha_{i_1 \ldots i_p}$ functions,

$$(5) \qquad \Box \alpha = \sum \Box \alpha_{i_1 \ldots i_p} dx^{i_1} \cdots dx^{i_p}.$$

Thus on Minkowski space, $\Box$ is the usual D'Alembertian

$$\frac{\partial^2}{\partial x_0^2} - \sum_{i=1}^3 \frac{\partial^2}{\partial x_i^2},$$

and on Euclidean R^n it is the Laplacian

$$\sum_{i=1}^n \frac{\partial^2}{\partial x_i^2}.$$

The proof depends on being able to find a right inverse $\Box_R^{-1}$ for $\Box$, i.e.

$$\Box \Box_R^{-1} f = f$$

for all functions f, with the additional property that $\Box_R^{-1}$ commutes with $\partial/\partial x_i$:

$$\frac{\partial}{\partial x_i} \Box_R^{-1} f = \Box_R^{-1} \frac{\partial f}{\partial x_i} \qquad \text{for all } f.$$

Appendix 2

For Minkowski space **M**, such a right inverse is given by the retarded integral 3.7(9):

$$\Box_R^{-1} f(t, \vec{x}) := \int_{R^3} \frac{f(t-|\vec{y}|, \vec{x}+\vec{y})}{|\vec{y}|} d^3\vec{y} \quad,$$

and similar right inverses are known for vector space spacetimes of other dimensions [Treves, p. 106] and for Euclidean R^n ([John, Section 4.1], cf. also the solution to Exercise 3.6). The proof below is purely algebraic given the existence of a right inverse for $\Box$ which commutes with d and δ and therefore applies to any space for which such a right inverse exists.

Extend $\Box_R^{-1}$ to a right inverse for the $\Box$ operation on p-forms in the obvious way: for a p-form α written as in (4),

$$\Box_R^{-1} \alpha = \sum (\Box_R^{-1} \alpha_{i_1 \ldots i_p}) dx^{i_1} \wedge \cdots \wedge dx^{i_p} \quad.$$

Let Λ_q denote the (infinite-dimensional) vector space of all C^∞ q-forms on R^n and

$$Q := \bigoplus_{q=0}^{n} \Lambda_q$$

their direct sum. The operators d, δ, and $\Box$ are extended in the obvious way to Q.† For example, if

$$\Phi = \sum_{q=0}^{n} \phi_q \in Q \qquad \text{with } \phi_q \in \Lambda_q \quad,$$

then

$$d\Phi := \sum_{q=0}^{n-1} d\phi_q \quad.$$

There is a similar obvious componentwise extension of $\Box_R^{-1}$ to a right inverse for $\Box$ on Q,

$$\Box_R^{-1} \Phi := \sum_{q=0}^{n} \Box_R^{-1} \phi_q \quad.$$

This extended $\Box_R^{-1}$ commutes with all partial differential operators $\partial/\partial x_i$, and hence with d and δ.

Now consider the problem of Section 2.9. Given a $(p-1)$-form α with $\delta\alpha = 0$ and a $(p+1)$-form γ with $d\gamma = 0$, we seek a p-form β with

(6) $\qquad d\beta = \gamma \qquad \text{and} \qquad \delta\beta = \alpha \quad.$

Define

(7) $\qquad \Psi := \alpha + \gamma \in Q \quad.$

First consider the easier problem of finding $\Phi \in Q$ with

(8) $\qquad (d+\delta)\Phi = \Psi \quad.$

---

† Since d actually maps n-forms into the zero-dimensional vector space of $n+1$-forms, for this to make strict logical sense one must specially redefine d as the zero linear transformation on Λ_n.

This is easier because if Φ happens to be a p-form (that is, if the q-form components $\phi_q \in \Lambda_q$ of $\Phi = \sum_{q=0}^{n} \phi_q$ vanish for $q \neq p$), then clearly $\beta := \phi_p$ is a solution of (6). Put slightly differently, a solution of (6) yields a solution of (8), but not obviously conversely. Since

(9) $$(d + \delta)^2 = d^2 + d\delta + \delta d + \delta^2 = d\delta + \delta d = \Box ,$$

for any $\Psi \in Q$,

$$\Phi := (d + \delta)\Box_R^{-1}\Psi$$

is a solution of (8). Moreover, for Ψ of the special form $\Psi = \alpha + \gamma$ with $\delta\alpha = 0 = d\gamma$, $(d + \delta)\Psi$ is a p-form, and hence

(10) $$\Phi := (d + \delta)\Box_R^{-1}\Psi = \Box_R^{-1}(d + \delta)\Psi$$

is also a p-form. In other words, (10) gives a $\Phi \in Q$ which solves (8) and is also a p-form, so that Φ solves (6) as well, and this completes the proof.

We remark in closing that the above proof can be expressed very elegantly and naturally in the language of Clifford algebras. For the case of Minkowski space $\mathbf{M}$, the associated Clifford algebra may be thought of as an algebra with four generators $\gamma_0, \gamma^1, \gamma_2, \gamma_3$ and relations

(11) $$\gamma_i\gamma_j + \gamma_j\gamma_i = 2g_{ij} \qquad 0 \leq i, j \leq 3 ,$$

so that γ_i and γ_j anticommute for $i \neq j$, and $\gamma_0^2 = -\gamma_1^2 = -\gamma_2^2 = -\gamma_3^2 = 1$. † Any given basis e_0, e_1, e_2, e_3 for $\mathbf{M}$ induces a vector space isomorphism from the Clifford algebra to the vector space Q defined above via the identification of a p-form

$$e_{i_1}{}^* \wedge e_{i_2}{}^* \cdots \wedge e_{i_p}{}^* \quad , \qquad i_1 < i_2 < \cdots < i_p \quad .$$

with the algebra element

$$\gamma_{i_1}\gamma_{i_2}\cdots\gamma_{i_p} ,$$

and under this isomorphism, the operator $d + \delta$ identifies with the Dirac differential operator (operating on Clifford-algebra valued functions)

$$\displaystyle{\not{\partial}} := \sum_{j=0}^{3} \gamma_j \frac{\partial}{\partial x_j} ,$$

which is a square root of the D'Alembertian operator: $\not{\partial}^2 = \Box$. Under these identifications, our problem is to solve

$$\not{\partial}\Phi = \Psi$$

for Ψ of the special form (7) under the side condition that Φ be a pure p-form, and a solution is

(12) $$\Phi := \not{\partial}\Box_R^{-1}\Psi .$$

---

† The algebra may be concretely realized as the algebra of 4×4 matrices generated by the Dirac matrices, which satisfy the relation (11).

Physicists may recognize (12) more readily in formally Fourier-transformed form, where $\Box$ becomes multiplication by the function $p \longmapsto <p,p>$, $p \in \mathbf{M}$, which has the formal inverse of multiplication by $p \longmapsto <p,p>^{-1}$. Then, denoting Fourier transformation by a tilde, we may write a formal (i.e. ignoring analytical questions of existence of Fourier transforms, etc.) version of (12) as:

$$\tilde{\Phi}(p) := \frac{\slashed{p}}{<p,p>}\tilde{\Psi}(p) := \frac{1}{<p,p>}\sum_{j=0}^{3}\gamma_j\, p^j\, \tilde{\Psi}(p) \quad .$$

Bibliography

Abraham, R., Marsden, J., and Ratiu, T., *Manifolds, Tensor Analysis, and Applications*, Addison-Wesley, Reading, MA, 1983

Adler, R., Bazin, M., and Schiffer, M., *Introduction to General Relativity*, McGraw-Hill, 1975

Baldwin, B. (Letter), Physics Today, January, 1975, 9-10

Barut, A. O., *Electrodynamics and Classical Theory of Fields and Particles*, Dover, N.Y., 1980

Bhabha, H. J. Classical theory of mesons, Proc. Roy. Soc. London **A172** (1939), 384-409

Bhabha, H. J., Classical theory of electrons, Proc. Indian Acad. Sci. **10**(1939), 324-332

Bhabha, H. J., and Corben, H. C., General classical theory of spinning particles in a Maxwell field, Proc. Royal Soc. London **A178**(1941), 273-314

Bhabha, H. J., and Harish-Chandra, On the theory of point particles, Proc. Royal Soc. London **A183** (1944), 137-141

Blewett, J., Radiation losses in the induction electron accelerator, Phys. Rev. **69**(1946), 87-95

Bonnor, W. B., A new equation of motion for a radiating charged particle, Proc. Royal Soc. London **A337** (1974), 591-598

Boyer, T., Random electrodynamics: the theory of classical electrodynamics with classical electromagnetic zero-point radiation, Phys. Rev. **D 11**(4) (1975), 790-808

Choquet-Bruhat, Y., Dewitt-Morette, C., with Dillard-Bleick, M., *Analysis, Manifolds, and Physics*, North-Holland, Amsterdam, 1982

Cohn, J., Hyperbolic motion and radiation, Amer. J. Phys. **46**(3)(1978), 225-227

Corson, D., Radiation by electrons in large orbits, Phys. Rev. **90**(5)(1953), 748-752

De Groot, S. R., and Suttorp, L. G., *Foundations of Electrodynamics*, North-Holland, Amsterdam, 1973

Dirac, P. A. M., Classical theories of radiating electrons, Proc. Royal Soc. London **A167** (1938), 148

Dirac, P. A. M., A new classical theory of electrons, Proc. Royal Soc. London **A209**(1951), 291-296

Dirac, P. A. M., A new classical theory of electrons II, Proc. Royal Soc. London **A212**(1952), 330-339

Dirac, P. A. M., A new classical theory of electrons III, Proc. Royal Soc. London **A223**(1954), 438-445

Dixon, W. G., *Special Relativity*, Cambridge University Press, Cambridge, England, 1978

Dodson, C. T. J., and Poston, T., *Tensor Geometry*, Pitman, London, 1977

Driver, R. and Hsing, D.P., in *Dynamical Systems* Proc. of a University of Florida National Symposium, Academic, N.Y., 1977, 427-430

Driver, R. and Hsing, D.P., Technical Report No. 61, University of Rhode Island, 1975

Eliezer, C. J., The hydrogen atom and the classical theory of radiation, Proc. Camb. Phil. Soc. **39/** (1943), 173

Eliezer, C. J., Radiating electron in a magnetic field, Proc. Camb. Phil. Soc. **42**(1946), 40-44

Eliezer, C. J., The classical equations of motion of an electron, Proc. Camb. Phil. Soc. **42**(1946), 278-286

Eliezer, C. J., The interaction of electrons and an electromagnetic field, Rev. Mod. Phys. **19**(1947), 147-184

Eliezer, C. J., On the classical theory of particles, Proc. Royal Soc. London **A194** (1948), 543-555

Erber, T., The classical theories of radiation reaction, Fortschritte der Physik **9**(1961), 343-392

Erber, T. Some external field problems in QED, *in Concepts in Hadron Physics* (Urban, P., editor), Springer-Verlag, New York, 1971

Feynman, R., Leighton, R., and Sands, M., *The Feynman Lectures in Physics*, Addison-Wesley, Reading, MA 1964

Flanders, H., *Differential Forms with Applications to the Physical Sciences*, Academic Press, N.Y., 1963

Frankel, T., Maxwell's equations, Amer. Math. Monthly, 1974, 343-349

French, A. P., *Special Relativity*, W. W. Norton & Co., New York, 1968

Friedlander, F., *The Wave Equation in Curved Space-Time*, Cambridge University Press, Cambridge, 1975

Godwin, R. P., *Sychrotron Radiation as a Light Source*, Springer-Verlag, Berlin, 1969

Hermann, R., *Energy-Momentum Tensors*, Math. Sci. Press, Brookline, MA, 1973

Herrera, J. C., Equation of motion in classical electrodynamics, Phys. Rev. D **15** (2) (1977), 453-456

Herrera, J. C., Equation of motion in classical electrodynamics II, Phys. Rev. D **21** (2) (1980), 384-387

Hestenes, D., and Sobczyk, G., *Clifford Algebra to Geometric Calculus*, Reidel, Amsterdam, 1984

Hicks, N. J. , *Notes on Differential Geometry*, Van Nostrand, Princeton, N.J., 1965

Hogan, P.A., The physical consequences of the Huggins term in classical electrodynamics, Il Nuovo Cimento **16B**(1973), 251-263

Hsing, D. K., and Driver, R.D., Radiation reaction in the two-body problem of classical electrodynamics, Technical Report No. 61, University of Rhode Island, October, 1975

Hsing, D. K., and Driver, R.D., Radiation reaction in electrodynamics, in *Dynamical Systems*, Proceedings of a University of Florida International Symposium, Academic Press, New York, 1977

Huschilt, J. and Baylis, W. E., Solutions to the "new" equation of motion for classical charged particles, Phys. Rev. D **9** (8)(1974), 2479

Huschilt, J. and Baylis, W. E, Numerical solutions to two-body problems in classical electrodynamics: Headon collisions with retarded fields and radiation reaction II, Phys. Rev. D **13** (1976), 3256-3268

Huschilt, J., and Baylis, W. E., Rutherford scattering with radiation reaction, Phys. Rev. D **17**(4)(1978), 985-993

Jackson, J. D., *Classical Electrodynamics*, John Wiley, N.Y., 1975

John, F., *Partial Differential Equations*, Springer-Verlag, New York, N.Y., 1982

Kasher, J. C., One-dimensional central force problem including radiation reaction, Phys. Rev. D **14**(1976), 939-944

Kerrighan, B., Arbitrary tensor concomitants of a bivector and a metric in a spacetime manifold, Gen. Relativ. Gravit. **13** (1981)

Kerrighan, D. B., On the uniqueness of the energy-momentum tensor for electromagnetism, J. Math. Phys. **23** (10) (1982), 1979-1980

Klemperer, O., *Electron Physics*, Butterworth, London, 1972

Kunz, C., Ed., *Synchrotron Radiation*, Springer-Verlag, N.Y., 1979

Landau, L. D., and Lifshitz, E. M., *The Classical Theory of Fields*, Addison-Wesley, Reading, MA, 1970

Landau, L. D., and Lifshitz, E. M., *Electrodynamics of Continuous Media*, Addison-Wesley, Reading, MA, 1960

Lang, S., *Algebra*, Addison-Wesley, Reading, Mass, 1965

Lang, S., *Real Analysis*, Addison-Wesley, Reading, Mass., 1962

Marion, J. B. , *Classical Electromagnetic Radiation*, Academic Press, N.Y., 1965

Lightman, A., Press, W., Price, R., and Teukolsky, S., *Problem Book in Relativity and Gravitation*, Princeton University Press, Princeton, N.J., 1975

Meystre, P., and Scully, M. (Ed.), *Quantum Optics, Experimental Gravity, and Measurement Theory*, Plenum Press, N.Y., 1983

Misner, C., Thorne, K., and Wheeler, J., *Gravitation*, W. H. Freeman, San Francisco, 1973

Mo, T. C., and Papas, C. H., A new equation of motion for classical charged particles, Phys. Rev. D **4** (1971), 3566-71

Moniz, E. J., and Sharp, D. H., Radiation reaction in nonrelativistic quantum electrodynamics, Phys. Rev. D **4**(1977), 2850-2865

Nelson, E., *Quantum Fluctuations*, Princeton University Press, Princeton, N.J., 1985

Panofsky, W., and Philips, M., *Classical Electricity and Magnetism*, Addison-Wesley, Reading, MA, 1961

Pollock, H., The discovery of synchrotron radiation, Am. J. Phys. **51**(3)(1983), 278-280

Pryce, M. H. L., The electromagnetic field of a point charge, Proc. Royal Soc. London **A168/** (1938), 389

Rindler, W., *Introduction to Special Relativity*, Clarendon Press, Oxford, 1982

Rohrlich, F., *Classical Charged Particles*, Addison-Wesley, Reading, MA, 1965

Rowe, E. G. P., Structure of the energy tensor in the classical electrodyamics of point particles, Phys. Rev. D **18**(10)(1978), 3639-3654

Rowe, E., and Weaver, J., Syncrotron radiation, Physics Teacher **15**(1977), 268-274

Sachs, R. K., and Wu, H., *General Relativity for Mathematicians*, Springer-Verlag, New York, N.Y., 1977

Schott, G. A., *Electromagnetic Radiation*, Cambridge University Press, Cambridge, 1912

Schwinger, J., On the classical radiation of accelerated electrons, Phys. Rev. **75**(12) (1949), 1912-1925

Shen, C. S., Magnetic *bremstrallung* in an intense magnetic field, Phys. Rev. D **6** (1972), 2736

Sokolov, A. A., and Ternov, J. M., *Syncrotron Radiation*, Pergamon Press, New York, 1968

Sternberg, S., *Lectures in Differential Geometry*, Prentice-Hall, Englewood Cliffs, N.J., 1964

Synge, J. L., Point-particles and energy tensors in special relativity, Ann. Mat. Pura e App. **84** (1970), 33-59

Synge, J. L., *Relativity: The General Theory*, North-Holland Publishing Co., Amsterdam, 1969

Synge, J. L., Relativity: The Special Theory, 2nd Ed., North-Holland Publishing Co., Amsterdam, 1965

Tabensky, R., Electrodynamics and the electron equation of motion, Phys. Rev. D **13**(2)(1976), 267-273

Taylor, E. F., and Wheeler, J. A., *Spacetime Physics*, W. H. Freeman, San Francisco, 1966

Teitelboim, C., Splitting of the Maxwell tensor: Radiation reaction without advanced fields, Phys. Rev. **1** (6)(1970), 1571-1582. (Also erratum, Phys. Rev. **2** (8) (1970), 1763)

Teitelboim, C., Villaroel, D., and Van Weert, Ch. G., Classical electrodynamics of retarded fields and point particles, Riv. Nuovo Cimento **3**(1980), 1-64

Bibliography

Teplitz, D., *Electromagnetism*, Plenum Press, N.Y., 1983

Thirring, W. E., *A course in Mathematical Physics*, vol 2., Second English edition, Srpinger Verlag, New York, 1986

Travis, S., Existence theorem for a backwards two-body problem of electrodynamics, Phys. Rev. **D 11**(2)(1975), 292-299

Treves, F. *Basic Linear Partial Differential Equations*, Academic Press, N.Y., 1975

Warner, F., *Foundations of Differentiable Manifolds and Lie Groups*, Scott, Foresman, & Co., Glenview, IL, 1971

Wheeler, J. A., and Feynman, R. P., Interaction with the absorber as the mechanism of radiation, Rev. Modern Phys. **17** (1945), 157

Notations

The notation mainly follows standard or commonly used conventions in electromagnetic theory or differential geometry. The chief exception is the use of of $v*$, where $v = v^i$ is a vector in Minkowski space, to denote the corresponding dual vector v_i. The adjoint of a linear transformation L is denoted $L^\dagger$ instead of the more usual L^* to avoid confusion with the star operation just described. The symbol " $:=$ " means "equals by definition", as in "$2x := x + x$ ".

Notations generally accepted in modern mathematics are used without comment; for example, $x \longmapsto x^2$ might be used to denote the function whose value at x is x^2. The convention that repeated upper and lower indices are summed is used throughout except in a few places where the opposite is specified. Thus,

$$v^i \frac{\partial f}{\partial x_i} := \sum_i v^i \frac{\partial f}{\partial x_i} .$$

Readers accustomed to classical tensor notation should keep in mind that we use the differential geometric definition of tangent vectors to a manifold, which identifies them with derivations (directional derivatives) on the ring of C^∞ functions. Thus if v is a tangent vector with components v^i with respect to a coordinate system $\{x^i\}$ and f a C^∞ function, then $v(f)$ denotes $v^i \partial f / \partial x_i$. Consistent with this viewpoint, the differential operator $\partial / \partial x_i$ represents a tangent vector in the x_i-direction.

Within a section, equations are numbered consecutively beginning with (1) and referred to by that number. Equations outside a section are referred to by chapter, section, and equation number in the format c.s(e). For instance, if equation (8) of Section 3 of Chapter 2 is referred to within that section it is called (8), but elsewhere 2.3(8).

For technical reasons, this table of notations had to be printed in a different font than the main text, so some of the symbols will vary slightly in appearance.

Symbol	Usage	Page
C^∞	infinitely continuously differentiable	44
M	arbitrary manifold, usually semi-Riemannian, always C^∞	43
$\mathbf{M}$	Minkowski space	4
$g = g_{ij}$	metric tensor on $\mathbf{M}$ or M, also denoted $<\cdot,\cdot>$	58
$v*$	The linear form dual to the vector v via g. Thus for all vectors w, $v*(w) = <v, w> = v^i g_{ij} w^j = v_i w^i$	33-34
Ω	the volume form on a manifold (the Levi-Civita tensor)	40
$\perp \alpha$	the Hodge dual of the alternating form α	40
$\alpha \otimes \beta$	the tensor product of two tensors α, β	35
$\alpha \wedge \beta$	the alternating (wedge) product of two alternating tensors α, β	38

V_q^p	the space of tensors $\alpha^{i_1 \cdots i_p}{}_{j_1 \cdots j_q}$ of contravariant rank p and covariant rank q on a vector space V	35
$\Lambda_p(V)$	the space of (alternating) p-forms on V	38
$v(f)$ or $v[f]$	where v is a vector and f a function denotes the directional derivative of f in the direction of v: $v(f) = v[f] = v^i \partial f / \partial x_i$. (See third paragraph above.)	46, 214
$\partial/\partial x_i$	either the usual partial differential operator or the corresponding tangent vector in the x_i-direction (see third paragraph above).	46
D	covariant derivative on M, usually induced by g;	48, 58
D_v (object field)	the covariant derivative of the object field (e.g. vector field, form field) in the direction of the vector v. Covariant differentiation is also sometimes denoted by a stroke, for example $v^i{}_{|j} := (D_{\partial/\partial x_j} v)^i$.	48
d	the differential operator on form fields	51
δ	the covariant divergence operator	59, 60
$\Box$	$\Box := \delta d + d \delta$; on Minkowski space this is the usual D'Alembertian (wave) operator $\partial/\partial t^2 - \sum_{i=1}^{3} \partial/\partial x_i^2$	102
τ	proper time on a particle's worldline $\tau \longmapsto z(\tau)$	16
$u = u(\tau)$	four-velocity, $u := dz/d\tau$	16
a	proper acceleration (four-acceleration) $a := du/d\tau$	17
$F = F_{ij}$	electromagnetic field tensor	97
$\hat{F} = F^i{}_j$	the field tensor F with an index raised, usually considered as a linear transformation: $\hat{F}^i{}_j = g^{ik} F_{kj}$, and similarly for other two-component tensors.	96
A	usually the four-potential: $dA = F$; also used for scalar proper acceleration in Chapter 5	102, 195
$T = T^{ij}$	energy-momentum tensor	110, 104
A_{ret}, A^{ret} A_{adv}, A^{adv}	respectively, the contravariant and covariant retarded potentials and the corresponding advanced potentials.	103, 128
$\tau_{ret} = \tau_{ret}(x)$	The "retarded (proper) time" at which a light signal would have to be emitted by a particle at a point $z_{ret} = z(\tau_{ret})$ on its worldline in order to reach the field point $x = (x^0, x^1, x^2, x^3)$	128

$r_{ret} = r_{ret}(x)$	The "retarded distance"; i.e. the spatial distance from $z(\tau_{ret})$ to x as measured in the particle's rest frame at $z(\tau_{ret})$	129-130
u_{ret}, a_{ret}	$u_{ret} := u(\tau_{ret})$; $a_{ret} := a(\tau_{ret})$	129, 131
$S_r(\tau_1, \tau_2)$	a "Bhabha tube" of "radius" r surrounding a particle's worldline between proper times τ_1 and τ_2	150
ϕ'	the derivative of a map $\phi: M \longrightarrow N$ between manifolds	145
$\phi^\dagger(\alpha)$	the "pullback" under ϕ of a form field α on N to a form field on M ; in essence, $\phi^\dagger$ is the adjoint of ϕ'	146
∇ or $\vec{\nabla}$	the usual operator of vector calculus: $\nabla = \vec{\nabla} := (\partial/dx, \partial/dy, \partial/dz)$	

Index

acceleration
  coordinate 18,
  proper 17, 18,
  relation between coordinate
    and proper 18
    and scalar proper 179
  scalar proper **179**, 195
addition of velocities formula 26, 249
adjoint
  of a linear transformation 36
  of differential 92
advanced
  distance, formula for 164
  field
    definition of 130
    formula for 134
  potential
    for a charged fluid 103
    for a point particle 130
  time, differential of 164
alternate fluid model 190 ff.
angular momentum, conservation of 123
anti-isometry 151
antisymmetrization 38
antisymmetry
  of differential forms 38
  of electromagnetic field tensor 97
basis
  dual 32
  formula for change of 86
  orthonormal 11
Bhabha and Harish-Chandra tensor 220
Bhabha tube **150**, 163
Bonnor equation 211-212, 228
boost 6, 9
bremsstrahlung **210**, 222
C^∞
  function 44
  tensor field 46, 47
caps 136
Cauchy problem for Maxwell-Lorentz
  system 172-175
causality, violation of 143
change of basis, formulas for 87
charge current
  for charged fluid
    classical 100
    relativistic 101
  for point particle 126
charge

conservation *see* conservation of charge
  definition of 94
  of electron 242
  units of 94, 124-125, **239-240**, 242
charged dust *see* charged fluid
charged fluid **64, 95, 114-115**, 189-190
Christoffel symbols 49
clock paradox 8, 27
clock, standard 6 ff.
collisions, conservation of energy-momentum
  in 20-22
commutator of vector fields 47
Compton wavelength **31**, 203
cone, light *see* light cone
conformal transformation 125
connection 48
  coefficients 49
  Riemannian 58
conservation
  of angular momentum 123
  of charge **64-66**, 174
  of energy-momentum of charged fluid
    and electromagnetic field 115 ff.
  of energy-momentum of uncharged
    particles 20-22
  of mass 184
constitutive relations 100
constraint equations
  for Maxwell-Lorentz system 174
  for specification of differential and
    codifferential 79
continuity equation 174
contraction
  length 7
  of a form with a vector 74, 75
  of tensor indices 36
coordinates
  local 1, **44**
  in which Γ^i_{jk} vanish 49
coordinatization **1-3**, 10, 43-44
  local 1, 44
  Lorentz 4
Coulomb
  field 134
  gauge 121
  law 178, 239
covariant derivative
  of a form field **49**, 89
    along a curve 130
  of metric tensor 90

of a vector field **48** ff.
    along a curve 96, 97
of a general tensor field **50, 51**, 90
covariant divergence
    of contravariant tensors **59**, 90
    of differential forms 60-61, 90
    sign in $\delta(\alpha) = \pm_\perp d_\perp \alpha$ 90
curl 57, **92**, 192
current vector see charge current
D'Alembertian operator **102**, 121-122
decomposable
    q-form **38**, 87-88
    tensor 35, 87
derivative
    covariant see covariant derivative
    of a map between manifolds 145
determinant
    abstract definition of 40
    inner product of forms in terms of 87
    of Lorentz transformation 9, 89
    of isometry 89
    of metric tensor 87
differential
    of a form field **51**, 52
    of a function **47**
    of a one-form 191
differentiation
    covariant see covariant derivative
dilation, time 7, 8
Dirac tube **150**
    integration over 167-168
distance 1
    units of **4**, 239-240
distribution 121-122, 126
divergence
    covariant see covariant divergence
    of alternate energy-momentum
        tensors 215-216
    of vector field on R^3 56-57, 92
    theorem 61-63
Doppler shift 27
doubly renormalized mass 140
dual
    basis 32
        formula for change of 86
    Hodge see Hodge dual
    of field tensor 119
    space 32
E = set of all events 1
Einstein summation convention 33
electric field **98**
electromagnetic field tensor
    components in classical notation 98
    definition 97

Eliezer equation 211-212, 225-228
energy flux vector **223** ff., see also Poynting
    vector
    for T_{L-D} 226
        calculation of 229-237
    for T_{pryce} 227
energy-momentum tensor **104** ff.
    for electromagnetic field
        alternate 219 ff.
        Pryce 220, 227
        usual **115-118**
            components of 117
            uniqueness of 204 ff.
    for a fluid 112-113
    for retarded fields expanded 155
    integration of
        over Bhabha and Dirac tubes
            compared 160
        over Dirac tube 167-168
        over light cones 163
energy
    kinetic 22 ff.
    potential 24
    -momentum 19
    relativistic 20, 23
    radiation see radiation, energy, also
        Larmor radiation law
event 1
    simultaneous _s 12-14
far-field approximation 224
Fermi-Walker transport 164-**165**
field point 128
field tensor **97**
    components in classical notation 98
    dual of 119
    matrix of **98**, 119-120
fields
    Coulomb 134
    electric and magnetic 98
    radiation 134
    retarded or advanced see
        retarded or advanced fields
fluid 112-115
    charged **64, 95, 114-115**, 189-190
force
    Lorentz see Lorentz force
    Newtonian 28, 238-239
form
    alternating 38
        inner product of _s **39**, 87
    differential 47
        integration of 54 ff.
    field 47
    linear 32 ff.

Index

four-acceleration 16, 17
four-momentum 19
four-velocity 16, 17
frame
  proper 17
  rest 17
function, C^∞ 44
gauge
  Coulomb 121
  Lorentz 102
  transformation 102
Gauss divergence theorem 61-63
Gaussian coordinates 70, 71
Gaussian units 238 ff.
geodesic
  coordinate system 90
  equation 115
global solutions of Maxwell-Lorentz equations 187-190
Green function for D'Alembertian 121-122
Herrera equation 211-212, 222, 225
Hodge double dual $\perp\perp\alpha = \pm\alpha$, formula for sign 41, **89**
Hodge dual 39 ff.
  component formula for 42
  of field tensor 119
index
  conventions 32-37
  lowering 33
  raising **34**, 86
induced orientation of a submanifold 56
inner product
  Minkowski 4
  of alternating forms **39**, 87
  of form fields 92
  of tensors 39
  space 32
integral curves 47
integration
  of vector field 46, 47
  of differential forms 54 ff.
isometry, determinant of 9, 89
isomorphisms, natural 36-37
kinetic energy **22-23**, 188
Lagrangian 206
Larmor radiation law 135
  generalized 170, **208**
length
  contraction 7
  units of **4**, 239-240
Lie derivative 72, 193
Liénard-Wiechert potential 130
light
  cone 5, 12-14

  Lorentz-invariant measure on 103, **121**
  ray or pulse 3-5
  velocity of 3, 4
    in cm/sec 239
local cordinatization 1, 43
Lorentz-Dirac equation **141**, 211-212, 222
  alternate energy-momentum tensor for 226
  in Gaussian units 241
  runaway solutions of 196

Lorentz
  boost 7, 9, 29
  coordinatization or frame **4**, 6 ff.
  force **96, 98-99**, 225
    term in energy-momentum tensor 159-160
  gauge 102
  metric 4
  rotation 6, 29
  transformation 6 ff.
Lorentzian manifold 58
lowering an index 33 ff.
$\mathbf{M}$ = Minkowski space 4
magnetic field **98**
  motion of particle in constant 123-124, 168-169
manifold 43 ff.
  Lorentzian 58
  Riemannian or semi-Riemannian 58
mass renormalization 127, **139-140**, 169
  for cone caps 163
mass
  conservation of 184
  of electron 242
  of proton 225
  relativistic 20
  renormalization *see* mass renormalization
  renormalized 139-140
  rest 19
  units of 19, 124-125, 239-240, 242
matrix of a tensor 35
Maxwell's equations **98-100**
  free-space 101
  Gaussian 242-244
Maxwell-Lorentz equations **101**
  Cauchy problem for **172-175**
  for a point particle 126 ff.
  global solutions for 187-190
  spherically symmetric solutions of **176-186**, 187-190
  two(1+1)-dimensional 181-183
metric tensor 58

Minkowski
  inner product (metric) 4
  space **4-5**
    3-dimensional subspaces of 25, 105
Mo-Papas equation 211-212, 222, 225
momentum
  conservation of **18** ff., 26, 115 ff.
  energy-_ (same as four-_ and relativistic _) 19
  Newtonian (three-momentum) 18 ff.
  radiation 169-171, 207-208
natural
  identification of vector space with tangent spaces 45
  isomorphisms 36-37
Newton's Third Law 238-239
non-degenerate inner product 4
normal, outward **55**, 60, 61
null
  relatively 12
  vector 5
orientation 10, 11
orientation (induced) of a submanifold 56
oriented
  basis
    for tangent space 53
    for Minkowski space 10
  manifold 53
orthogonal transformation, determinant of 4, 89
orthonormal basis 11, 24
outward normal vector **55**, 62, 63
photon 5, 29-31
Poincaré Lemma 73, 74
point electrodynamics 126
positivity of local energy density 221
potential
  classical vector or scalar _ 120
  energy **23**, 188
  Liénard-Wiechert 130
  one-form **102** ff., 120
  retarded or advanced
    for a charged fluid 103
    for a point particle 128-*130*
  scalar and vector 120
Poynting vector **117-118**, 135, 212, 223-224
  for retarded field **135**, 223
  radiation _ **135**, 224
  unphysical 210-211
preacceleration 143
pressure 113
proper
  acceleration 17
  acceleration, scalar 29, **179**, 195

frame 4
time 16
Pryce tensor 220
  angular dependence of 227
pullback of a form field 146-147
radiation reaction 142
  term in Lorentz-Dirac equation, size of 166
radiation
  Coulomb 134
  energy 169-171, 207-211, see also Larmor radiation law
    in lab frame 168
  fields 134
  law see Larmor radiation law
  momentum 169, 207-208
  reaction 142
  synchrotron 208-210
  term in energy-momentum tensor 158, 160
raising an index **34** ff., 86
relatively
  null 12
  spacelike 12
  timelike 12
relativistic
  energy, mass, and mass-energy (all the same) 20, 23
  momentum 19, 115 ff.
renormalization see mass renormalization
renormalized mass **139-140**
retardation condition for Maxwell-Lorentz equations 162, 185, 189-190
retarded
  distance r_{ret} **129-130**, 198, 214
    differential of 133
  field
    definition of 130
    formula for 133
  point 128
  potential
    for a charged fluid 103
    for a point particle 128-**130**
  sphere 150
  time τ_{ret} 128
    differential of 132
Riemannian
  connection 58
  manifold 58
rotation
  spatial 6
  of a fluid 192
runaway solutions of Lorentz-Dirac equation 197
scalar potential 120

Index

scalar proper acceleration 29, **179**, 195
semi-Riemannian manifold 58
shell-crossing 187
simultaneous events 12-14
spacelike
  relatively 12
  subspaces of spacetime 1 ff.
spatial rotation 6
sphere
  laboratory 169
  retarded 150
  zero-acceleration 180, 184
spherically symmetric solutions of Maxwell-Lorentz equations **176-186**, 187-190
Stokes' Theorem 55
subspaces of Minkowski space 25
summation convention 33
symmetry
  of energy-momentum tensors 110, **122-123**
syncrotron radiation 208-210
T_{EZR} 226
$T_{EZR\,2}$ 227-228
T_{L-D} 226
T_{miz} 223 ff.
tachyons 14, 28
tangent
  vector 45
  space 45
temporal order of light signals 200
tension 114
tensor 34 ff.
  associated array or matrix 35
  contravariant or covariant 34, 35
  decomposable 35
time 1
  dilation 7, 8
  proper 16
  units of 124-125, **239-240**
timelike
  relatively 12
  vector 5, 10
trace of a linear transformation 36
transformation
  Lorentz 6 ff.
  orthogonal **9**, 89
tube 136
  Bhabha or Dirac *see* Bhabha or Dirac tube
twin paradox 27
unit vector 11, 24

units
  in electromagnetic theory **238-244**
  length 4
  of time, mass, and charge,
    relation between 124-125
vector
  backward or forward timelike 11
  field 46
    along a curve 96-97
  null, timelike, spacelike 5
  potential 120
  tangent
  unit 11, 24
volume form **40**, 87
wedge product
  of alternating forms 38
  of 1-forms, formula for 89
zero-acceleration sphere 180, 184

Errata

Negative line numbers are counted from the bottom. Numbers in parentheses are equation numbers for the relevant section rather than line numbers. Lines counted forwards or backwards from equation (E) are denoted $(E)+n$ or $(E)-n$, where the positive integer n is the count; e.g. (19)–1 is the line just before equation (19) and (19)+2 the second line after. The statement "$a \succ\!\!\longrightarrow b$" means "replace phrase a by phrase b".

Page	Line	Error
vi	7	two oppositely charged point particles $\succ\!\!\longrightarrow$ two equally massive but oppositely charged point particles
24	14	Synge $\succ\!\!\longrightarrow$ [Synge, 1965]
59	-2	indiviual $\succ\!\!\longrightarrow$ individual
62	(13)-2	spanned by tangent vectors $\succ\!\!\longrightarrow$ spanned by positively oriented collections of tangent vectors
102	-1	and usually denoted Δ $\succ\!\!\longrightarrow$, which is usually denoted Δ
121	(1)-1	inhomogeneous, $\succ\!\!\longrightarrow$ inhomogeneous, linear,